AUTOMATED PEOPLE MOVERS 2009

Connecting People, Connecting Places, Connecting Modes

PROCEEDINGS OF THE TWELFTH INTERNATIONAL CONFERENCE

May 31–June 3, 2009
Atlanta, Georgia

SPONSORED BY
Committee on Automated People Movers
The Transportation & Development Institute (T&DI)
of the American Society of Civil Engineers

EDITED BY
Robert R. Griebenow, P.E., S.E.

Published by the American Society of Civil Engineers

Library of Congress Cataloging-in-Publication Data

Automated people movers, 2009 : connecting people, connecting places, connecting modes : proceedings of the twelfth international conference, May 31-June 3, 2009 : Atlanta, Georgia / sponsored by Committee on Automated People Movers, The Transportation & Development Institute (T&DI) of the American Society of Civil Engineers ; edited by Robert R. Griebenow.
   p. cm.
   Includes bibliographical references and indexes.
   ISBN 978-0-7844-1038-7
   1. Personal rapid transit--Congresses. I. Griebenow, Robert R. II. American Society of Civil Engineers. Committee on Automated People Movers.

TA1207.A995 2009
388.4'1--dc22   2009015799

American Society of Civil Engineers
1801 Alexander Bell Drive
Reston, Virginia, 20191-4400

www.pubs.asce.org

Any statements expressed in these materials are those of the individual authors and do not necessarily represent the views of ASCE, which takes no responsibility for any statement made herein. No reference made in this publication to any specific method, product, process, or service constitutes or implies an endorsement, recommendation, or warranty thereof by ASCE. The materials are for general information only and do not represent a standard of ASCE, nor are they intended as a reference in purchase specifications, contracts, regulations, statutes, or any other legal document. ASCE makes no representation or warranty of any kind, whether express or implied, concerning the accuracy, completeness, suitability, or utility of any information, apparatus, product, or process discussed in this publication, and assumes no liability therefore. This information should not be used without first securing competent advice with respect to its suitability for any general or specific application. Anyone utilizing this information assumes all liability arising from such use, including but not limited to infringement of any patent or patents.

ASCE and American Society of Civil Engineers—Registered in U.S. Patent and Trademark Office.

Photocopies and reprints.
You can obtain instant permission to photocopy ASCE publications by using ASCE's online permission service (http://pubs.asce.org/permissions/requests/). Requests for 100 copies or more should be submitted to the Reprints Department, Publications Division, ASCE, (address above); email: permissions@asce.org. A reprint order form can be found at http://pubs.asce.org/support/reprints/.

Copyright © 2009 by the American Society of Civil Engineers.
All Rights Reserved.
ISBN 978-0-7844-1038-7
Manufactured in the United States of America.

Preface

APM 2009 is the 12th in the series of international automated people mover conferences and continues the tradition of transit professionals meeting together to share the latest technology, projects and planning of automated transit.

The conference theme, "Connecting People, Connecting Places, Connecting Modes – APMs" reflects the conference's focus on APMs and driverless transit systems as connectors for people, places and modes in cities, airports and private developments. Around the world automation in transit is accepted as safer and less costly that manual operation. Full automation has been applied to a wide range of transit technologies, from rapid transit systems to small circulator systems. APMs have matured, and this conference has recorded the experience and lessons learned from recent projects and developing technologies.

Special recognition goes to the organizers of the first APM conference: Murthy V.A. Bondada, Conference Chair, and Edward S. Neumann, Program Chair, who through their creative and persistent efforts on the APM Committee crystallized the first conference in 1985. Now every two years, this conference brings together the planners, inventors, designers, suppliers, builders, owners and operators of automated transit systems to share their experiences, technologies and innovative ideas. Past conferences, held around the globe, have built an international fraternity of profession and friendship.

APM 1985 – Miami Florida
APM 1989 – Miami, Florida
APM 1991 – Yokohama, Japan
APM 1993 – Irving, Texas
APM 1996 – Paris, France
APM 1997 – Las Vegas, Nevada
APM 1999 – Copenhagen, Denmark
APM 2001 – San Francisco, California
APM 2003 – Singapore
APM 2005 – Orlando, Florida
APM 2007 – Vienna, Austria

The editor acknowledges the significant efforts of the many authors who through their papers have created the record in these APM proceedings in both time and place.

Robert Griebenow

Acknowledgments

Conference Steering Committee
- Hal Lindsey, A.M.ASCE, Lea+Elliott, Conference Co-chair/Exhibits and Sponsorships
- Robert Griebenow, P.E., S.E., M.ASCE, Berger/ABAM, Conference Co-chair/Technical Program
- Janice Li, Lea+Elliott, Track Chair
- J. Sam Lott, P.E., M.ASCE, Kimley Horn, Track Chair
- Martin Lowson, FREng, Automated Transport Systems Ltd, Track Chair
- Larrence L. Smith, P.E., F.ASCE, Consultant and ASCE APM Standards Committee Chair, Track Chair
- William Sproule, Ph.D., P.E., F.ASCE, Michigan Technological University, Track Chair

Conference Co-Sponsors
- Advanced Transit Association (ATRA)
- Airports Council International North America (ACI/NA)
- Institution of Civil Engineers
- City Mobil
- INRIA

ASCE Staff
- Jon Esslinger, P.E., F.ASCE, Director, Transportation & Development Institute of ASCE
- Debra Tucker, CMP, Conference Manager
- Sean Scully, ASCE, Exhibits
- Joanna Colbourne, Sponsorships
- Donna Dickert, Publications

Contents

Airport Experiences

Landside APM Planning at Seattle-Tacoma International Airport 1
  Dave Tomber, Harley Moore, and Bob Griebenow

NDIA APM: Building an APM in a Terminal Building 11
  Steve Beebe, Jackie Yang, and Harley L. Moore

The Dallas Love Field People Mover Connector 20
  Terry Mitchell, Diego Rincon, and Scott Kutchins

The Sacramento International Airport APM 34
  Jenny Baumgartner and Harley L. Moore

DFW Skylink: Tracking Success 44
  David Taliaferro

Operations and Maintenance at Atlanta Airport 50
  Melvin Redd, Russell Woodley, and Jerome Page

O'Hare ATS—The Teenage Years 56
  Dennis Gary and Mark Piltingsrud

APM Systems: The Key to Atlanta Airport Expansion 69
  John Kapala

Planning and Integration—MHJIT at Atlanta Airport 80
  Sambit Bhattacharjee, John Kapala, and Mike Williams

Expanding the APM System at Atlanta Airport 91
  Gregory A. Adams, Thomas Sheakley, and Frank Jeffers

The Impact of APMs on Property Value 105
  David D. Little and Margaret Picard

London Heathrow Terminal Five APM Project 116
  Jon Brackpool and Glenn Morgan

MIA Mover APM: A Fixed Facilities Design-Build Perspective 128
  B. M. Schroeder

MIA Mover Procurement 141
  Sanjeev N. Shah, Larry Coleman, Margaret Hawkins Moss, and Franklin Stirrup

Major Activity Centers

APM Feasibility Study for the Vienna Central Station Development Area 151
  Heimo Krappinger

Las Vegas People Mover Integration Potential 164
  Wayne D. Cottrell

Planning and Procurement of the Doha, West-Bay APM 179
  Kamel-Eddine Mokhtech, Sanjeev Shah, Hassan Eisa M. Al-Fadala,
  Ghanim Hassan Al-Ibrahim, and Hassan Qaddoura

PRT Case Study at the Village West Development in Kansas City, Kansas 190
  Stanley E. Young, Peter Muller, Moni El-Aasar, Dean Landman,
  and Steven Schrock

Project Funding Opportunities 201
  Sanjeev N. Shah and Larry Coleman

Evolving Clark County APM Code Requirements 212
  David Mori and Eric Troy

A Campus Transportation System for Michigan Tech 219
  William H. Leder, Frank W. Baxandall, and William J. Sproule

California University of Pennsylvania (CALU)—Maglev Sky Shuttle 233
  Thomas E. Riester and Husam (Sam) Gurol

New System Developments

Market Trends and Comparative Study of Economic and Technological
Parameters of APM Systems 245
  E. Todt, A. Gehlen de Leão, L. A. Lindau, E. Bortolini, and B. M. Pereira

Financing Transit Usage with Podcars in 59 Swedish Cities 257
  Göran Tegnér and Elisabet Idar Angelov

Sustainability, PRT, and Parking 297
  Shannon Sanders McDonald

The Impact of PRT on Army Base Sustainability 309
  Peter J. Muller

Ride Sharing in Personal Rapid Transit Capacity Planning 321
  John Lees-Miller, John Hammersley, and Nick Davenport

Wireless Communication Based Computer Simulator to Assess
the Operational Scenarios for the PRT Systems 333
  Jun-Ho Lee and Yong-Kyu Kim

Extending PRT Capabilities 343
  Ingmar J. Andreasson

Open-Guideway Personal Rapid Transit Station Options 350
  Peter J. Muller

Introducing PRT to the Sustainable City 361
  Robbert Lohmann and Luca Guala

The Need for High Capacity PRT Standardization 379
  Raymond MacDonald

Vectus PRT Concept and Test Track Experience 389
  Jörgen Gustafsson and Svante Lennartsson

Plenary

Defining the Right Roles for Automated Guideway Transit Systems 403
Hal Lindsey

Somewhere in Time—A History of Automated People Movers 413
William J. Sproule

System Improvements

Eco-Industrial Design for Cityval, Siemens' New AGT 425
Marc Zuber

Guideway Design and Construction 428
Brian K. Adams, John A. Heath, and Gary B. Lineback

How to Design a PRT Guideway 436
J. Edward Anderson

Heathrow PRT Guideway, Lessons Learned 450
A. D. Kerr and R. J. Oates

New Technology Integration for Older System Technologies 471
Steven M. Castaneda

Optimizing APM Failure-Mode Capacities 484
James W. Green

Orlando APM Running Surface Rehabilitation: Airsides 1 and 3 492
Sambit Bhattacharjee, Dan McFadden, and Tuan Nguyen

System Demonstration: Preparing for Success 499
Matthew Sturgell

Energy-Efficient APM Using High Performance Batteries 507
Masaya Mitake, Hiroshi Ogawa, and Katsuaki Morita

Standards for Successful APM Implementation 511
Frank Culver and Mario Nuevo

Advanced Composite Carbody Systems 525
Takaomi Inada, Genichirou Nagahara, Seung-Cheol Lee, Dae-Hwan Kim,
Masaaki Kuwabara, and Tsutomu Hoshii

Advances in Passenger Convenience and Comfort 534
Kunihiro Tatecho, Masafumi Kawai, Yuji Koike, Motoaki Tanaka,
Masaaki Kuwabara, and Tsutomu Hoshii

Urban Metro

Objectively Assessing Automatic vs. Manual Control for Transit Systems 546
John E. Joy

Rubber Tired APM—A Better Solution for Honolulu Rapid Transit 555
J. David Mori

Guangzhou APM: First Urban APM in China 562
Rob DeCostro

Simulation Analysis of APM Systems in Dense Urban Environments— Part 1: Transit User Experience 574
    J. Sam Lott, Douglas Gettman, and David S. Tai

Simulation Analysis of APM Systems in Dense Urban Environments— Part 2: System Operations 588
    J. Sam Lott, Douglas Gettman, and David S. Tai

An Enhanced Bombardier CX-100 APM Vehicle 603
    Jack Galanko and Scott Moore

Indexes

Author Index 621
Subject Index 623

Landside APM Planning at Seattle-Tacoma International Airport

Dave Tomber[1], Harley Moore[2], and Bob Griebenow[3]

[1] Aviation Planning Manager, Port of Seattle, Seattle-Tacoma International Airport, P.O. Box 68727, Seattle, WA 98168-0727; PH (206) 444-4368; FAX (206) 431-4496; email: tomber.d@portseattle.org
[2] Senior Principal, Lea+Elliott, 785 Market Street, 12th Floor, San Francisco, CA 94103, PH (415) 908-6450, FAX (415) 908-6451, email: hlmoore@leaelliott.com
[3] Vice President, Berger ABAM, 33301 9th Ave S, Suite 300, Federal Way, WA 98003-2600, PH (206) 431-2323, FAX (206) 431-2250, email: bob.griebenow@abam.com

ABSTRACT

Increasingly, large-hub airports in the United States (U.S.) are developing rental car facilities remote from their terminals in a facility that consolidates the operation of all rental car companies. In many U.S. large-hub airports, the percentage of passengers using rental cars is a significant percentage of overall traffic. At Seattle-Tacoma International Airport (SEA), almost 20% of all passengers use rental cars, creating a need for a transportation system that can move large volumes of people and baggage in peak periods with reliable frequencies and both reliable and minimal transit times. This paper is a case study for a landside automated people-mover (APM) system at SEA. It explores issues related to the feasibility of installing an APM system at SEA between the main terminal and a future consolidated rental car facility remote from the terminal. It analyzes the trade-offs between busing and different APM systems.

BACKGROUND

Seattle-Tacoma International Airport (SEA) served 32 million passengers in 2008. Forecasts indicate that this passenger activity could grow to almost 60 million annual passengers over the next 20 years. SEA is predominantly an origin-destination airport: in 2008 about 80% of the passengers started or ended their journey at SEA. This high concentration of origin-destination activity places heavy demands on landside facilities, such as airport roadways, curbsides, parking, and various ground transportation modes.

Increasingly, large-hub airports in the United States (U.S.) are developing rental car facilities remote from their terminals in a facility that consolidates the operations of all rental car companies. SEA began construction of a remote consolidated rental car (CONRAC) facility in 2008. The CONRAC site is located approximately 1 mile north of the existing terminal. An APM guideway connecting the existing terminal to the remote CONRAC needed to be planned to cross over a regional light rail system, also under construction in 2008 and scheduled to be completed by early 2010. A further planning constraint is that envelope of the APM vehicle height needs to be under airspace navigational ceilings determined by the Federal Aviation Administration (FAA) for interaction with arrival and departure aircraft flows from nearby King County Airport.

SYSTEM CONFIGURATION

A diagram of the APM system configuration is shown in Figure 1. Adjacent to the terminal is an existing 8-story parking garage. Midway between the existing terminal and the CONRAC is the site that potentially could be used for long-term parking in a multi-story structured facility. The regional light rail alignment is shown in a dashed line and there are two stations at or adjacent to the airport. The APM alignment is shown in a solid line for a 3-station configuration. Station 1 is at the existing terminal. Station 2 is at the future parking garage. Station 3 at the future CONRAC. Several options for the APM alignment were explored. The interior of Station 1 at the existing terminal is shown in Figure 2. The exterior of Station 1 is shown in Figures 3 and 4.

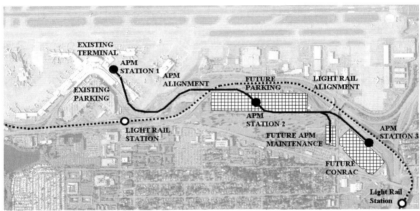

Figure 1

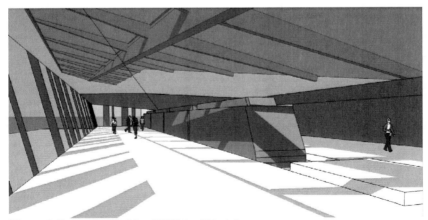

Figure 2 (Image by Perkins+Will Architects)

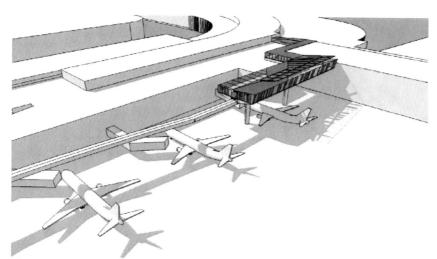

Figure 3 (Image by Perkins+Will Architects)

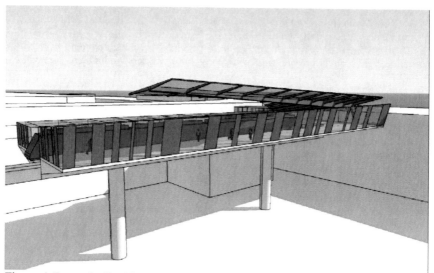

Figure 4 (Image by Perkins+Will Architects)

SYSTEM OPERATION

At SEA, almost 20% of all passengers use rental cars, creating a need for a transit system that can move large volumes of people and their baggage in peak periods with reliable headways and reliable and minimal transit times to and from the terminal.

The 20-year activity forecasts result in an airport-wide design hour arrival flow of approximately 8,300 passengers per hour and a departure rate of about 7,500 passengers per hour. Origin-destination (OD) passengers are about 80% of these values, and about 20% of the OD passengers are expected to use the CONRAC. Consequently, the estimated passenger flow to the CONRAC would be 8,300 x 80% x 20%, or 1,328 riders. Adding a factor for the future parking garage users in this hour and a surge of riders within the design peak hour results in an estimated APM passenger design flow rate of 1,500 passengers per hour per direction (pphpd).

For simplicity, only larger vehicle APM products were used in the analysis. These cars, like those of an existing airside APM system at SEA, are 40 feet long. Passenger carrying capacity varies with the type of passenger carried. For recent analyses at SEA for the airside APM system on which riders have only carry-on bags at most, up to 50 passengers typically ride in each vehicle. For passengers using the landside APM system serving the CONRAC and future long-term parking garage, passengers were assumed to have all their baggage, some of which (skis, golf clubs), can be large. Thus each rider requires more space than on the airside APM system. Assuming no baggage carts are allowed on the trains, the car capacity was assumed to be 40 riders. Smaller vehicles, such as those provided by some suppliers of cable-propelled APMs, could provide similar capacity, but more cars would be required.

The same operations were assumed for all seven days of the week. Three operating periods were assumed over a typical day. Peak period operations with both lanes operational would be provided 16 hours per day. Off-peak would have only one lane operating in a shuttle manner for three hours per day. Night operations for the remaining five hours would be with one lane operating in an on-call mode, much like an elevator.

Both shuttle and pinched loop routing scenarios were analyzed. Shuttle routing would involve one train going back and forth between the end stations on each guideway; thus two operating trains during peak periods. Pinched loop routing would have switches at each end station so that the trains could change guideway lanes, allowing more than two trains to operate on the system. Shuttle routing is much simpler, but with a two train limit that typically has longer trains for a given capacity. Cable-propelled technologies, which often are less costly, have, in the past, been more suited for shuttles without intermediate stations; however recent technology advances by suppliers of two such systems make pinched loop configurations possible. Pinched-loop systems usually have shorter headways (time between trains), and fewer cars per train for a given capacity requirement; however, pinched-loop systems require self-propelled APM technologies.

OPERATIONAL RESULTS

The operational modeling results from analyzing the train operations are summarized in Figure 5 for shuttle operations and Figure 6 for pinched loop operations. APM system options depict operational results for both cable- and self-propelled.

Operating Results – Shuttle System								
Option	Round Trip Time (Sec)	# of Operating Trains	Average Headway (Min)	# of Cars/ Train	System Capacity (pphpd)	Load Factor (%)	Operating Fleet	Total Fleet
Cable	590	2	4.9	4	1,953	77	8	12
Self-Propelled	576	2	4.8	3	1,500	100	6	9

Figure 5

Operating Results – Pinched Loop System								
Option	Round Trip Time (Sec)	# of Operating Trains	Average Headway (Min)	# of Cars/ Train	System Capacity (pphpd)	Load Factor (%)	Operating Fleet	Total Fleet
Cable	590	3	3.3	3	2,197	68	9	12
Self-Propelled	576	3	3.2	2	1,500	100	6	8

Figure 6

Several conclusions can be drawn from these train capacity analyses, in addition to the previous shuttle versus pinched-loop discussion:

1. Shuttle operations would work for capacity.
2. The levels of service for a shuttle system would be slightly different from those for a pinched-loop system, as the latter reduces the average wait time from about 3 minutes to 2 minutes.
3. Shuttles could be a reasonable starting operational approach.
4. Shuttle operations are often less expensive and it would be possible to use cable as a well as self-propelled APM technologies.
5. Pinched-loop routing would result in shorter trains (typically one car less), and stations (about 40 feet less due to the shorter trains). The fleet size would remain the same in most cases; the trade-off is more operating trains of fewer cars each.
6. Possibly the best, most flexible approach would be to be able to operate in either mode. This would not be the least expensive approach.

7. Short inter-station distances and a curvy alignment can result in higher cruise speeds not paying off in terms of fleet or service levels.
8. For a self-propelled system, the fleet could require fewer cars. For a shuttle system, there would be no difference. This is because of the curves and station spacing: there are relatively few locations that a train could take advantage of the increased speed. Thus a somewhat slower and possibly less expensive APM technology (including cable-propelled or some of the existing airside APM fleet) could be used.
9. Adding an extra station at the future parking garage improves service to the users of that garage, but reduces the level of service for everyone else: the vast majority of the riders who are going between the terminal and the CONRAC. Adding a station increases system and facility costs.
10. The station platform would be about 200 ft. long, with vertical access at the ends of the center platform station, the access points from the garage would be about 400 ft. apart. This helps reduce the amount of walking to a single station. Having only one centrally located station in the future parking garage and moving walkways on one of the levels to access that station is worth consideration.

SYSTEM REQUIREMENTS

System requirements used in the cost modeling are given in Figure 7. For the cost model, the system was assumed to operate as a dual lane shuttle. There would be three stations: one each at the existing terminal, future parking garage, and future CONRAC.

System Cost Elements		
Cost Element	Quantity	Comment
Fleet	9	40 foot cars
Operating guideway (linear feet of dual lane)	6,500	Guideway equipment
Maintenance guideway (linear feet of single lane)	730	Guideway equipment
Guideway switches	10	Guideway equipment: 8 main line and 2 maintenance
Automatic train control	1	Communication-based for pinched loop operation
Power distribution and substations	2	DC distribution using power rails along the guideway
Stations	3	Station equipment, including platform doors, communications, and CCTV
Maintenance equipment	1	All tools and equipment to outfit the maintenance facility

Figure 7

FACILITY REQUIREMENTS

Facility requirements are summarized in Figure 8. Detailed descriptions for the guideway, stations, maintenance facility, and power substations are greater described below.

Facility Cost Elements		
Cost Elements	Quantity	Comment
Operating guideway (linear feet of dual lane)t	6,500	Guideway structure
Maintenance guideway (linear feet of single lane)	730	Guideway structure, some within the maintenance facility in the future parking garage
Power distribution substations (number)	2	750 or 1,500 VDC distribution
Power distribution substations (sq.ft., each)	2,800	
Stations (number and layout)	3	Center platform
Stations (square feet each)	6,000	Platform 30 feet wide by 200 feet long
Station access at each end (square feet)	1,000	Plus two escalators, one elevator and one stairway
Maintenance facility (square feet)	70,000	Three levels within the future parking structure

Figure 8

1. <u>Guideway.</u> The main line guideway is dual-lane and elevated at least 20 feet above the ground (including roadways and parking lots) below. It would be a stand-alone structure not connected to or supported by other structures. It would be a concrete structure about 25 to 30 feet wide (10 to 12 foot wide track areas and a 5 foot wide emergency walkway). Columns would be at 60 to 100 foot intervals. The beams would be nominally 6 feet deep. Going over some roadways and building could require a greater depth of structure if column spacing must be over about 120 feet. The lead tracks to the maintenance facility would be similar, with adaptations to support switches. Guideway within the maintenance facility would be supported on the future parking garage building structure.
2. <u>Stations.</u> Each station would be a stand-alone center platform with appropriate vertical transportation cores (elevators, escalators, and stairs) connecting them to the airport facility they serve. The landside terminal station would be built at the north end of the existing terminal, and it would be connected to the rest of the terminal by a series of moving walkways in an enclosed corridor built above the ticketing level of the terminal. The future parking garage station would be next to the parking garage. The CONRAC station would be similarly located next to that building. Each station would have connections from one or both ends. The terminal and CONRAC stations would have connections at only one end. The future parking garage station would have connections at both ends to reduce the

walking distances within the garage. These connections would consist of a pedestrian bridge either above or below one of the guideways serving the station with a vertical core connecting it and the station platform. The area would be about 1,000 sq. ft. including the bridge and the base at the end of the platform. Two escalators (one up and one down), one elevator, and a stairway would connect the station platform and the bridge. The CONRAC station at the end of the line could have a walkway directly into the building at the counter or return level, depending on the design of that facility.

3. Maintenance Facility. This would occupy three levels on the northern end of the Doug Fox Parking Garage. The main area would be the full height (nominally 25 feet) for the vehicle maintenance and storage and the open shop areas. This area would be about 300 ft. by 140 ft. and include two lanes of guideway for vehicle maintenance and storage. Part of the facility (100 ft by 140 ft.) would be two levels with a nominal 11 ft. height. The lower level (an extension of the main maintenance floor) would be for shops and spares/equipment storage; the upper level for offices, central control, and personnel facilities.

4. Power Substations. There would be two buildings, each about 80 ft. by 35 ft. One would be located near the maintenance facility and the other near the existing terminal station. The sites would depend on available land and power grid feed locations.

CAPITAL COSTS

APM system costs were estimated for a parametric model based on data from similar projects within the last 5 years. The goal of the parametric model is to develop a rough order of magnitude estimate for hard and soft costs by system for project scoping. Figure 9 summarizes the estimated direct costs for systems and facilities totaling $394 million. All costs are stated in 2008 USD (United States dollars). Since many APM suppliers are European and Japanese, it should be noted that these costs can easily fluctuate due to currency valuation of the U.S. Dollar compared to the Euro or Yen.

Summary of System and Facility Costs		
Cost Elements	Quantity	Comment
APM system & equipment	$70,000,000	Fleet, Controls, Power distribution, Station equipment, Guideway equipment, Maintenance equipment, System verification
APM facilities	$187,000,000	Maintenance facility, Guideway, Stations, Terminal integration
Soft costs	$137,000,000	Sales tax, Hazardous materials, Contingencies, Permits, Management
Total costs	394,000,000	Direct hard costs, soft costs, and markups

Figure 9

OPERATION & MAINTENANCE COSTS
Operation and maintenance (O&M) costs for APM systems are comprised of five major categories:
1. Labor. This category consists of management and administrative staff, central control operators, and vehicle and equipment maintainers.
2. Utilities. Primarily for electricity for propulsion power, but also housekeeping power in the APM facilities, plus minor costs for other utilities.
3. Spare Parts and Consumables. Parts, consumables, and the like used in routine maintenance.
4. Administration. Insurance, uniforms, and office items.
5. Other. All other O&M costs, including contracts for janitorial and engineering support.

Although O&M costs for APM systems can vary widely with different assumptions, there is a reasonable correlation with annual APM vehicle miles traveled (VMT). This provides a quick method of estimating such costs. The factors used are shown in Figure 10.

Cost Elements	Cost (dollars per vehicle mile traveled)
Labor	$4.50
Utilities	$1.60
Spare parts/consumables	$0.80
Administration	$0.30
Other	$0.70
Total	$7.90

Figure 10

System annual VMT (vehicle miles traveled) was estimated to be about 650,000 miles. At a cost of $7.90 per vehicle mile traveled, the total annual O&M cost for the APM system would be $5,135,000. Additionally, the average annual O&M and renewal costs for the APM facilities (stations and guideway) would be approximately 1.5% of the initial capital cost of construction averaged over the life of the facilities, or $2,805,000 (1.5% x $187,000,000). Consequently, total annualized cost of ownership for the APM system and facilities would be $7,940,000 ($5,135,000 + $2,805,000).

DEVELOPMENT TIMEFRAME
The development timeframe was analyzed for two options: Option 1, shown in Figure 11, was for a typical 24 month environmental process, and Option 2, shown in Figure 12, was for an overlap of environmental with APM system procurement and facility design. Option 1, a typical 24 month environmental process, resulted in a development timeframe of 9 years. Option 2, an overlap of environmental with APM system procurement and facility design, resulted in a development timeframe of 7 years.

AUTOMATED PEOPLE MOVERS 2009

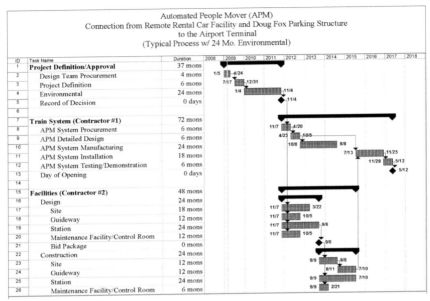

Figure 11

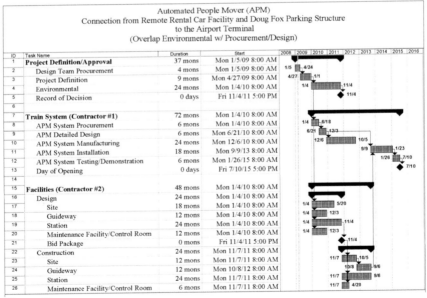

Figure 12

NDIA APM: Building an APM in a Terminal Building

Steve Beebe*, Jackie Yang**, and Harley L. Moore***

* Senior Associate, Lea+Elliott, Inc., sbeebe@leaelliott.com
** Manager of Planning Projects, Lea+Elliott, Inc., jyang@leaelliott.com
*** Senior Principal, Lea+Elliott, Inc., hlmoore@leaelliott.com
785 Market Street, 12th Floor, San Francisco, CA 94103, (415) 908-6450; (fax) 908-6451

Abstract

The New Doha International Airport (NDIA) APM system will have two parallel single-lane shuttles operating on the second floor within the "Y" shaped Concourses C, D, and E of the new terminal building. Due to the terminal facilities design and construction schedules, the APM infrastructure was designed for a generic technology that could include self-propelled rubber tire, low-speed maglev, and cable-propelled rubber-tire and air-levitated systems. The stations were located, and the trains designed, to serve both secure and non-secure passengers with different passenger peaking characteristics and times. The maintenance facilities were designed to fit inside the building on the same level as the system, yet not constrain the passenger flows in or around the gate areas. Providing access for inserting vehicles into the system, initially and in the future for replacement, was particularly problematic.

This paper describes the system, discusses challenges and solutions for the APM design due to its interior location, provides information on the procurement process and results, and summarizes issues that have arisen since the start of the implementation period.

Overview

NDIA in Doha, the State of Qatar, is envisioned to be a major international gateway to the region and a transferring hub. The 67-gate terminal has a "Y" configuration (Figure 1) and is expected to serve 48 million annual passengers (MAP) in 2020. An APM was included in the terminal design to reduce passenger walking distances and to improve connecting times.

Figure 1 - NDIA Terminal Building and Concourses
Source: Doppelmayr Cable Car

Design Issues

Interior APM Location

An interior APM system was planned to avoid problems with wind, blowing dust and sand, extreme temperatures, and other outside elements. This was for both more reliable system operations and to seal the building from these outside elements. The interior location led NDIA and its architect, HOK, to require close control of the design of the APM facilities and the visible system elements, as well as implementing stringent noise and vibration requirements. Many configurations were developed and evaluated given the constraints of the building, ridership estimates, and level of service requirements. The selected design is a pair of nominally parallel single shuttle guideways, connecting the North and South Nodes, centered along the Concourse C spine, and on Level 2 of the building (see Figure 2).

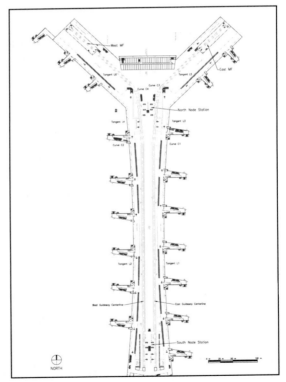

Figure 2 – NDIA APM Alignment Configuration
Source: HOK

Generic APM Facilities Design

Originally, construction of Concourses A and B was to be completed in 2010 and it was anticipated that the airport would be open for about a year before Concourses C, D, and E, and the APM, would begin operation.

The terminal and APM facilities needed to be designed before the APM system was procured. The NDIA Steering Committee, (the organization responsible for the airport project and overseeing its design and construction), wanted to maximize the potential for competition among APM suppliers for cost and technology selection reasons. Thus the APM facilities (guideway, stations and maintenance facilities) were designed to accommodate a wide range of APM systems.

NDIA decided that it would provide the guideway and other terminal building facilities for the APM supplier to install its system equipment. This gave NDIA control of the design and construction of all civil works, which was important to the building and APM design and their construction schedules. A generic, concrete "bathtub" (or channel) guideway design accommodated all known self-propelled, cable-propelled, and low-speed maglev APM technologies. This opened up the competition for the APM system to at least six APM suppliers and technologies. The stations were designed to accommodate the various train widths and lengths to achieve capacity requirements. The maintenance facilities were configured to permit the various train sizes and access and equipment for both self-propelled and cable-propelled systems.

Operational Requirements

Secure and Non-Secure Ridership

Transferring and origin/destination passengers will use the APM. The secure (security screened – typically departing and some transferring) and non-secure (non-screened – typically arriving) passengers must be physically separated on the system. The ridership demand of each group will vary over the day given anticipated flight arrival and departure times. On the typical design day, about 50% will be origin/destination passengers and 50% transferring passengers; however, during the peak hours, 70% will be transferring passengers. The APM ridership and capacity simulation modeling was based on the expected future demands over a 24-hour design day, and the system operations were developed to serve the peak hour changes in ridership over time, as shown in Table 1.

SYSTEM PEAK HOUR RIDER FLOW RATE ESTIMATES		
	Southbound	Northbound
Peak Arrival Demand		
Non-Secure Riders (pphpd)	2,100	0
Secure Riders (pphpd)	2,550	2,350
Peak Departure Demand		
Non-Secure Riders (pphpd)	250	0
Secure Riders (pphpd)	2,550	4,950

Table 1. Ridership Demand Requirements

Operating Configuration

Because of the short length of the system (approximately 500 meters station-to-station), pinched loop configurations were determined not to be viable. Various shuttle operating configurations (single lane, bypass, and dual lane) were analyzed. Alternatives included shuttles with two sets of stations, one each at the North and South Nodes, and shuttles with three sets of stations, which would require expansion into Concourses D and E when they were built. The two-station alternative was selected for level of service, building spatial constraints, and because of the complexity of expanding to a three-station system, complete with maintenance facilities that would work for both.

The APM will operate 24 hours a day. Two shuttle trains will operate during peak hours. Only one will operate during off-peak hours, either as a scheduled shuttle or in an on-call mode, depending on the passenger volumes. Non-secure passengers on the non-operating side will walk and use moving walkways when one of the two single shuttles is closed, for maintenance or failure modes. As this is not a "must–ride" system and there are moving walkways, this was deemed acceptable, although the level of service (trip times) will not be as good as with the APM in normal double shuttle operation.

Split stations and trains will be operated to accommodate both secure and non-secure passenger groups. Trains were planned to have secure (north end), non-secure (south end), and "swing" (middle) cars to shorten their length and provide more efficient train and station designs and operations (Figure 3). The swing car would carry secure passengers at some times and non-secure passengers at other times; a security sweep would be done before switching passenger types. Train control will use a double set of train berthing positions for the arrival and departure peak periods to allow the doors of these three types of cars to open in the appropriate station areas.

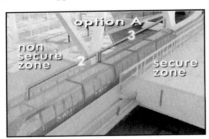 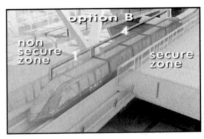

Figure 3 – Swing Car Concept (at North Node)
Source: Doppelmayr Cable Car

Station / Platform Configuration

The stations and platform configurations provide different boarding and deboarding areas to separate the two passenger groups (Figure 2). The South Node station has a center platform configuration with secure passenger boarding/deboarding at the north end, and non-secure passenger boarding/deboarding at the south end. The South Node station secure/non-secure platform areas will be separated by a wall. The North Node station has two side platforms for non-secure passengers and one center platform for secure passengers. With the center platform configuration for secure passengers, operations are similar to a simple dual lane shuttle where passengers can use either lane to go either direction (i.e., north or south). Non-secure passengers will use the North Node station platform closest to their gate and operations will be similar to a single lane shuttle, as there is no east-west connection between these stations or the areas they serve.

The emergency walkway is located along the inside edge of the guideways. Thus they will exit into the secure station platforms. Any evacuation will result in mixing secure and non-secure passengers on this walkway, so special procedures will be used to control their entry into the stations and deal with the security issues.

Maintenance Facility Issues

The two APM maintenance facilities, one for each lane, are inside the passenger terminal building in the "winglet" areas of Concourses D and E. These locations were selected to minimize the impacts on all three levels of the terminal. Other alternatives, including combining the facilities and bringing the guideways to grade, were found to be infeasible.

The maintenance facilities are primarily on the ground and second floor levels, with spaces provided through the first floor for cable and other chases and for the vehicle access shafts. The minimal footprint on the first floor also allows for better passenger movement on that level.

The ground floor areas include the electrical substations and system electrical equipment rooms, apron access, emergency backup generators, and some storage areas. For this cable-propelled system, they also include the main drive rooms, the tensioning equipment, and the deflection towers. The facilities in the east winglet include the Central Control Room and the administrative offices; those in the west winglet include the main power distribution substation with the two independent 11 kV line feeds from the airport power grid, transformers, switchgear and other electrical equipment.

As the guideway is on the second level of the terminal building, access hatches will be provided in both maintenance facilities to allow for the initial installation of the trains (one car at a time) and to allow for their removal if required for major overhaul and/or replacement. Adequate building entry facilities (a large enough floor area on the ground floor level) were planned to allow the vehicles to be brought into the building, aligned, and then hoisted by crane to the second floor level.

An employee access corridor between the two winglets allows easy access for O&M personnel to travel between both maintenance facilities. Access to other areas of the APM (e.g., along the guideway and at the stations) requires special considerations and procedures to meet the terminal security provisions.

Procurement Issues

Procurement Process

The procurement process consisted of a request for interest and information (RFI), a request for qualifications (RFQ), a request for proposals (RFP) from those selected as qualified for this project, and ultimately, best and final offers (BAFOs) from the finalist proposers. Six suppliers with varying technologies (self-propelled, cable, and maglev) responded to the RFQ: Bombardier - Innovia; Doppelmayr Cable Car (DCC) - Cable Liner; IHI/Niigata - NTS; Leitner-Poma MiniMetro (LPMM) - air-levitated cable MiniMetro; Mitsubishi - Crystal Mover and HSST Maglev; and Siemens - Airval. Five of these suppliers were deemed qualified to receive the RFP. Four provided proposals: two with self-propelled systems (Bombardier and Siemens) and two with cable-propelled systems (DCC and LPMM). After a thorough "best value" evaluation of the proposals, DCC and LPMM were asked to provide a BAFO. Based on that, the NDIA Steering Committee selected DCC. Notice to Proceed (NTP) was given in July 2007 and the system completion date was originally scheduled for December 2009.

Best Value Procurement

In the best value procurement evaluation, price was not the ultimate factor. Price items, including capital, operations and maintenance, and life cycle costs, accounted for 30% of the total score. Technical qualifications (40%) and commercial items (30%) were also considered to determine the best value by the NDIA Steering Committee.

Pricing

There was considerable, and close, competition for this project, in part because the suppliers saw it as a good initial project in the region, where many APMs were being planned. Several items had a significant effect on pricing in addition to competition and the potential for other projects in the region: exchange rate, shipping, and local contractual requirements. The Qatari riyal was pegged to the US dollar, so the relatively strong Euro and Yen added pricing risk to those suppliers. Given the location of the project and the location of the potential suppliers' factories (US, Europe, and Japan), most project materials had to be shipped to Doha. The standard contract requirements developed by NDIA and its PM/CM (Overseas Bechtel, Inc.) were based on construction in Doha. Although this was modified somewhat for this supply, install, operate, and maintain procurement, vestiges added to the effort required by, and prices from, the proposers.

Detailed Design and Implementation Issues

The APM contract and specifications identify performance requirements, but the detailed design by the contractor further defines the specific system, and to an extent, the APM-related facilities. The generic design of these facilities accommodated all candidate technologies. With the selection of DCC and its cable-propelled system, some aspects of the APM facilities needed refinement, and adjustments have been ongoing. There have been some unusual changes that have made the project more challenging, some of which are discussed in the following sections.

Terminal Design Changes

The NDIA Steering Committee made major changes to the North Node and Concourses D and E to add significantly more retail and amenity space. There have been changes, albeit relatively fewer, to the South Node and Concourse C, which have had only a minor impact on the APM facilities and system. The primary impact has been on project schedule, particularly the facility access dates for the APM contractor to start installing and testing the system equipment. The contractual access dates are shown in Table 2. Final dates and contractual aspects of these changes are in negotiation as of this writing.

Area	Original Date
Concourse C, South Node Station	January 2009
Concourse C, South Guideway	February 2009
Concourse C, Middle Guideway	March 2009
Concourse C, North Guideway	April 2009
North Node, All Levels	April 2009
Substantial Completion	April 2010
Final Acceptance	February 2011

Table 2. Access and Milestone Dates

Changes to the APM facilities as reflected in the contract and DCC's proposal are also being made. Most of these are based on architectural concerns in this very public terminal space. These include the design of the cover for the return wheels, painting most of the guideway equipment black so it is relatively unobtrusive, minimizing the impacts of the station platform walls and doors, and the location and design of passenger information and security devices, including the station dynamic signs and CCTV cameras.

The design of the DCC trains, with five individual cabins, affected the lengths of the stations, and station locations and lengths were adjusted accordingly. The train is a departure from DCC's previous individual cabin design: the vehicle structure and bogies have been redesigned to allow "walk-through" cars. The three northern cars are "walk-through" to serve secure passengers. The fourth car is an individual cabin, which will be the swing car. The southernmost car is an individual cabin for non-secure passengers only.

There have been changes to the maintenance facilities as well. Because the trains are passive (i.e., cable drive instead of self-propelled), maintenance activities are simpler and therefore less space is needed in the Level 2 maintenance rooms. Making these rooms smaller allowed for space to be returned to terminal uses. Changes were also made to the location of the drive equipment for terminal space and architectural reasons.

System Equipment Changes

The major change to system equipment was to the vehicles. As part of the system design reviews, NDIA determined that it wanted a different shape to the noses at the ends of the vehicles. Other nose designs were proposed by DCC. The one that NDIA selected is less rounded and adds passenger space to the end cars (Figure 4). This required a redesign of the vehicle structure, and affected dwell times and system capacity.

Aesthetic changes to both the interior and exterior of the trains have also been made. The concept has been to maximize the "transparency" of the train so passengers can more fully appreciate both the terminal building and train riding experiences. The glass area of the train has been maximized (there already was a glass roof), and the interior seats and stanchions have been designed for elegance and minimum obtrusiveness.

Figure 4 - Proposed Vehicle Design (left) and Final Vehicle Design (right)
Source: Doppelmayr Cable Car

Lessons Learned

Although it is too early to list definitive lessons learned, the following can be made from the project to date. These are based on the authors' observations, and are not necessarily those of the owner, architect, PM/CM contractor, or APM contractor.

1. Procurement
    a. Owners often direct the designers and consultants to do whatever is reasonable to foster good competition. This usually results in a lower system price. It can, however, add to the APM facility price, particularly if there is a wide variation in APM technologies under consideration. It would be helpful to select an APM supplier and technology early in the facility design process, but often this is not possible, as with this project.
    b. The procurement process should be structured to give the Owner a good basis to choose the APM system and contractor. A multi-step, best value approach can accomplish this while maintaining an open and fair competitive process.

c. Many contractual requirements developed for building construction are inappropriate for a system design, supply, and install contract. They can also cause problems for the operations and maintenance period. The entire package should be developed for the approach and purpose intended and not cobbled together from different types of projects or contracting approaches.
2. An APM Inside a Building
   a. Having an APM within the building adds many requirements and challenges. Among these are fire code, noise and vibration, and, most difficult for this project, system and facility designs to meet architectural and other desires and requirements.
   b. Not all APM technologies are appropriate for a system inside a building, but eliminating some *a priori* limits competition and can bring challenges. For example, the low speed maglev would be good from a noise standpoint (assuming shuttles and no switches) but the speed for which it is designed is not useable.
3. Impacts of Redesigns
   a. Any major redesign to the facilities or system equipment will affect schedule and price, possibly very significantly, which must be recognized when it is being contemplated.
   b. Changes to the schedule for such design changes will also affect equipment manufacturing, shipping and storage, and warranties. The later the changes, the more significant these impacts.
4. Design Process Control
   a. Adding people and organizations to the design process, particularly those that were not involved in the planning or contractual requirement development, can make the process more difficult, lengthy, and expensive. All who will participate in the detailed design should participate in the earlier aspects of the project as well.

The Dallas Love Field People Mover Connector

Terry Mitchell[1], Diego Rincon[2], Scott Kutchins, P.E.[3]

[1]**Terry Mitchell,** City of Dallas, Department of Aviation; Love Field Terminal Building, 8008 Cedar Springs, LB16, Dallas, TX 75235; Phone: (214) 670-6086, Fax: (214) 670-6051, terry.mitchell@dallascityhall.com
[2]**Diego Rincon,** City of Dallas, Department of Aviation; Love Field Terminal Building, 8008 Cedar Springs, LB16, Dallas, TX 75235; Phone: (214) 670-6149, Fax (214) 670-6051, diego.rincon@dallascityhall.com
[3]**Scott Kutchins, P.E.,** Member ASCE, BSCE, Lea+Elliott, Inc; 1009 W. Randol Mill Rd., Arlington, TX 76012; Phone: (817) 261-1446, Fax (817) 861-3296, skutchins@leaelliott.com

Abstract

In anticipation of long-haul flights in 2014, Dallas Love Field is underway with a modernization program to maximize the efficiency of the airport operation. The program consists of the design and construction of a new terminal building consisting of twenty (20) gates in a dual loaded pier configuration and other terminal improvements such as baggage, security and check-in facilities. Dallas Area Rapid Transit (DART) is currently underway with the expansion of the light rail transit (LRT) system to the northwest including a station adjacent to Love Field less than one mile away. One option considered in the DART environmental impact statement was to route the LRT alignment through the airport in a tunnel however it was determined that this did not meet the cost-effectiveness guidelines of the Federal Transit Administration and was not further pursued.

In the interest of regional mobility, the City of Dallas has determined that a direct connection via an Automated People Mover (APM) to the LRT is desired. This report discusses the steps that have been taken to date to determine the feasibility of providing an APM link between the LRT system and the new terminal facility. This paper will discuss tunneling and construction methods, system performance requirements, system alternatives, the recommended alternative, facilities considerations, technology assessment and program schedule. Illustrations of the proposed passenger stations, the new terminal complex and the proposed alignment will be included as well.

INTRODUCTION

Dallas Area Rapid Transit (DART) is in the process of expanding its Light Rail Transit (LRT) System to the northwest suburbs of Dallas on a new segment called the Green Line. The Green Line will run on current and former railroad rights-of-way owned by DART including a rail corridor adjacent to Dallas Love Field (DAL). During the planning phase of the Green Line, many alignments were considered to provide direct rail access to DAL. Numerous discussions were held between the City of Dallas, DART, the Federal Aviation Administration and others about the potential to tunnel underneath the DAL airfield, roadways and terminal building to construct a passenger station in the basement of or adjacent to the terminal building.

At the time of these discussions, direct flights from DAL were limited to the states directly adjacent to Texas. Based on this fact, the annual passenger load at DAL was limited. It was determined through the Federal Transit Administration (FTA) environmental process that the added cost to construct the tunnel through DAL jeopardized the Green Line's cost effectiveness for federal funding. As a result of these findings, the recommended Green Line alignment would remain on the existing railroad right-of-way and the DAL tunnel and station within the terminal building would not be constructed.

In late 2004, talks began regarding the repeal of the legislation which prohibited long-haul flights out of DAL. These talks evolved into a spirited discussion and negotiation involving the cities of Dallas and Fort Worth, Dallas/Fort Worth International Airport, Southwest Airlines and American Airlines through 2005 and into 2006. On September 29, 2006, the 109th United States Congress passed the bill which would allow for a staged repeal of the legislation restricting flights out of DAL. The complete repeal will take effect in the fall of 2014. President Bush signed this bill into law on October 13, 2006.

Once the legislation was completed, discussions once again focused on the possibility that DART could revise the design of the Green Line to include the previously studied tunnel and station at DAL. The FTA informed DART and the City of Dallas that this would jeopardize the $700 million in funding that had been granted to DART for the Green Line and the DAL LRT tunnel concept was not developed further.

The City of Dallas (COD) intends to provide a transportation connection between the Love Field Station and the Love Field Terminal Building that will connect Dallas Love Field to the regional rail network. As part of the legislation mentioned above, the parties agreed to study a "people mover" connector (PMC) between the DART Love Field LRT Station (on the Green Line) and the DAL passenger terminal building. The Dallas Love Field People Mover Connector Feasibility Study was performed to investigate how this connection would interface with the regional rail network and the modernized Love Field terminal building. During the course of the feasibility study, coordination meetings with stakeholders including DART,

Southwest Airlines and other city departments were held to develop a needs assessment for the system.

The PMC would need to provide an equal level of service to the direct rail connection and be "seamless" to the passenger changing modes of transportation to access the airport. Exhibit 1 below provides an overall view of the study area for the DAL PMC.

Exhibit 1 - Dallas Love Field People Mover Connector Area Map

TECHNOLOGY ASSESSMENT

A Technology Assessment was prepared for the purpose of identifying technology categories and assessing their characteristics for applicability for the DAL PMC System (the "System"). This technology assessment was intended to identify and evaluate a range of current technologies that could potentially transport passengers between the Dallas Area Rapid Transit (DART) - Love Field Station of the Northwest Corridor (Green Line) Light Rail Transit (LRT) project and the Dallas Love Field (DAL) Terminal. This assessment defined and recommended a group of representative technologies that meet the System design requirements and performance criteria. This group of representative technologies has been used to establish generic design criteria to be used to advance the PMC facilities design until such time that a System Supplier is under contract and technology specific criteria are available.

Information such as the alignment, ridership forecasts, budget, and integration with existing or planned development and transportation services will assist in choosing the most appropriate technology. It was recommended that the final selection of a technology supplier be achieved through a competitive basis using an approved procurement process.

The technology assessment evaluated all transit technologies that were potentially applicable to the DAL PMC. This evaluation provided an initial screening for the project and information on various modes of transit to assist in the evaluation and selection of a technology.

From this initial screening the following technology alternatives were evaluated:

- Moving Walks including: Conventional and Accelerating
- Buses including: Conventional, Bus Rapid Transit (BRT) and Guided Buses
- Streetcars including: Historic and Modern
- Light Rail Transit (LRT)
- Automated People Movers (APM) including: Self-propelled APMs, Cable-propelled APMs, Monorail, and Low-speed Maglev technologies
- Personal Rapid Transit (PRT)

The technologies that appear to be applicable for this project will be evaluated against factors grouped into four categories:

- Performance
- Level of Service (LOS)
- Environmental Impacts
- Cost Effectiveness

Performance Factors

Performance factors used in the technology assessment included: capacity, speed, geometry/configuration, expandability, automation and technological maturity. At a minimum, the technology must be able to provide sufficient capacity to meet the estimated peak hour ridership demand in passengers per hour per direction (pphpd). A technology should have the flexibility to meet a range of capacities over a daily operating schedule and over the life of the system. The technology must be able to operate at a reasonable speed to generate desirable travel times. For this project, the estimated operating speed is in the range of 25-35 miles per hour (40 – 56 kph). Technologies must be able to fit physically and to operate over the alignment envisioned between the DART Love Field LRT station and the terminal without undue disruption to current and planned development. Alignment requirements are considered to identify technologies that could not physically operate within the people mover connector criteria. These requirements consider a technology's performance capabilities and constraints with regard to the geometry of the baseline alignment(s).

Expansion should be possible without significant disruption to the operating system. This also refers to the ability to expand the system by increasing the fleet size. Some technologies considered can operate fully-automated without drivers. Other technologies considered are manually-driven and require drivers. A fully-automated system may reduce Operating & Maintenance (O&M) costs, provide operating flexibility and increase safety. Automated operation can be especially important in systems with long hours of operation such as airports. The technology must be developed to a state that it can be implemented with minimum technological, budget, and schedule risks. In selecting a technology for a new system, it is important to assess the developmental and implementation risk associated with the technology. Risk can be determined by examining such factors as the years of proven service in similar transit applications, the number of systems currently in operation, the reliability and safety records of the operational systems and the experience of the technology supplier. For the DAL PMC, a period of two years in operating service was recommended.

Level of Service Factors

The Level of Service (LOS) provided by the System depends on planning and design considerations, including but not limited to: ride quality, passenger trip times, walk distances, ease-of-use, frequency of service and passenger wait times. The technology category selection should provide the optimum level of service in terms of minimizing passenger trip times and providing the best ride quality possible.

Level of Service factors are used to measure the passenger's experience. "Performance" measures, such as trip times and headways, combined with "perception" measures, such as the degree of seamless connection and the perception of where they "enter" the airport, all contribute to the LOS provided to the passenger.

Environmental Impacts

The interaction of the PMC with the surrounding environment was considered to identify technologies that emit unacceptable or undesirable levels of noise or other pollutions into the environment. The PMC must be compatible with the Airport and surrounding area and not induce objectionable noise or other emissions. The technology should not cause any significant impact to environmentally sensitive areas or cause air or water quality issues. Electricity or clean fuels should be used to propel the vehicles.

The technology should not create unacceptable noise or vibration levels in the surrounding areas, especially in residential neighborhoods which exist directly adjacent to DAL. The technology must physically fit into the urban/airport fabric, along rights of way, and into/next to specific developments. It should not create unacceptable physical and visual impacts in vehicle design, guideway design, stations, and the maintenance facility.

Cost Effectiveness

The capital cost of the initial system needs to be within the available budget, and any expansion must be at a reasonable cost. The capital cost of the alternative alignments will vary with elevated/at grade guideway, specific site conditions, system length, use of alternative structures, fleet size, and many other variables. The O&M costs must also be cost effective. Automated systems are able to meet the goals of the criterion due to reduced operator labor costs. Operator costs can be significant when a system is operating seven days per week for three shifts per day.

One of the biggest technical challenges will be integrating the DAL PMC into the operating airport, both during construction and after completion. Some technology categories may have an impact on the design of the PMC facilities and the adjacent airport facilities, both existing and those being planned. This could result in increased cost and increased sizes of facilities such as the stations or tunnel envelopes. Technologies that have requirements significantly different than the other candidate technologies could require special accommodations to allow integration into the airport facilities. This could result in the technology being unsuitable for the project.

Based on the results of the assessment, three classes of driverless, Automated People Movers (APMs) were retained as potential candidates for the DAL PMC (hereafter known as the DAL APM) System – Self-Propelled APM, Cable-Propelled APM and Low-speed Maglev.

TUNNELING METHODS ASSESSMENT

The most direct access between the DART Love Field LRT Station and the Love Field Terminal Building will be via a tunnel underneath the runways, taxiway and apron around the terminal building. An at-grade solution was considered at the onset of the feasibility study. The route that was considered paralleled runway 13R/31L along Denton Drive towards Mockingbird Lane and then proceeded into Love Field on Cedar Springs Road. This alignment was not further developed as it was determined that this alignment would infringe on the airfield clearances. Based on this fact and an initial assessment of the other site conditions (streets, existing buildings, parking facilities, future development and utilities) as well as discussions with the City of Dallas, a tunnel configuration was determined.

Two alignment corridors were initially selected. For each corridor a deep and a shallow tunneling option, as well as a combination with an aerial alignment, were investigated and evaluated. The recommended alternative for the DAL APM will have a tunnel that gradually slopes to a maximum depth of approximately sixty-five (65) feet (20 m) under the runway.

Tunnel Construction Methodologies that may be implemented include Shielded Pressure-Face Tunnel Boring Machines; the New Austrian Tunneling Method (NATM) or Sequential Excavation Method (SEM); Cut and Cover Construction; and the Doorframe Slab Method. Station Construction Methodologies identified include aerial construction (DART side only), Cut-and-cover construction, NATM tunneling and Doorframe Slab Method.

SYSTEM ALTERNATIVES

The connection of Dallas Love Field to the DART Love Field LRT station will offer patrons of the airport an alternative means to access the terminal building and its ancillary facilities. The level of service for this new transportation connection needs to compare favorably to the level of service experienced by the air traveling public at airports throughout the world.

The design of the system must not only take into consideration the means by which to connect these two locations but also must appear seamless to the passenger utilizing the connection. The APM Station adjacent to the DART Love Field Station is a "new entrance" to Love Field and should therefore provide an experience to the passenger similar to that of entering the main terminal building at Love Field. The patrons of the APM should not only have a sense of arrival at Love Field but should feel secure in their surroundings in the people mover stations and on the system itself.

Two options for the DART Love Field APM Station and two options for the Love Field Terminal Building APM Station were determined to be most feasible. These four station location options are shown in Exhibits 2 – 5 below:

DART Love Field APM Station location options:

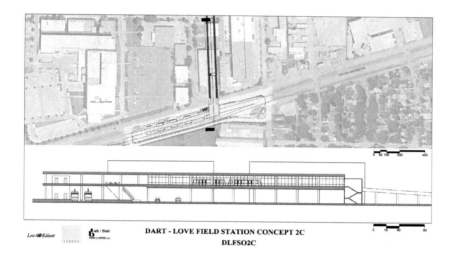

Exhibit 2 - Aerial APM Station at DART

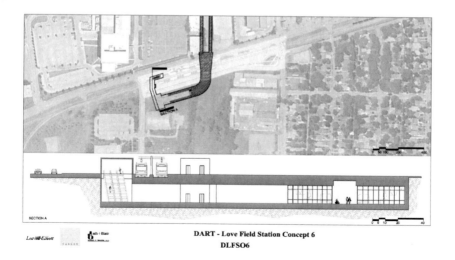

Exhibit 3 - Underground APM Station at DART

Love Field Terminal Building APM Station location options:

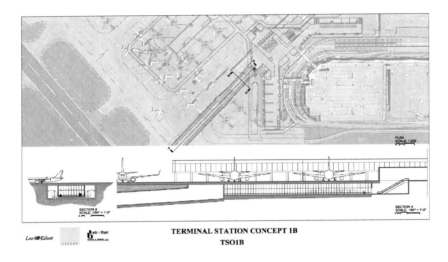

Exhibit 4 - Underground Airside APM Station at Terminal

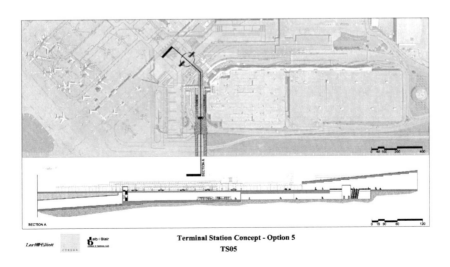

Exhibit 5 - Underground Landside APM Station at Parking Garage

RECOMMENDED ALTERNATIVE

These short-listed alternatives were quantitatively and qualitatively evaluated to compare their strengths and weaknesses. Each of the short-listed alternatives were analyzed using Lea+Elliott's LEGENDS© family of analytical tools. Computer simulations of the alternatives were performed to precisely describe the performance of each study system. Output of these system analyses provided a complete description of each alternative.

After completion of the Alternative Evaluation, the recommended alternative was the Aerial APM Station at DART to Underground Airside APM Station at Terminal. This alternative is shown in Exhibit 6.

This recommended alternative should be considered a feasibility level concept and not a final design concept. During the next phase of design of the DAL APM, the concept will be refined to optimize the interface between the DAL APM and the modernized terminal. Coordination with the terminal designers to create the best possible passenger orientation and experience is paramount to the ultimate success of the DAL APM and the modernized terminal.

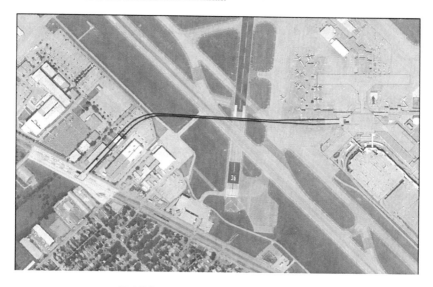

Exhibit 6 - Recommended Alternative

Refinement of Recommended Alternative

Subsequent to the selection of the recommended alternative, a more detailed analysis took place to determine if an alternative operating mode could be developed that would result in a reduction in the order of magnitude cost for the DAL APM program. The alignment for these alternatives would remain the same however the operating condition of the APM would vary from the recommended dual-lane shuttle. This analysis focused on the highest cost construction components of the project (the tunnel and associated guideway and the APM System).

The initial ridership forecasts that have been prepared to date indicate that a single operating APM in a shuttle has the required capacity to convey the passengers between the DART Love Field LRT Station and the Love Field terminal building. The dual lane shuttle was recommended to provide a level of redundancy in the APM that a single operating APM shuttle does not provide. The travel time between these two locations is estimated to be less than three (3) minutes. The recommended alternative was the shortest alignment of those that were studied in the feasibility study at total length of 3,400 LF (1,036 m). The Maintenance Facility is included in the passenger station adjacent to the DART LRT Station.

The first option to the recommended alternative that was studied was what savings could be realized if both tunnels were constructed and only one lane was implemented at the onset. The cost savings for this alternative would be limited to the APM System only. The tunnel and guideway fixed facilities cost would remain the same. The initial capacity would be reduced and the redundancy does not exist with this option. Exhibit 7 shows a schematic plan and section of this operating scenario.

Single Lane Shuttle - Build Both Tunnels   *Tunnel Cross Section*

Exhibit 7 – Option One to Recommended Alternative

The second option that was studied involved the construction of a single tunnel with a bypass area in the center of the tunnel where two trains would pass one another. This alternative provides an opportunity for cost savings in the tunnel and guideway, station and APM System. A similar capacity to the recommended alternative can be achieved with this option. A level of redundancy is also provided with this option. Exhibit 8 shows a schematic plan and section of this operating scenario.

Single Lane Bypassing Shuttle in Single Tunnel   *Tunnel Cross Section*

Exhibit 8 – Option Two to Recommended Alternative

The third option that was considered as part of this additional analysis was the construction of only a single tunnel and the installation of a single APM operating in a shuttle. Capacity and redundancy issues are the same for this option as the initial phase of the first option that was studied. Elements of work which would potentially realize cost savings include the tunnel and guideway, station and APM System. Exhibit 9 shows a schematic plan and section of this operating scenario.

Single Lane Shuttle in Single Tunnel   *Tunnel Cross Section*

Exhibit 9 – Option Three to Recommended Alternative

These three options were evaluated based on the level of service issues that were used to determine the original recommended alternative. At a minimum, the bypassing APM shuttle (Option Two) in lieu of a dual lane shuttle was recommended as a revised recommended alternative. This option provides a similar level of capacity and redundancy to the original recommended alternative as well as potential cost savings opportunities to the fixed facilities and APM Operating System. The added capacity provides flexibility to the City of Dallas to serve other airport functions (airport parking, consolidated rental car, etc) should the need arise without an immediate expansion the System. The procurement process for the APM System may allow an alternative to the bypassing shuttle.

This option reduced the amount of tunneling from 3,400 LF (1,036 m) to 1,900 LF (579 m). As a result of this reduction, the amount of APM System equipment was also reduced. Each station platform can be reduced since only one APM will be berthed at any one time resulting in a reduced platform width. Exhibit 10 shows an aerial view of the revised recommended alternative.

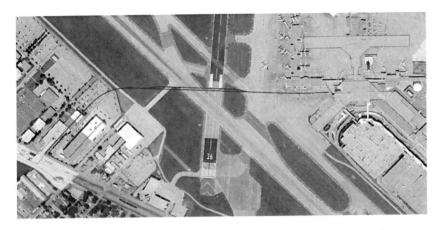

Exhibit 10 – Single Lane Bypass Revised Recommended Alternative

PROJECT SCHEDULE

A capital development project such as the DAL APM will take place over a number of years and include design, construction, implementation and commissioning phases. It is estimated that the overall project duration of the DAL APM is seventy-two (72) months. The phases of the project are listed below.

- Schematic Design
- Final Design
- APM System Procurement
- Facilities Construction Procurement
- Civil/Sitework Construction
- APM System Manufacturing
- Tunnel Construction
- Station Construction
- APM System Installation
- APM System Testing And Demonstration

FUNDING SOURCES AND OPTIONS

For a project of this nature, there are several funding alternatives available to airports. Since the primary mission of the project is the safe and efficient handling of passengers into and out of the airport, the project becomes eligible for funding under the Airport Improvement Program (AIP) and the Passenger Facility Charge (PFC) Program. This eligibility would extend to all elements of the program with the exception of operation and maintenance costs. Other known sources of funding include a $20 million commitment from DART and a portion of a $100 million commitment for rail access into airports in the North Texas Region from the Regional Transportation Commission.

PROJECT FEASIBILITY

The DAL APM project is currently developing the schematic design (SD) package of the revised recommended alternative shown in Exhibits 8 and 10. The SD package will be completed by November 2009. Based on the findings of the SD phase, the Dallas City Council will determine whether or not the project will proceed through procurement and implementation phases. It is anticipated that if the project proceeds to these steps that an APM System Procurement will take place in 2010 to maintain the overall scheduled completion by the end of 2014. The DAL APM will provide a high level of service to air travelers and employees connecting between Dallas Love Field and the DART Love Field Station and will provide access to Dallas and the region via the regional rail network. The APM will create a new entrance to Love Field that will reflect a high quality of service consistent with the new image of Dallas Love Field.

THE SACRAMENTO INTERNATIONAL AIRPORT APM

Jenny Baumgartner, P.E., AICP* and Harley L. Moore**

* Senior Engineer, Lea+Elliott, Inc., 785 Market Street, 12th Floor, San Francisco, CA 94103, (415) 908-6450; (415) 908-6451 (fax); jbaumgartner@leaelliott.com
** Senior Principal, Lea+Elliott, Inc., 785 Market Street, 12th Floor, San Francisco, CA 94103, (415) 908-6450; (415) 908-6451 (fax); hlmoore@leaelliott.com

Abstract

The Sacramento International Airport (SMF) APM is a dual lane shuttle that will connect the new landside Central Terminal B and airside Concourse buildings. Planning started in August 2003 and preliminary design of the system and the buildings in October 2006. The procurement process started with an RFI (interest and information). An RFQ sent to six potential suppliers resulted in a short list of four firms and technologies. The APM contractor was selected in June 2008. The terminal and APM are scheduled to open in early 2012.

This paper describes the system and discusses why an APM was essential to the terminal concept and construction process. It also provides insight into the procurement process and results, including the effect of local requirements for terms and conditions and typical construction processes. It looks at the implications of a very constrained project site and the resulting challenging system configuration. It discusses system expansion, both for the capacity of the initial dual lane shuttle and the possible addition of two similar systems in the future. Finally, it uses this project to discuss some of the challenges foreign companies have in competing in the US APM market and providing innovative APM technologies.

Background

The Sacramento County Airport System (SCAS) is planning and implementing an airport APM system at the Sacramento International Airport (SMF) as part of the Terminal Modernization Program (TMP). SMF currently has two terminals: Terminal A, completed in 1998, and Terminal B, opened in 1967. An interim international arrivals building, located between Terminals A and B, was completed in 2002. The existing passenger terminal facilities have a capacity of about 12 million annual passengers, compared to 10.3 million served in 2006. Projected increases in the number of passengers will cause the capacity of the existing terminals to be reached in 2013.

SCAS has embarked on a four-phase airport development process to identify and implement the vision developed in the Final Sacramento International Airport Master Plan to modify existing airport infrastructure and develop new facilities through the year 2020. The Master Plan was approved by the Sacramento County Board of Supervisors in February 2004 and is Phase I of the Airport Development process.

Phase II, completed concurrently with the Master Plan, was called the Terminal Modernization Program. The TMP included preliminary facilities requirements, terminal complex alternatives, and evaluation of four alternatives, two of which included an APM system. The Board approved the selection of the preferred terminal development concept which was ranked highest with respect to long term strategic, operational, environmental, feasibility / constructability, and customer service. A key decision factor was that this allowed the existing Terminal B to continue in operation while its replacement was constructed.

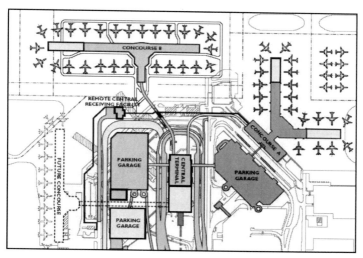

Figure 1: Master Plan Concept E2 with APMs

Phase III was completed in two steps. Step 1 was the preliminary design and design development of Terminal Concept E2 to support the EIR/EIS and also involved discussions with the Airport Airline Technical Committee. The EIR/EIS was completed and approved in the summer of 2007. Step 2 then resulted in the preparation of design and construction documentation and the procurement of the major contractors: the design-build contractor for the landside Terminal, the design-build contractor for the airside Concourse and APM guideway, and the APM system contractor.

Phase IV is the construction of Terminal B. The estimated opening date is early 2012.

Terminal and APM Design

One initial and two potential future APM systems will transport passengers between the landside terminal and the gates. The initial Terminal-Concourse B APM system is considered a "landside" or non-secure system, as passengers will be processed through TSA security screening just after they exit the APM at the Concourse. A future APM would operate between the new landside Central Terminal B and the existing Terminal A, which would then become Concourse A. A very long term future airside Concourse C might also have an APM connecting it to the landside Terminal B. All would be dual lane shuttles with non-secure trains. Document check points at the landside Terminal B stations would limit APM riders to ticketed passengers and airport employees.

The initial Terminal B APM is a two-station dual lane shuttle, as shown in Figure 2. The Terminal station will be located on level three; the Concourse station will be located on level two. The guideway will extend from the north end of the Terminal, across airport roadways and apron and terminate at the center of the south side of the Concourse. This alignment has a significant grade change and "S" curves. The maintenance facility will be under the airside Concourse station. Additional information is given in Table 1.

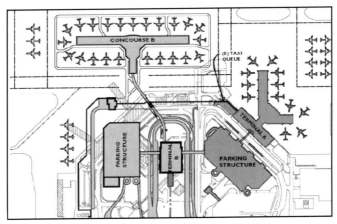

Figure 2: Terminal B / Concourse B APM

Table 1: APM System Characteristics	
Length	1100 ft.
Min. Horizontal Curve Radius	270 ft.
Min. Vertical Curve Radius	3300 ft.
Maximum Grade	5%
Number of Flow-through Stations	2
Number of Trains	2
System Capacity: Initial	2300 pphpd
Ultimate	3000 pphpd
Operational Mode: Peak Period	Dual Lane Shuttle
Off-Peak	Single Lane Shuttle
Night	On-call Single Lane Shuttle
Average Round Trip Time / Lane	~ 3.0 minutes
Cruise Speed	~ 20 mph
Operating Hours	24 / 7 / 365

Procurement Process

SCAS opted to use a typical Design, Build, Operate, Maintain (DBOM) approach for the APM procurement. Because it is only for the system equipment, and to make sure there was no confusion with the provisions of the California Design-Build law that governs facility construction, this approach was defined as: Design, Supply, Install, Operate, and Maintain (DSIOM). The procurement process included a Request for Interest and Information (RFI), Request for Qualifications (RFQ), Request for Proposals (RFP), and Best and Final Offers (BAFO).

In March 2007, six APM suppliers responded to an RFI that was released in February. Information provided in these responses was used to forward the preliminary, generic design of the APM system facilities so as not to preclude any of these APM technologies.

In September 2007, the Sacramento County Board of Supervisors (Board) approved the DSIOM procurement process for the APM system and the release of the RFQ. A "best value" approach was selected by the County as it offered the flexibility to consider all aspects of the APM suppliers and systems, not just price. The RFQ was advertised publicly and provided to all suppliers that requested it. Representatives of seven APM suppliers attended a pre-submittal meeting October 2007. Six Statements of Qualifications (SOQ) were received; these were (in alphabetical order):

1. Bombardier Transportation (Holdings) USA, Inc., from Pittsburgh, PA, with self-propelled APMs.
2. DCC Doppelmayr Cable Car GmbH & Co., of Austria with a cable-propelled APM.
3. IHI California / IHI Inc. / IHI Corporation, a Japanese supplier with a self-propelled APM.
4. Leitner-Poma Mini Metro, a French-Italian company with a cable-propelled APM.
5. Sumitomo Corporation of America / Mitsubishi Heavy Industries, a Japanese supplier with a self-propelled APM.
6. Schwager-Davis Inc., from San Jose, CA, with a self-propelled APM.

An initial evaluation of the six SOQs was conducted by the Technical Review Team, which rated the SOQs and listed key advantages and disadvantages of each. Based on this information and its own evaluation, the Selection Committee recommended the following pre-qualified APM suppliers be invited to respond to the RFP:

1. Bombardier Transportation (Holdings) USA, Inc.
2. DCC Doppelmayr Cable Car GmbH & Co.
3. IHI California / IHI Inc. / IHI Corporation
4. Sumitomo Corporation of America / Mitsubishi Heavy Industries

The Board approved this recommendation and SCAS released the RFP to the four firms in January 2008. Proposals were submitted in April 2008. The initial evaluation resulted in requesting a BAFO from each of the four firms.

Under the best value approach, proposals and BAFOs were evaluated with respect to system technical aspects; the suppliers' experience and qualifications, and project management strategy; and price. Although the technical / management scores were similar, the proposal prices had a differential of $13 to $24 million between that of the domestic supplier and those of the foreign suppliers. The BAFOs had a differential of $5 to $20 million. After an extensive evaluation of the proposals and BAFOs, the Selection Committee recommended that the Board authorize the Design, Supply, and Install Contract and the separate, but interrelated, Operate & Maintain Agreement with Bombardier for a CX-100 system. The Board authorized these agreements on June 18, 2008. Their scope included detailed design, supply, installation, and acceptance testing of the APM, then operation and maintenance services for an initial period of five years. SCAS retained the option to renew the O&M period for an additional five years.

APM Preliminary Design

As the terminal building facility preliminary design process occurred before selecting the APM contractor, that design provided for sufficient and appropriate space, and guideway structure, to suit the variety of APM technologies and their operations and maintenance needs.

The nominal space and structural requirements for both the self-propelled and cable-propelled systems, and airport function- and cost-based space constraints in both the Terminal and Concourse buildings required the location of the APM facilities (central control, maintenance facility, PDS substation, etc) be split between the two. Preliminary design efforts resulted in the maintenance facility being under the Concourse station, while central control, offices, and the PDS substation were in the Terminal building, as shown in Figure 3. Although the preferred space for the APM facilities in the Terminal was on level 1, the cable-propelled technology cable return equipment required space directly adjacent to and under the station, which is located on level 3 in the Terminal building. To meet these requirements, two scenarios were provided in the RFP reference drawings for the spaces in the Terminal (Figure 3): one for the self-propelled technology (APM facilities located on level 1) and another for the cable-propelled technologies (APM facilities located on level 2 directly under the station).

Due to the passenger demand as well as the length and the vertical and horizontal differential between the stations in the Terminal and Concourse, a dual lane shuttle configuration was the optimal choice for the system. The challenge was to develop a preliminary design that would not preclude any technologies as well as developing a vertical and horizontal geometry that would meet ride quality needs and minimize the impacts to an existing building and future airport facilities.

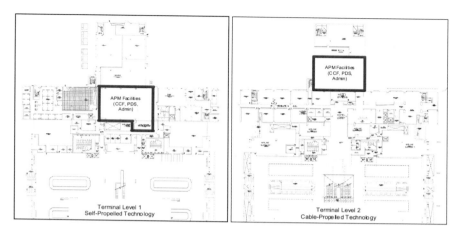

Figure 3: Preliminary Design - APM Spaces

The APM guideway goes over the existing interim International Arrivals Building (IIAB), which must remain operational during the construction of the new buildings and the guideway. The guideway construction method initially was segmental to address the issues of the span over the IIAB. As this option was quite costly, two other methods were reviewed: pre-cast and cast-in-place. The alignment length and geometry did not allow for economies of scale with the pre-cast method. Cast-in-place was challenging due to the formwork needed and vertical clearance available between the guideway soffit and the top of IIAB. The savings in construction time and access of a pre-cast guideway was reviewed to determine if it would offset the cost of material, making it as economical as or more economical than a cast-in-place solution. The review resulted in cast-in-place being the preferred method.

The guideway is also above the future elevated roadway system adjacent to the Terminal and the roadway and the entrance to the tug tunnel in the vicinity of the Concourse. These brought vertical clearance requirements for the guideway design. To accommodate these clearances, the guideway structure was designed in three sections (bridges). The end bridges, 1 and 3, included the spans over the roadways and tug tunnel entrance. These were designed as cast-in-place slabs and are shallower than bridge 2, which was designed as cast-in-place box girder. Figure 4 shows the transition from bridge 1 to bridge 2.

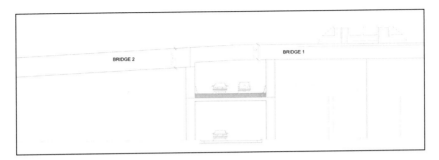

Figure 4: Section - APM Guideway Spanning Elevated Roadways

The alignment geometry was a challenge due to the vertical and horizontal differences between the Terminal and Concourse stations. The alignment has a grade change of 30 feet between level 3 of the Terminal and level 2 of the Concourse. The horizontal shift between the stations is approximately 350 feet. The alignment needed to accommodate those distances in under 900 linear feet. The resulting alignment consists of overlapping horizontal curves (radii ranging from 270 to 365 feet) and vertical curves (radii of 3300 feet) and fairly short tangents for transitions, shown in Figure 5.

APM Final Design

Once the APM Contractor was selected in June 2008, the final design phase began. Facilities could then be detailed to the specific requirements of the CX-100 system. Bombardier's requirements were coordinated with, and incorporated into, the overall Airport facility design at monthly design workshops. This coordination resulted in an adjustment to the APM spaces given in the RFP reference drawings. The central control facility and associated equipment room, which were originally located on level 1 of the Terminal, were relocated to the Concourse building within the APM maintenance facility (Figure 6). This resulted in a more efficient use of space for Bombardier (particularly for a system with a small, cross-trained O&M staff) and opened up space in the Terminal for other Airport needs (Figure 7).

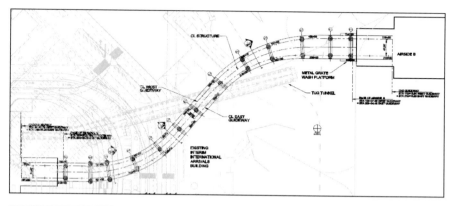

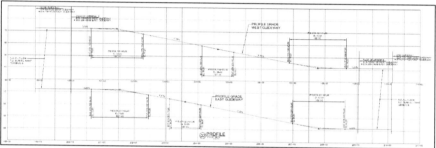

Figure 5: Alignment Plan and Profile

The guideway design was also refined to better fit the CX-100. Although the geometry did not change, the guideway was made narrower, providing a cost savings.

Bombardier and Lea+Elliott worked with the designers, Corgan Architects, Inc. and Fentress Architects, to finalize the facility construction drawings during late 2008 and early 2009. At the same time, Bombardier also finalized the operating system design, through numerous design submittals and design review meetings.

Schedule

Bombardier's implementation schedule is being coordinated with, and incorporated into, the overall Terminal Modernization Program schedule. Schedule items of particular importance are the APM facilities access dates. Once given beneficial occupancy, Bombardier will finish out and install equipment in/on the guideway, maintenance facility, central control, PDS substation, and equipment rooms.

The program remains on schedule with substantial completion of the APM system in November of 2011 and the new Central Terminal B opening in early 2012.

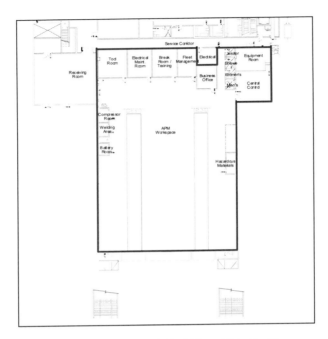

Figure 6: Final Design –APM Facilities within the Concourse

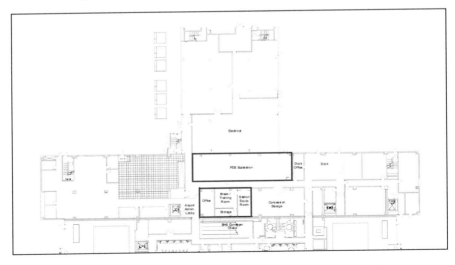

Figure 7: Final Design –APM Facilities within the Terminal

Lessons Learned

Although as of this writing the project is still in the detailed design phase, there are useful lessons to be learned from issues that were encountered in the procurement and early design processes. The following are the opinions of the authors and not necessarily those of SCAS, the facility designers, or the APM contractor.

1. The sooner the APM contractor is selected, the sooner the APM facility designs can be finalized to meet the needs of the system and the client, potentially resulting in significant design, schedule, and construction cost savings.
2. Local building construction contract terms and conditions, when superimposed on an APM design, supply, and install contract, can add significant costs and complications. When developing APM system contract terms and conditions, the implications of local "standard requirements" need to be considered. Changing such standard terms and conditions is not easily done; local jurisdictions can be reluctant to understand and accept variations to what they are used to.
3. Good competition, with a flexible yet defensible procurement approach, can lead to a cost-effective set of proposals and system design. Confidentiality in the process is imperative to obtaining good prices.
4. A clear, documented qualifications and proposal evaluation process can effectively mitigate any post-proposal challenges. Being able to explain in a clear and concise manner, using the evaluation criteria and process, to the suppliers that were not selected assisted greatly in avoiding formal challenges to the outcome in this procurement.
5. When the proposals were submitted, the US dollar had a relatively low value compared to the Euro and Yen. This has a significant effect on the prices of the non-US suppliers. Adjusting for the exchange rate would have made all four price proposals very close. Although a best value approach was used, in this case it turned out that price was the deciding factor, as the technical and management proposals were rated relatively equal. The exchange rate, and possible need for currency hedging, can greatly affect the potential for foreign suppliers to compete in the US market. As the exchange rates have changed significantly since the proposals were submitted, the situation at the time of the procurement can have a telling effect on prices and supplier selection.
6. One supplier proposed a "green" system that was highly regarded by the Owner. This new technology was priced significantly higher than the winning "off-the-shelf" technology. The "technology differential" could not overcome the price differential in the evaluation process. This has implications for the introduction of new technologies in the competitive APM market.
7. Having a set of APM products that can be applied to relatively simple shuttle systems and more complex systems can offer a significant cost advantage. Several suppliers had complex train control, power distribution, and/or maintenance subsystems and approaches that could not be cost-effectively simplified, thus were less cost-competitive.

Sources: Drawings are courtesy of Corgan Architects, Inc. and Fentress Architects.

DFW Skylink: Tracking Success

David Taliaferro

Skylink Manager, Dallas/Fort Worth International Airport, Energy & Transportation Management Department, P.O. Box 619428, D/FW Airport, TX USA 75261; PH 972-574-8009; dtaliaferro@dfwairport.com

Abstract:

The Dallas/Fort Worth International Airport (DFW a.k.a. the owner) Skylink Automated People Mover, (APM), launched passenger service on May 21, 2005, replacing Airtrans, the LTV Aerospace Corporation APM system in service since 1973. Skylink transports secure-side passengers to any of five separate airline terminal facilities.

As the system develops, valuable experience is being gained that has led to enhanced passenger service, improved operational efficiency, and preferred maintenance business practices in various ways. After the design, construction, testing and commissioning phases reached completion, the continual operation and maintenance phase vigorously began. The lessons gleaned from the areas of quantifiable passenger ridership, passenger wayfinding, training of operations staff, operating procedures, implementation of the maintenance plans, data collection, contract oversight and administration have paved the way with new opportunities to gain valuable insights for projects yet to come.

The Skylink team gained experience and was presented with opportunities to assess data to best adapt to the realities of specific application of knowledge in dynamic and flexible ways in order to garner the most advantageous impact for passengers and all stakeholders. Challenges, success, and the lessons derived from them provide the pathway for greater achievement short-term and on future projects.

Introduction: Innovative Partnership

The Skylink Automated People Mover entered passenger service on May 21, 2005, after completing construction "on-time and on-budget." [1] It replaced the aging Airtrans APM (supplied by LTV Aerospace Corporation), which had been in continuous service since 1973.

The launching of the Skylink system introduced an entirely new elevated guideway structure approximately 15.24 meters (50 feet) high at all points in the revenue system. The 64-vehicle Innovia fleet is a bi-directional, dual-loop circulator system with each loop measuring approximately 8 kilometers (5 miles) in length. One loop moves along in a clockwise direction while the other loop runs in a counter-clockwise path.

Passenger service is provided to 10 stations at any of five separate airline terminal concourse facilities. Skylink was built to transport 5,000 passengers per direction, per hour with expansion capacity to 8,500. The gate-to-gate connection speed increased from 27.4 kilometer/hour (17 mph) on Airtrans to 59.55 kilometer/hour (37 mph) on Skylink in two-minute headways with an average passenger ride time of 5 to 8 minutes.

Skylink's O&M is an innovative hybrid partnership between DFW and Bombardier Total Transit Systems, (BTTS a.k.a. the supplier). Uniquely, Skylink operation is the sole function of the owner, while the supplier delivers the maintenance service. Train recovery staff is supplier sub-contracted.

Quantifiable Passenger Ridership

To enable the most efficient use of owner asset and passenger service delivery, quantifiable passenger ridership data, (a way to measure passenger service level demand), are used to adjust operating schedules, the number of trains in service, headway, and vehicle frequency to provide the capacity to meet demand during peak and off-peak periods. A real-time count provides more accurate data than projected input can. It is used to accurately analyze the impact on service.

DFW planned Skylink during a time when airline industry growth was expected to increase from 650 million enplanements to almost 1 billion enplanements and international air travel was projected to reach almost 250 million enplanements by 2010.[2] After the events of September 11, 2001, the growth projections were revised. Without a precise counting application with which to measure and therefore alter operations to fit actual demand, Skylink often was operating at a level higher than actual demand required. A schedule of routine manual counts during designated times was undertaken using a sizable sampling of passengers. The data was examined and an adjustment to the number of trains in service during peak times was reduced. Additionally, the start times for peak service each morning were adjusted to thirty minutes later.

The owner learned that it is necessary to have a process in place to quantify actual passenger service demand to be able to be dynamic and more responsive to changes in service-level demands on the system. Therefore, real-time measurements and trends in the service delivery can be monitored and service adjusted more expediently.

Passenger Wayfinding

The modern Skylink stations give the first impression to Skylink passengers in a grand way. The floors of each terminal's stations are decorated in unique artistic designs by local artists and fabricated in terrazzo. The stations feature glass and steel architecture measuring 146.3 meters (480 feet) long and soaring ceilings of 23.16 meters (76 feet) high. The spacious, well-lit stations with signage plus announcements provide passengers a comfortable, secure atmosphere. However, beyond the stellar appearance, passenger satisfaction surveys revealed the need to enhance passenger service with a few improvements to station electronic display signage to add to the ease of use by passengers from gate-to-gate.

During Skylink's testing and commissioning phase, DFW planners used airport employees and airline volunteers to perform a passenger wayfinding exercise to evaluate the effectiveness of signage and directions. The personnel who were familiar with the airport were able to find their way easily from terminal to terminal.

Later with actual passengers, the DFW marketing department performed two passenger satisfaction surveys during the first six months of service. Some comments from both surveys involved signage and directions in Skylink stations. The surveys revealed that some passengers were unsure about the service options available. When passengers enter the station, they view two sets of boarding platforms on the right or left side of the center platform station. With the dual-loop circulator system configuration, either side serves all stations, but the option for the most direct path to reach their destination may not have been immediately apparent to all passengers.

Skylink utilizes two types of electronic graphic panels to display information to the passenger at each loading platform door area. One type is flush mounted above each door-set and the other is extruded from the wall displaying messages on both sides. Combined, they provide a dynamic information stream to the passenger approaching the boarding area from all directions.

DFW hired a consultant for recommendations for improvements to the messages displayed on the dynamic graphic signs in the stations. As a result, the message script of the electronic graphic signs received modifications to provide intuitive and more immediate destination information to passengers upon entering the stations. Additional static signs, maps, and information were placed in the stations at strategic and centrally-located terminal concourse areas.

The owner learned that intuitive graphics and additional signage strategically placed can benefit passenger wayfinding. Also, it appears that surveying people who were unfamiliar with the layout of the airport may have produced wayfinding results more in-line with actual first-time passengers as well as those who are returning travelers.

Training Operations Staff

DFW Airport has a 35-year foundation of training its own APM operators. The initial BTTS training on the new Skylink system successfully enabled the operators to perform daily tasks. Together, as a new hybrid O&M partnership, the initial basic BTTS technical training plus the long-tenured foundation and experience of the Airtrans/Skylink operations staff resulted in a seamless transition from the former system to the new. System experience has reinforced the advantages of the unique

O&M partnership.

With the symbiotic partnership, the owner assumed the responsibility for ongoing training and testing programs, while BTTS continues to provide technical input. A major advantage of the owner's role as trainer is the inclusion of the DFW philosophy, which focuses on providing exceptional customer service. With that in mind, the owner began monthly formal "Lessons Learned" sessions to share information and analysis of system-delay events. The central operations staff and BTTS technical staff meet to discuss the details and the significance of recent events, insights of the individual team members involved, and the lessons learned from all input. The discussions provide insights and exposure to the reactions to system events and how to apply those lessons to future scenarios.

Formerly, testing of DFW central operations staff members had been routinely performed by the application of paper tests that lacked sufficient comprehensive feedback to the test candidate and to the management team. The owner purchased a testing software package for annual recertification tests. The software tracks student progress, provides immediate printed feedback to the student, statistical information to the instructor of the candidate's progress, and records the results and characteristics of the test.

Productive results can be realized by taking advantage of previous years of owner-trained staff experience plus ongoing monthly event-sharing sessions and flexible electronic testing software to provide student progress and historical data. The number of downtime events related to repeated or similar scenarios is reduced.

Operating Procedures

The BTTS procedures were a combination of O&M procedures that were recovery-specific. However, the owner realized the need to have a unique set of protocols with a strict focus on operations for the central control operations staff to handle circumstances that occur from multiple-scenario events such as weather issues, medical emergency, safety or security alerts, routine and unscheduled maintenance.

Skylink schedules trains at two-minute intervals during peak hours to each airport terminal station. Consistency of APM service at the posted times is critical to efficiency. If a disruption occurs, contingency procedures provide a comprehensive set of instructions to accommodate passengers and to return the system to normal operations quickly. Many prior procedures would not adapt easily to the Skylink system except in concept. Some experience that was based on the previous Airtrans system's 35-year history needed to be adapted to Skylink. Different technology and system behavior dictated the creation of operating procedures specific to the unique characteristics of Skylink's design and configuration, while using experience gained from the previous APM system.

Pre-existing procedures and protocols, event history, and backup contingency plans should be considered as parts of the overall failure management process, especially if a rare event causes prolonged system downtime. The result has been successful avoidance of lengthy delays for occasional recurring events, which benefits the owner, supplier, and especially the Skylink passenger.

Implementation of the Maintenance Plans

In the BTTS and DFW partnership, the supplier is responsible for the maintenance with direct daily engagement and oversight by the owner. Maintenance plans usually precede project launch.

To discuss ongoing issues, the owner holds daily meetings with the supplier to encourage and support preferred maintenance business practices to ensure that the Skylink system is maintained at the level to sustain the owner's asset value. Joint review using data collection processes to monitor periodic and corrective maintenance with daily O&M feedback from all areas keeps the partners informed of system status.

The implementation of solid maintenance plans will provide requirements and targets to maintain DFW's asset value in the system. Site-specific system maintenance requirements should be specified by the owner and written into the plans at the onset of the project with active owner oversight.

Data Collection

A properly specified, designed, and installed APM system should be accompanied by a data collection system capable of tracking and reporting on each area requiring oversight by the owner. The original Skylink system specifications called for a maintenance management information system to interface with the owner's existing database application. BTTS provided a package called Site Information Management System, (SIMS).

DFW and BTTS Information Technology departments struggled to enable integration between the two independent computer networks on which they existed. The hurdles were not overcome and the two systems remained isolated from each other by electronic security firewalls. Skylink management received four PCs with an installed SIMS package. The SIMS terminals are located in the owner's administrative offices and can be used to access all SIMS portions related directly to the Skylink site.

Research into data collection systems should be explored prior to implementation to ensure integration to a level satisfactory to both the owner's and supplier's requirements to provide full functionality in the reporting process, preferably on the owner's existing system. The ideal data collection system will present the chosen data in dynamic formats that enable analysis easily. The system should be onsite and fully operational at the beginning of system testing and commissioning.

Maintenance Contract Oversight and Administration

Skylink maintenance oversight administration involves tracking periodic and corrective maintenance documentation to ensure that the specifications of the maintenance plan are being followed. Having an experienced, knowledgeable, and motivated staff to follow-up on the maintenance plan keeps the focus on the goals and targets of the owner.

The DFW Skylink Management team ensured that experienced staff was ready prior to system testing and commissioning. When knowledgeable personnel are in place, oversight can begin immediately.

With the APM-experienced personnel (who could anticipate requirements and issues) providing oversight in the early stages of the project, the owner could monitor safety-related issues or premature parts failures, while maintaining owner's asset value.

Conclusion

Tracking success through the lessons learned during the Skylink team's experience has strengthened the innovative hybrid O&M partnership in the areas of quantifiable passenger ridership, passenger wayfinding, training of operations staff, operating procedures, implementation of the maintenance plans, data collection, contract oversight and administration. Experience (DFW) plus technology (BTTS) form dual loops of opportunities to assess, adapt, and apply principles for greater future impact and achievement in the APM industry enhancing passenger service and improving operational efficiency and business practices.

---

[1] Dallas/Fort Worth International Airport Press Release, May 23, 2005
http://www.dfwairport.com/mediasite/pdf/05/05/050521-skylink-opens.pdf

[2] Dallas/Fort Worth International Airport 1997 Updated Airport Development Plan D/FW Planning Department

Operations and Maintenance at Atlanta Airport

Melvin Redd SDC Director
Russell Woodley O&M Manager
Jerome Page Technical Compliance Manager
Bombardier Transportation
Systems Division
Hartsfield-Jackson Atlanta International Airport

Abstract

The Bombardier-supplied Automated People Mover (APM) at Hartsfield-Jackson Atlanta International Airport is one of the busiest and most complex systems of its kind in the world. Bombardier Transportation is operating and maintaining the system that operates underground with 20 hours of pinched loop service daily over 4.4 miles with an additional 30-minute 2-train shuttle at the end of normal loop service nightly.

This paper provides details of how the Atlanta APM system is operated and maintained and how it has grown over the past 28 years to provide an essential transportation service around one of the world's busiest airports.

Since opening to service in 1980, the customer has increased the number of APM vehicles to meet rising passenger demand, as well as the replacement of the original fleet of Bombardier-provided C-100 vehicles. This paper details the growth of the APM fleet, the number of stations and concourses that the APM services, as well as full loop service travel time.

To maintain the system, approximately 90 employees perform various duties to maintain system performance. This paper describes the maintenance facility and the day-to-day maintenance and inspections of the vehicles, track and guideway and the engineering group's technical guidance to ensure compliance to established specifications. Also described are the major subsystems and upgrades of the APM system.

In 2005, Bombardier received a 10-year contract with two 5-year options to operate and maintain the *BOMBARDIER* CX-100** APM system. The paper shows the benefits of a long-term operations and maintenance contract with the City of Atlanta customer.

Hartsfield Atlanta International Airport Automated People Mover System

Introduction

Hartsfield-Jackson Atlanta International Airport people mover system is one of the busiest and most complex systems of its kind in the world. It is operated and maintained by Bombardier Transportation. The system operates 20 hours of underground pinched loop service daily on a 2.5-mile full loop system. The system provides an additional 30- minute 2-train shuttle at the end of normal loop service nightly. To accommodate late flights, the shuttle service is extended upon the request of the airport.

System Operation

The system operates seven days a week and moves an average of 255,000 passengers daily between the landside terminal and five airside concourses. In 2007, the APM system moved 89,409,237 million passengers through the airport. The peak hour capacity is approximately 10,000 passengers per hour and more than 8 million passengers monthly.

Bombardier and the customer continuously modify the APM schedule to accommodate the volume of passenger traffic through the airport. This modification includes providing additional train service to match the peak periods of airport travel.

Atlanta APM History

The people mover system began operation in September 1980 with 17 C-100 vehicles and expanded to 24 C-100 vehicles by 1993. In 1994, the system was expanded to service the new international terminal at concourse E. This expansion included 25 additional *CX-100* vehicles for a total fleet of 49 vehicles in service today.

In 2001, the original 24 C-100 vehicles placed in service in 1980 had accumulated more than 1,000,000 miles per vehicle and were replaced with 24 new *CX-100* vehicles.

CX-100 Vehicle

CX-100 Vehicle

The CX100 vehicle is 39 feet in length and weighs 26,000 lbs. It has a 100 standing passenger capacity. The vehicle is equipped with two 100 HP series wound DC traction motors for propulsion and a pneumatic braking system. The vehicle maximum speed is 25 miles per hour. The vehicle is supported by eight rubber tires mounted on two drive axle assemblies. There are four door openings per vehicle, four double seats and seven stanchion poles.

Passenger information is provided by static and dynamic graphic displays and digital voice announcement units.

System Performance

The system is currently operating above 99.8% availability which equates to high passenger satisfaction ratings. System availability is determined by the system reliability and fleet availability. System availability, delays and adherence to maintenance schedules is the basis of the contracted management fee. The system delays are categorized as incidental delays (1-3 minutes), significant delays (3-10 minutes), major delays (10-59 minutes) and catastrophic delays (>60 minutes). The contractual guidelines are that the system operates above 99.50% system availability with no more than 60 incidental delays, 15 significant delays, 2 major delays and zero catastrophic delays.

Station Door System

Station Door System

The APM system services 14 stations with automatic platform doors. Our normal daily service provides 8 to 10 trains consisting of four cars during peak travel periods. The station doors cycle approximately 1200 times per day. The average full loop service time is 12 minutes while operating ten trains.

Power Distribution System

The major subsystems of the APM system consist of a relay-based fixed-block signaling system, a 600Vac 3 phase power system; a guideway mounted power rail, four main traction power stations and two sub-stations.

Typical Guideway Switch

Guideway Switches

The Atlanta guideway system is comprised of a series of 20 rail switches to accommodate storing of trains in two bypasses, dispatching trains onto the mainline loop service, running bypass shuttles and manual driving of trains into the maintenance facility for inspections, repairs and routine maintenance.

The system utilizes three types of switches (pivot, rotary and turntable) to perform train movements. There are sixteen (16) pivot switches, three rotary switches and one turntable switch.

The three rotary switches are used at the end of the pinched loop to turn back the trains. Normal cycle rate is approximately 850-970 cycles daily.

Four pivot switches are used for bypass shuttle mode operation and for storing trains in the bypass area. During normal 30 minute bypass shuttle mode these switches cycle 6-8 times.

The remaining twelve pivot switches are used during dispatching, maintenance shop vehicle movement and turn back operations. They cycle approximately 12-20 times a daily depending on the work load or service requirement.

The one turn table switch is used during train dispatch from the maintenance shop and alternate loop operation.

Maintenance Shop

Maintenance Operations

To maintain the system, approximately 9 employees perform various duties to maintain system performance. During a typical month the 49 vehicles undergo 600 bi-daily inspections, 20 periodic inspections at 7,500 mile intervals, 5 periodic inspections at 30,000 mile intervals and 5 periodic inspections at 60,000 mile intervals. The 14 stations undergo a total of 420 daily inspections and ten 20,000 cycle door PM inspections per month. The operations and maintenance group performs day-to-day maintenance and inspections of the vehicles, track and guideway while the engineering group provides technical guidance to ensure compliance to established specifications. The maintenance and administrative tasks are performed in a 70,000 sq. ft. facility with a 21-car storage capacity.

APM Contract

In 2005, Bombardier received a 10-year contract with two 5-year options to operate and maintain the Bombardier-built *CX-100* APM system. The benefit of this type of contract is that it provides a 20-plus year customer relationship and the ability to provide long-term system performance enhancements. This was due to the efficiency, professionalism and high level of service required to provide on-time customer service.

Bombardier administrative offices and the customer's office are both centrally located within the maintenance facility. This allows the customer to have a clear vision of the day-to-day requirements of the operation which is vital to the success of the site. It provides opportunities for partnering and coordinating with the customer develop strategies to improve shop performance and employee morale. It also provides a one-on-one relationship with the customer, which is an integral part of Bombardier running a complex operation of this size.

BOMBARDIER and CX-100 are trademarks of Bombardier Inc. or its subsidiaries.

O'HARE ATS – THE TEENAGE YEARS

Dennis Gary* and Mark Piltingsrud, **
AECOM Technical Services (previously Earth Tech)

* Vice President – Technical, AECOM Technical Services, 10 South Riverside Plaza, Suite 1900, Chicago, IL 60606-3728, (312)777-5572; dennis.gary@aecom.com
** Senior Program Director, AECOM Technical Services, 10 South Riverside Plaza, Suite 1900, Chicago, IL 60606-3728, (312)777-5575; mark.piltingsrud@aecom.com.

Abstract

In the fifteen years since it opened, the O'Hare Airport Transit System (ATS) has performed well, and is generally ready to grow with the airport. A number of ATS challenges have arisen or have become evident over the years. The range of these issues include system capacity, a need for improved de-icing facilities, inefficient access to the maintenance bays, and the distances between the ATS, the Chicago "L" and Metra's commuter rail stations. This paper describes some of the larger current issues facing the ATS in terms of both its internal operations and its interaction with the airport environment that it serves. The extent to which these issues will be addressed is currently under discussion as the city plans for a $16 B expansion of the airport and potentially for the 2016 Olympics, should Chicago be selected in October 2009.

Introduction

Chicago O'Hare International Airport (ORD) has been one of the world's busiest and best known airports since President John F. Kennedy dedicated new terminals opening the airport in March 1963. Thirty years later, a major expansion was completed (May 1993) including a new international terminal and the opening of the O'Hare Airport Transit System (ATS). The ATS was conceived as an "intra-airport" transit system providing an alternate landside connection between the core area airport terminals and remote parking facilities. Featuring quick and convenient service, it was intended to relieve the access roadway system of the many buses that had previously labored through the dense ground traffic around O'Hare. It also provided a critical landside connection between the new international terminal and the three domestic terminals for international-domestic transferring passengers.

The current ATS system is an automated guideway transit (AGT) system, or an automated people mover (APM). It is based on the first application of AGT technology to a line-haul urban transit system, the

French Matra (now Siemens) VAL technology, which opened in Lille, France in 1983. The O'Hare ATS opened in 1993 and uses a wider VAL 256 vehicle. Those wider vehicles were also originally delivered as part of VAL systems built in Jacksonville and Taipei. The current ATS fleet is 15 vehicles which operate on 2.71 miles of double guideway with passenger stations at three domestic terminals, the international terminal and remote parking facilities.

The original 1983 request for proposals to design and build the ATS specified a calculated initial capacity of 2,400 passengers per hour per direction (pphpd). The system was designed to handle up to 6,000 pphpd through the use of a pinched loop guideway configuration, three-car train consists, and a fixed-block automatic train control system capable of providing 90-second headways between trains. ATS passenger stations were designed to handle up to a three-car train. The system was also planned to accommodate above ground expansion to the northern boundaries of the airport and west below the tarmac to future access/parking stations. Provisions were also made for an additional station between the international terminal and the remote parking stations for a rental car center. However, inserting that station would require some system modifications.

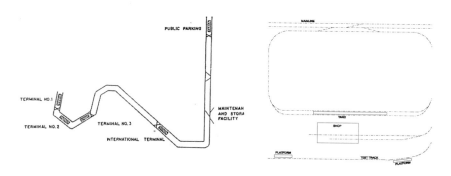

Existing ATS Guideway Configuration   **MSF Existing Facility Layout**

The O'Hare ATS has an exemplary record of airport AGT speed, safety and availability. The ATS is currently maintained and operated for the Chicago Department of Aviation (DOA) by a private firm, O'Hare Airport Transit System, Inc. (OATS), under contract with the City.

However, successful operation of the ATS system during its 14 years of service and the continuing growth of O'Hare have resulted in significant increases in ATS ridership to the point that the capacity of the system as built is now being severely taxed. Furthermore, the City has undertaken an O'Hare Modernization Program (OMP) which will increase the capacity of the airport, thereby further increasing the number of travelers using the ATS. Because of these growth factors, the Department of Aviation of the City of Chicago (City) is initiating expansion and improvement of the ATS. The driving forces behind these ATS growth changes are varied and generally fall into four technical categories of ATS needs, as discussed below.

Capacity Improvements: The most immediate and highest priority need of the ATS is added capacity. This added capacity will be achieved in part simply through an increase in the fleet of vehicles. Added vehicles would allow maximum length 3-car trains to operate at closer to the minimum operational headways of the existing technology. To achieve this, the existing fleet needs to be expanded by seven or eight vehicles from the current fleet of 15.

Further capacity improvements would require not only additional fleet, but also a shortening of operating headways as discussed below under technology upgrades and added off-peak storage capacity for the extra vehicles.

Physical Expansion: A number of airport expansion projects have been considered. Some of those that have a direct impact on the ATS include:

- Expansion of the line north to a new integrated rental car facility, remote parking garage and intermodal connection to the Metra O'Hare transfer station.
- Relocation of the current Remote Parking ATS station to be out of the runway safety area (RSA) of a new east-west runway.
- Relocation of the ATS guideway north of the current international terminal to accommodate a new second international terminal. To make room for this new terminal the current ATS Maintenance and Storage Facility (MSF) would also need to be relocated.
- Western access to the airport including a new terminal and connection with the current core area.
- Expansion of the maintenance and storage facility to serve the additional fleet vehicles.

These plans are still in formative stages. However, flexibility to accommodate future expansions such as these is an underlying theme for all current ATS planning.

Dependability Improvements: Other needs of the ATS relate to the more mundane, behind-the-scenes, but critically important activities of daily operations and maintenance (O&M) functions. When these needs are addressed, the system will run or be repaired more efficiently and provide more reliable service. These include items such as:

- Sufficient, flexible yard storage capacity for vehicles
- Indoor vehicle storage, cleaning and light maintenance facilities
- Flow and access to maintenance positions within the shop building
- Improved provisions for winterizing vehicles.

Technology Updates: A final category for ATS improvements is driven by the age of the technology. While the ATS is a very robust and well-built transit system that operates with high reliability, it is based largely on technology from the 1970s and is fairly unique hardware in the transit industry. Since the ATS is facing major expansions as discussed above, a change or updating of the basic technology of the system may be warranted early in the expansion schedule. Waiting until later in the program to change specific technologies would risk the added costs of "go-back" modifications where recent updates would have to be re-engineered for a second change within a short time period.

The technologies of most concern in this area are the rubber tire suspension and guidance of the vehicles, and the fixed-block automatic train control (ATC) system. In particular, a moving-block ATC might double line-haul capacity but would also require a near doubling of both the fleet size and the capacity of the maintenance and storage facility to provide that new capacity.

The Issues

There are several current challenges facing the ATS that the City of Chicago is beginning to address.

1. **Growth of O'Hare Airport compared to the ATS designed line haul capacity**

    The successful performance of the ATS system during its 14-year operation, the "must-ride" nature of international-domestic transferring passengers, and the continuing growth of O'Hare have resulted in significant increases in ATS ridership to the point that the capacity of the system as originally built is now being severely taxed.

    Currently ATS capacity expansion is required to meet the immediate needs of the airport. However, as the airport grows there will also be a need for physical and additional capacity expansion of the ATS. Airport and ATS improvements may also be considered should Chicago be selected for the 2016 Summer Olympics. Chicago has been chosen as the US city to bid to the International Olympic Committee, and is one of four worldwide cities being considered. The winner is to be announced in October 2009.

    The long-term plans of the ATS are embodied in the O'Hare Modernization Program (OMP). Phase I of the OMP is scheduled to be completed in 2012, at which time the airport will have been reconfigured to provide four parallel runways in an east-west orientation and any improvements required for the Olympics, and will include the expansion of the ATS to new parking and/or perhaps a rental car center to the north. The remainder of the OMP will be accomplished in Phase 2 and includes construction of a western terminal complex on the opposite side of the airport from the current central terminal area. Western access to the airport will need a transportation link to the current eastern terminals and parking facilities. That link could be an ATS extension from the current last station in the core terminal area. The extension would loop around under itself and go underground. Alternately, the western access could be reached by an independent underground line from the core area of the existing terminals. If the link is an extension of the current ATS, allowing for longer trains, a total fleet of perhaps 100 vehicles could be needed.

2. **The need for additional storage space for spare parts beyond the space originally provided in the Maintenance and Storage Facility (MSF)**

    The existing spare parts room on the ground floor of the MSF is conveniently located to the maintenance floor area of the shop. It is also managed appropriately with a computerized stock control system. However, a shortcoming of the original design is that there is insufficient room for the volume of all of the varied parts needed for the ATS. As a result, temporary outbuildings have been located around and near the shop building and 20,000 ft^2 of warehouse storage are used close to the MSF in a currently unused airport building. The temporary buildings are manufactured shelters similar to conventional shipping containers that are brought in (or removed) by truck. The warehouse storage is in a building that is scheduled to be razed for the future T6 terminal.

The outbuilding and warehouse approach to parts storage is a workable solution to a shortage of space for spare parts. In fact, some infrequently needed spare parts are readily able to be stored in these less convenient locations without disruption to routine maintenance. However, more generally, this approach to spare parts storage is inefficient and risks long-term damage to parts in infrequently accessed locations. It also provides less security for valuable spare parts than is possible in a single storage area. The clear preference would be to store all spare parts in one secure area under one roof.

This is not a critical problem that needs immediate attention. Rather, if there are plans for expanding the service areas of the shop building to accommodate a larger fleet of vehicles, those building plans could include relocation of the receiving dock further to the north of its current position and expansion of the MSF parts storage area. This expansion should accommodate not only the proportionally larger number of spares needed for a larger fleet, but also bring all appropriate spare parts storage inside one conveniently located facility.

Although there are some spare parts related to stations (platform door actuators), traction power substations and other infrastructure elements along the line, the majority of spare parts are related to the vehicles. Therefore, whatever the reason for a fleet expansion, whether expansion of capacity on the current system, or extension of the line to other stations, in general, the ultimate anticipated fleet size should be the dominant factor in determining the size of the spare parts area. Selecting an appropriate planning horizon and estimating the fleet size at that date may be difficult and represents a risk area that must be addressed to provide for future system expansions.

3. **The need for improving the vehicle de-icing facilities to make the process more efficient and not interfere with normal maintenance**

The 2.71-mile long double-track ATS mainline along with the MSF storage, test and access tracks represent a total of 6.3 miles of single-lane trackwork composed of running rails, power/guidance rails and negative return rails. To keep those rails free of ice and snow in winter conditions, the system uses over 75 miles of heating cables which consume over 6 MW of power. Much of that melted snow is thrown by the rotation of the vehicle primary suspension and guidance wheels into areas of the vehicle undercarriage around those rubber tires where it refreezes. That ice creates a maintenance issue that, under worst case conditions, requires vehicles to be brought in for de-icing.

OATS currently brings iced vehicles into the shop to the position shown in the photo, for example. There the ice on the vehicles is allowed to thaw in the warmth of the ambient air of the building. However, this melting is a time-consuming process that cannot be accelerated through mechanical chipping processes because of the risk of damage to ice-encrusted cables or parts. Nor could hot air, chemical spray or steam cleaning be used effectively in the open environment of the shop without shutting down normal repair activities in adjacent areas.

Any expansion of the shop building, covering of the yard storage tracks, or construction of additional or replacement MSF facilities should consider including specific enclosed facilities for de-icing cars to facilitate and speed the de-icing process so as to minimize the operational risks associated with winter operations. While it might be desirable to test various de-icing techniques (hot forced air, steam, infrared, chemical, etc.) and the positioning of the tools associated with those processes relative to the vehicle before designing a facility, in general, simply having an enclosed separated area with access to utilities will probably be sufficient to try the various approaches.

Furthermore, the provisions for washing vehicles in the original system design was to use long handle brushes from the single platform in the yard area shown in the photo. This is an outside operation which limits the cleaning to the ends and one side of the vehicle above the platform level. An indoor full-vehicle washing facility would allow for faster, more efficient and more complete cleaning of the  vehicle exteriors. This is a function that can readily be included in the design of any new enclosed de-icing facility. Therefore, any plans for a new de-icing facility should also consider including provisions for car washing within the same facility.

Finally, the cleaning and light maintenance of vehicles in the yard can be generally enhanced and the vehicles can be prepared to provide improved climate control and service for passengers if at least some part of the yard allows for storage of the vehicles under the protection of a building.

4. **The stub-end maintenance building restricts access to shop maintenance positions.**

A shortcoming of the current shop layout is that it was designed and built with entrance tracks at only the north end of the building. One track does exit the south end of the building, but it stub ends a short distance from the building and is used primarily for shipping or receiving complete vehicles on flatbed trucks. As noted earlier, because the shop service tracks end in the building, their "stub-end" design can "trap" vehicles inside the building after work on them has been completed but work on an adjacent vehicle to the north is still in progress. The inefficient layout of the stub-end tracks also work in the reverse direction when a new vehicle may not be brought into a northern working position because a vehicle south of it is nearly completed and the track must be kept open to get the southern vehicle out when work is complete. So generally, the two northern positions, with their immediate access to the building exit/egress tracks, are inherently more flexible than the southern vehicle maintenance positions with their potential for trapping vehicles.

The shop was designed with four vehicle service positions, two on each of the two tracks. Track 1 is on the west side of the shop area adjacent to the offices, while Track 2 is on the east side. On Track 1, the fixed vehicle jacks are located on the northern service position. South of the two Track 1 positions is an area which is not quite long enough to accommodate a third separate vehicle, unless the spacing between positions were reduced to about seven feet. That would severely constrain movements in the area. Alternately, a coupled pair of vehicles can be parked and serviced in that area as shown in the photo. But generally OATS uses that area for temporary storage and parts layout. On Track 2, the fixed jacks are positioned for the southern service position. South of it is a cleaning cubicle that is enclosed on three sides and prevents positioning of a third vehicle on that track under any circumstances. Because of the limited space south of the current positions and the trapping potential discussed above which would be worsened if three individual vehicles were serviced on any one track, there is little opportunity to increase the number of vehicle service positions inside the current shop.

This inefficiency is workable for the current vehicle fleet size, but any expansion of the fleet size will only increase the frequency of the trapping and, hence, the inherent inefficiency. Furthermore, if it is decided to expand the shop area, the creation of a third or fourth service position on the current tracks would exacerbate the trapping problem. The risks associated with this inefficiency should be addressed in future system expansions.

5. **The need for a rapid recovery vehicle to supplement the current diesel powered maintenance vehicle which is limited by its speed and flexibility**

The VAL technology was designed and built to be a robust system. It has proven to be reliable throughout its 14 years of service. But failures and outside events do and will, of course, occur and will lead to trains being stranded on the system. Normally when individual cars of a train have propulsion or brake failures, any one car of even a 3-car train has sufficient propulsion and braking power to get the train to a station where passengers can disembark. Some brake failures require that maintenance personnel must reach the vehicle to release the brakes manually for the train to proceed in a recovery mode. Other failures can be addressed by utilizing push recovery of the following train.

Nevertheless, the ATS must have provisions for responding and recovering from a worst case scenario that disables the system. That worst case scenario is an area-wide utility power outage that would cut off all power to the trains and leave them stranded on the guideway, unable to reach a station to unload passengers or to return to the yard. This has happened once in the first 14 years of ATS operation. The cars do have one-hour battery back up power sufficient for running auxiliary equipment to maintain the train control system and the passenger interior environment. After that hour, the brakes are automatically set such that they can be released only by a technician on board, but more importantly the interior environment will begin to degrade. So when there is a major power outage event trains stranded on the guideway need to be moved to a station to unload passengers as expeditiously as possible and many of them also need to be moved to the yard to clear the guideway so that other trains can be reached for similar recovery.

To make these moves of stranded trains under this worst case scenario, the current procedures require use of the ATS maintenance of way (MOW) vehicle, a diesel-hydraulic propelled vehicle which can operate either on paved roads or the ATS guideway. The vehicle is regularly used for heavier maintenance activities on one track along the guideway while the system continues to operate, alternating directions, on the second track. The vehicle is approximately the same length as one of the passenger vehicles but where a passenger vehicle can be operated manually from either end, the MOW has an operating cab at only one end and, therefore, requires a second crew member positioned for safety at the opposite end during reverse moves. The MOW vehicle is normally stored mixed in with other passenger vehicles in the yard or shop.

Using the MOW vehicle for train recovery can be a slow process. Activating the MOW for use in train recovery requires assembly of a crew of two. But more importantly, at the time of an event that requires a recovery vehicle, the MOW may already be involved in active repair work on other parts of the system and may need time to disengage itself from that work. Or worse yet, it may be undergoing repairs that cause it to be temporarily disabled. Also, in the worst case scenario of an area-wide utility power outage, many trains may be stranded on the line such that it may be preferable to apply two independently powered (i.e. diesel) vehicles in getting passengers to stations and recovering the system.

In addition to the above limitations on the MOW vehicle itself for emergency recovery, the current enhanced security environment of airports suggests that a faster acting and redundant system be available for the worst case failure scenario.

6. **Timing and extent of the upcoming vehicle mid-life overhauls**

Compared to other major hardware systems of any fixed guideway transit property, the vehicles usually have the shortest projected life because they are exposed to the wear and tear of mechanical operations at speed, extremes of climate, and abuses of the public. The Federal Transit Administration (FTA), for example, assumes the useful life of a rail transit car is 25 years, compared to 80 years for elevated structures and yards, 70 years for stations, 50 years for maintenance and administration buildings, and 30 years for train control, traction power and central control equipment.[1] Only rubber tire buses running in the aggressive environment of public streets have a significantly lower projected life of 12 years. Even though the VAL vehicles run on and are guided by rubber tires, because they run on an independent, well maintained, isolated guideway, their useful life is expected to be equivalent to that of rail vehicles and they should be treated that way.

Typically, rather than waiting for rail vehicles to begin to have increased failure rates near the end of their original useful careers, operators of public transit vehicles frequently go through their fleet and do a complete overhaul of each vehicle, a few at a time. These are usually referred to as "mid-life" overhauls, are performed once in the life of a vehicle, and are timed to occur in the third or fourth quarters of the projected lives of the vehicles. The timing is usually triggered by increasing failures rates of various parts, difficulty with parts supplies, and perhaps a parallel procurement of additional fleet. Performing the overhaul in parallel with purchase of new cars has the advantages of extra (new) vehicles starting to become available to replace the older ones, the efficiency and convenience of replacement parts being obtained as additions to the procurement of parts of the new vehicles, and the opportunity to achieve greater commonality of spares by retrofitting the older vehicles with the same improved parts of the new vehicles.

Mid-life overhauls are preceded by a careful inspection and inventory of all vehicle subsystems and a determination of the need of each subsystem for upgrading the fleet technology, full fleet or individual vehicle part replacement, or individual vehicle part repair. The overhauls themselves typically disassemble the vehicles as much as practical and reassemble them with the planned new technology, parts (e.g. seats), and redesigned parts (e.g. interior lighting and lenses), frequently resulting in what the public may perceive as a new vehicle.

The ATS vehicles are in the third quarter of what might be considered their nominal useful life (14 years old in a useful life of 25 years). However, they are well maintained and generally are in very good condition. OATS has performed 350,000-mile major overhauls of the vehicles roughly every five years. These are the most detailed overhauls performed on the ATS vehicles to date, but still are short of the extensiveness of a mid-life overhaul. The regularity of those 350,000 mile overhauls along with the excellent performance of the vehicles so far in their careers suggests that the useful life of these vehicles may be longer than the 25 years suggested above. That also argues strongly for performing mid-life overhauls on the vehicles since it is a re-investment in a proven design. In contrast, when a vehicle design is experiencing significant maintenance problems, transit agencies will frequently not do mid-life overhauls and instead simply replace the problem cars through procurement of new cars.

So, eventually, the DOA is likely to face a decision on when to do mid-life overhauls on the current fleet of vehicles. The timing of that decision would seem to be somewhere in the 5 to 10 year horizon.

7. Advances in train control and impacts on the ATS

---

[1] Standard Cost Categories for Major Capital Projects," Planning & Project Development, Federal Transit Administration, Rev.3, November 2005.

The basic technology of the existing automatic train control (ATC) system originated with an automatic train operation (ATO) fixed-block train control system originally designed for the Paris Metro. It was first used in public service on April 13, 1952, but was not fully applied to a major line until 1969. The technology was advanced to a fully automated (driverless) ATC system and opened for public service for the first time in Lille, France in April 1983. So the current ATC technology is well proven, but it is also well dated. Since the Lille and O'Hare systems opened, in addition to using advancements in basic technology, the rail transit control industry has seen the development of two major trends that have relevance to the O'Hare system. These trends point to the current ATC system being the limiting factor or shortcoming of the ATS in adapting to future airport growth.

First, "moving-block" control systems have become much more common. The first moving-block AGT system in North America was the Vancouver Skytrain system which opened in 1985. A more recent communication based train control (CBTC) program has further advanced moving-block technology. From a system capacity viewpoint, the net difference between fixed-block (Lille and O'Hare) and moving-block systems is that moving-block systems allow the trains to operate at closer headways thereby increasing the system line-haul capacity.

Specific values for the minimum operating headway (MOH) depend on the vehicle speeds, electronic cycle times, mechanical reaction times (for brake applications), braking rates and other variables. Typical MOH values of 60 seconds or better are achievable with current moving-block systems. For pinched-loop systems, the shortest headways are achieved when trains stop at end-of-line (EOL) stations, discharge passengers, and then continue on past the station before making the crossover move to the other track to reenter the station for boarding passengers before starting their return trip. A more conservative 75 or 80 second target headway is reasonable for the O'Hare ATS because the crossovers at O'Hare EOL stations are in front of the stations, not after them, and passenger loading with luggage requires longer vehicle dwell times in stations than in conventional urban transit. In general, replacing the O'Hare ATC with a moving-block control system would roughly double the system line-haul capacity.

In addition, because current ATC systems use much more modern technology than the 1970's technology of the O'Hare ATC, modern moving-block systems have much less wayside equipment, higher reliability, and lower capital and maintenance costs. For example, in September 2007 single-track ATS operations were required over progressive sections of the entire guideway to permit construction crews to change the aging wiring of the ATC and communications transmission line assembly (TLA), which is mounted between the running rails of the guideway. With the most current moving-block systems (i.e. CBTC) there is no TLA or complicated guideway wiring, so there is no need for a recurring replacement program that requires extensive track occupancy by work crews.

The second major trend in the rail transit control industry that relates to O'Hare has been a movement toward "open architecture" ATC systems. The components of the initial fully automated ATC systems used proprietary components. In addition, the interface between trackwork and vehicle had to be significantly more complex for the added operational and safety functions, so the interface effectively also became proprietary through being driven by highly technical proprietary components. In contrast, the transit agency driving the CBTC program has placed a heavy emphasis on establishing a non-proprietary interface between major components of the new ATC systems, and particularly between the vehicle and wayside. In this open architecture approach, a vehicle with the functional equivalent of a vehicle on board control system (VOBC) from one company would be able to operate without any service degradation in areas with wayside signaling installed not only by the same company, but also by other companies. This means that whole transit systems can be expanded using competitive bidding among train control companies for the wayside equipment, potentially resulting in lower capital costs and faster construction schedules. Furthermore, on-board or wayside problem equipment could be replaced with equipment from another supplier, allowing a transit agency to escape from the confines of a poorly performing vendor in a procurement that would otherwise necessarily be sole-source.

In summary, the current ATC of the O'Hare ATS is very reliable and well supported by the system supplier and microprocessor industry. However, over the long term, its advanced age and older technology may raise concerns about the risks of reliability, maintainability, and operating cost beyond the near-term period.

1. It may become increasingly difficult to find ATC replacement parts over the next 10 to 20 years.

2. The fixed block nature of the system places a limitation on the minimum operating headway possible between trains, whereas other moving-block technologies are available that would allow that headway to be lowered, thereby increasing the system capacity.

3. Newer technology with "open architecture" may provide greater competition among train control vendors for both future extensions of the system and, perhaps, even the supply of replacement components for the current ATC system.

8. **Three "rail" transit systems serve O'Hare in various capacities, the ATS, CTA's "L" (heavy rail) and Metra commuter rail, yet no two of them meet in a common station.**

A catch-phrase that has recently become popular in transit planning is "seamless transfers." As a design goal, it represents making the transfers from one system or line to another as short, convenient and comfortable as possible. Currently O'Hare is served by both the CTA and Metra which consider the airport an origin or destination, and not a transfer point within or between those two rail systems. In contrast, the ATS is an internal circulator for the airport, so seamless transfers with both the CTA and Metra are desired. The Metra O'Hare station is northeast of the current ATS Remote Parking station and relies on bus service to connect to the ATS. Northern extensions of the ATS are being considered, which, among other goals, would make that transfer more convenient.

The CTA's service comes right into the heart of the O'Hare core terminal area with a single end-of-line, multi-track, stub-end station below the core area parking garage. However, although the CTA station is surrounded by three ATS stations, there is no convenient connection between the CTA and the ATS. None of the pedestrian routes between the two systems is an obvious direct path. All routes between them require pedestrians to walk long distances and the vertical circulation along all connections requires an up, over and down pattern among several levels.

In addition, others are working on long-term plans to provide express CTA service with downtown Chicago luggage check-in and for extending the CTA under Terminal 2 to reach the western access proposed in the OMP. Those conceptual studies are being coordinated with ATS infrastructure expansion plans.

9. **Station platform crowding issues.**

The platforms of each of the five ATS passenger stations is centered between the two guideway tracks so vertical circulation is required for all access and egress. In the airport core area the three domestic terminal stations (T1, T2 and T3) are accessed from the ATS by separate pedestrian bridges to each terminal. A pair of escalators and an elevator at one end of each platform creates a "muzzle loading" configuration. The international and remote parking stations provide platform access from both ends of their platforms.

In the core area, emergency egress stairs are located at the opposite ends of the platforms from the escalators and elevator, providing an emergency exit path to grade level. Access to and from the guideway emergency walkway is provided at the platform ends as well.

Passenger stations will experience increasing platform crowding as traffic increases. Consistently using three-car train consists will improve the distribution of passengers on the platform. But given the need for increased line-haul capacity of the ATS, station vertical circulation may also need to be expanded accordingly. Finally, since National Fire Protection Association (NFPA) design guidelines for rail transit systems[2] did not yet apply to AGT systems when the ATS was built, emergency

---

[2] National Fire Protection Association, NFPA 130 Standard for Fixed Guideway Transit Systems, Quincy MA.

egress capability may also need to be expanded to ensure conformance with that standard.

Solutions

A very wide variety of specific solutions to the above major challenges and other lesser issues are beginning to be discussed by the City and its team of consultants. However, none are as yet sufficiently well developed or officially selected to allow specifics to be addressed at this time.

APM Systems
The Key to Atlanta Airport Expansion

John Kapala
APM Project Manager, International Aviation Consultants, 1255 South Loop Road, College Park, GA 30337, Tel 404-530-5707, *John.kapala@atlanta-airport.com*

ABSTRACT

The Hartsfield-Jackson International Airport has always been at the forefront of Automated People Mover innovation and implementation from its opening day in 1980 to continuing plans for current and future expansion. The Atlanta Airport continues to expand to accommodate passenger growth from about 40 million in 1980 to over 90 million in 2008 and a projected 110 million in 2020. This paper examines the APM's role in the Atlanta Airport continuing growth and expansion through past, present, and future projects.

INTRODUCTION

The Hartsfield-Jackson Atlanta International Airport (H-JAIA) is the busiest airport in the world. In 2008 it welcomed a record number of passengers – more than 90 million – exceeding every other airport in the world.

This great achievement could not have been possible without Atlanta Airport's innovative design, extensive use of Automated People Movers (APM), and continued dedication to growth and customer service.

ATLANTA AIRPORT DESIGN

When it first opened in 1980 it was referred to as the Central Passenger Terminal Complex because it was centrally located between the runways. It consisted of the two main terminal buildings and 4 concourses (A, B, C, and D), each with about 26 gates, all connected with an underground Automated People Mover (APM) System.

This central terminal complex layout allowed planes to get to and from the runways and gates very quickly and efficiently. It also provided for relatively easy future expansion.

The key to this airport design is the Automated People Mover System, which connects the terminal and the concourses (which are 1,000 feet apart) with a quick and reliable transportation system.

ATLANTA AIRPORT AND APM SYSTEM EXPANSION

As the number of passengers grew, the airport and the people mover system expanded to meet the demand. And the passenger growth has been dramatic, from about 35 million in 1980 to over 90 million in 2008.

The initial APM system in 1980 included 17 cars, which could run individually or in 2 or 3 car trains. The stations were configured to handle up to 3-car trains but were expandable to 4 car positions. The system originally ran six 2-car trains.

The high demand on the APM system required an increase in the fleet to 24 cars in 1983 and six 3-car trains were run in normal service.

The fleet was again increased in 1992 by 4 cars concurrent with the addition of Concourse E and 7 more cars were added in 1995 bringing the fleet total to 35 cars.

In preparation for the Olympics, the stations were expanded from 3 to 4-car train berths in 1993-94, five cars were added and the fleet was reconfigured to run 4 car trains. Eight 4-car trains were operated during the 1996 Olympics.

The fleet size was again increased by 4 cars in 1997 and 5 cars in 2000 to end up with a fleet of 49 cars, which is where it is today. In 2001, 24 new cars were purchased to replace the original 24 cars.

Currently the system can operate up to ten 4-car trains with a headway of 108 seconds.

H-JAIA APM Cars in the Maintenance Facility

AUTOMATED PEOPLE MOVERS 2009 71

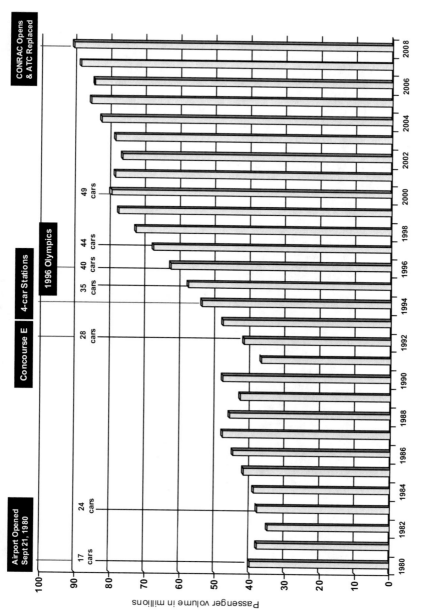

CURRENT APM PROJECTS AT HJAIA

There are currently five APM projects underway at the Hartsfield Jackson Atlanta International Airport. These projects are part of the Hartsfield Jackson Development Program which is an ongoing $6 Billion plus program which includes:
- A new Fifth Runway which went into operation in May, 2007
- Consolidated Rental Car Agency (CONRAC) see below.
- Maynard H Jackson International Terminal (MHJIT) see below.
- And numerous upgrade and refurbishment projects in the Central Terminal Complex and Concourses (CPTC).

Automated Train Control (ATC) Replacement Project – This project replaces all the original ATC equipment with new computer based Train Control technology from the original supplier (currently Bombardier).

Old Interlocking Equipment being replaced

This project has been quietly underway since 2006 and the significant aspect of this project is that the ATC equipment is being replaced without impact to the APM system operation. All the work is done during system shutdown (1:00am to 5:00am). This project is in its last phase and substantial completion is scheduled for June 2009.

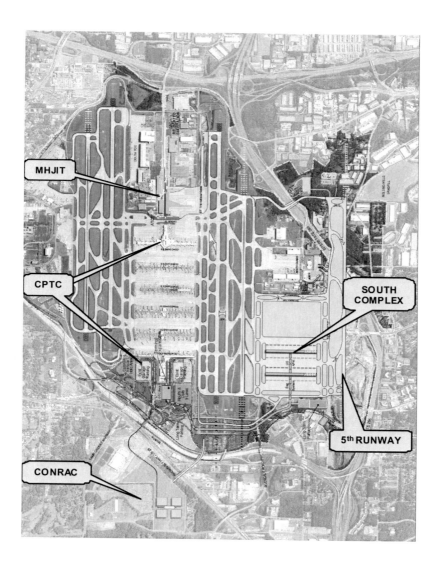

Hartsfield-Jackson Atlanta International Airport Master Plan Development

CONRAC – Consolidated Rental Car Facility APM System – This new stand-alone APM system connects the Airport Terminal to a new remote consolidated rental facility a mile and a half away. Moving the rental cars to an off-airport location frees up valuable space for future airport expansion and eliminates all the rental car busses and associated congestion and pollution.

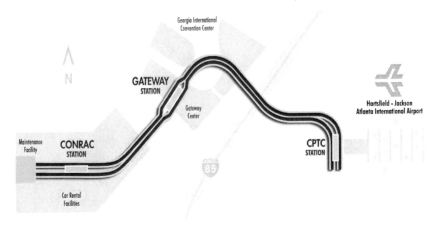

CONRAC APM System Alignment

This project is a DBOM (Design Build Operate Maintain) project and includes 1.5 miles of elevated guideway, platform doors and controls for 3 stations, 12 cars (6 married pairs), Automatic Train Control System, Power Distribution System, a fully equipped Maintenance and Storage Facility, and a 5-year Operations and Maintenance Contract.

Once in operation the system can handle 2700 passengers per hour per direction (pphpd) initially with 2.5 minute headways and is expandable to handle over 5000 pphpd.

The DBOM contractor is Archer Western/ Capitol Contracting and the APM system supplier is Mitsubishi Heavy Industries.

This project is nearing completion and is scheduled to open in November, 2009.

CONRAC APM Vehicle

CONRAC APM Vehicle Interior

APM Expansion to the Maynard H. Jackson International Terminal (MHJIT) –
This new terminal will provide another entrance to the Atlanta Airport from the east and will be the international gateway to Atlanta.

MHJIT Terminal Concept

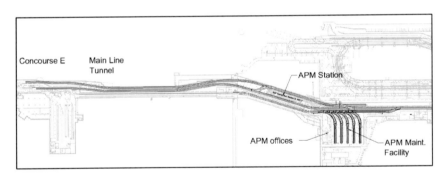

MHJIT APM System Alignment

The current H-JAIA APM System is being expanded to the east to provide service to this new terminal. The expansion includes approximately 1,200 feet of the dual lane system (mostly in a tunnel), ten new cars, ATC expansion, PDS expansion, a new station, and an additional light maintenance and storage facility for 20 cars. The APM contract for this expansion was awarded to Bombardier.

The project is currently under construction and is scheduled to open in 2012.

Automated Couplers Replacement – The current system uses mechanical couplers with two jumper cables for the electrical controls. This makes it very difficult and time consuming to couple and uncouple cars and change train consists. When the fleet was small, this was not a big issue, but as the system expanded it became more and more of a problem and would only get worse as the fleet continued to grow.

The new cars for MHJIT will have fully automated couplers and this project provides coupler kits to replace the couplers on the existing fleet of 49 cars with new automated couplers.

This project is under way, with coupler kits starting to arrive summer of 2009 and should be completed by the end of 2010 before the new MHJIT cars arrive.

H-JAIA APM Mechanical Coupler Being Replaced

ATC at E Replacement – This project replaces the relay interlocking equipment which was installed in 1992, when the system was expanded to Concourse E, with new computer based interlocking equipment from Bombardier. This will make all the Automatic Train Control equipment the same new technology. In addition, the Central Control computer ATS (Automatic Train Supervision) equipment will be upgraded as well.

This project is underway and is scheduled to be completed before MHJIT goes into operation.

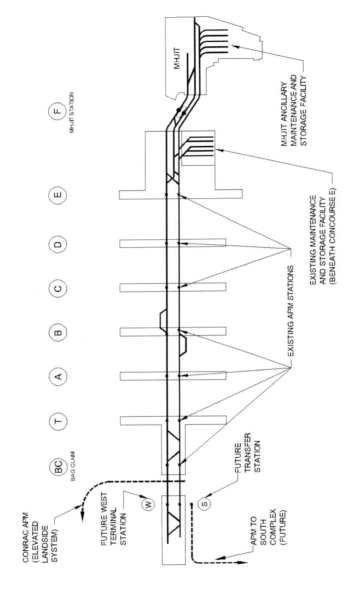

H-JAIA APM System Alignment and current and future projects

FUTURE APM PROJECTS AT H-JAIA

As the number of flights and passengers continues to grow, plans are being developed to expand the airport in the future. The following projects are still in the concept/planning phase and their actual timing is dependent on future passenger and airlines growth:

APM Expansion to the West Terminal Expansion – This project would expand the current Main Terminal to the west to provide for more facilities for passenger check in and baggage.

The APM system would be expanded to the west approximately 1,200 feet (all underground). Expansion would include a new station, up to 10 additional cars, and expansion of associated subsystems.

APM System Capacity Upgrade – Currently the system operates at 108 sec. headways at peak periods, which equates to 10,000 passengers per hour per direction (using 75 passengers as the car capacity). The system is capable of running 95 sec headways which would be an increase of 13.7% in the system capacity.

The limiting factor at this time is the turnback at the Baggage Claim station. In order to increase the capacity, the track would have to be extended to the west to have a turnback beyond the station. This change would be included if the APM system is expanded to the West Terminal. The new train control system is designed to handle the shorter headways. Additional trains and an upgrade of the Power distribution system would be needed.

APM System to the South Complex – If additional gates are needed beyond what is being provided at MHJIT, the potential exists for developing an additional group of concourses near the new fifth runway.

This new complex of concourses would be connected to the main terminal by a new stand alone APM system. This system would share a new station with the existing APM system which would be expanded to the west to provide simple and efficient passenger transfer.

CONCLUSION

As can be seen from the numerous Airport expansion projects presented above, the Hartsfield-Jackson Atlanta International Airport continues to be in the forefront of APM planning and innovation to support the Airport's continued growth into the future.

Planning and Integration - MHJIT at Atlanta Airport

Sambit Bhattacharjee [1], P.E., M.ASCE (Primary Contact)
John Kapala [2], M.ASCE
Mike Williams [3]

(1) Senior Associate, Lea+Elliott, Inc., 5200 Blue Lagoon Drive, Suite 250, Miami, FL 33126; PH 305 500 9390; Fax 305 500 9391; sambitb@leaelliott.com
(2) APM Project Manager, Hartsfield Jackson Development Program, 1255 South Loop Road, Atlanta, GA. 30337; PH (404) 530 5707; John.Kapala@atlanta-airport.com
(3) Director, MHJIT, 1255 South Loop Road, Atlanta, GA. 30337; PH (404) 530-5528; Mike.Williams@atlanta-airport.com

Abstract

Successful implementation of APM expansion requires a comprehensive understanding of the existing design features and a real-time evaluation of facility interfaces. The design process of Atlanta Airport's new East Terminal, officially named the Maynard Holbrook-Jackson Jr. International Terminal (MHJIT), occurred concurrently with the finalization of the APM Supplier Contract for Bombardier. The paper examines the development and integration of APM aspects in the MHJIT facility planning ahead of the APM operating system contract award. This includes the project's APM alignment, station location and facility interfaces that require balancing of a complicated set of existing parameters (tunnel under existing terminal, tie-in to an existing system guideway located immediately adjacent to arguably the busiest APM maintenance facility. The Project team has committed to maintain the operation and maximize the available capacity of most heavily traveled Airport APM. In addition, the City has fast-tracked the terminal space planning to support an accelerated construction schedule.

Background

Atlanta airport's existing APM System also called the Automated Guideway Transit System (AGTS), started service in 1980. The APM System was planned to be the backbone of the airport's passenger movement and transportation. The success of the Atlanta APM system in enhancing passenger level of service has become a model for other airports and translated into the APM being a critical element of the Atlanta airport expansion planning. A significant extension of the originally constructed APM System to Concourse E was undertaken in the year 1992. The system currently operates with Bombardier's CX-100 vehicles. The CX-100 vehicles were originally purchased in the 1980s. An additional fleet of 24 CX-100 vehicles were most recently procured in 2001. Atlanta Airport has programmed and planned for the new MHJIT to provide an additional landside access to the airport. The new terminal is proposed to have 12 gates and is called the Maynard Hartsfield-Jackson Jr. International Terminal (MHJIT).

Initial planning for the MHJIT was completed by Hartsfield Planning Collaborative (HPC) during 2000 - 2002. Subsequently, the project went into design in 2004, but was stopped in 2005. In 2007 the project was re-initiated by the Department of Aviation (DOA) with Atlanta Gateway Designers (AGD) as the design architects for the project. The project is being managed by Hartsfield-Jackson Development Program (H-JDP). The scope of design for AGD included development of a schematic terminal layout. The task included the layout of the APM at the basement of MHJIT with certain site constraints. The design team was encouraged to use already constructed elements of the terminal. This included some underground utilities and masonry retaining wall.

Figures 1 and 2 illustrate the progression of the project from programming document to the layout developed in 2005. In addition to the project programming layout, shown in Figure 1, the design layout from 2005 also included a mainline extension with a station at the new terminal. The station was added into the project based on the input from stakeholders and approval by DOA. The location of the station in the terminal as well as some site constraints, such as the location of existing smoke evacuation fan room, location of the piles and structure for Concourse E forced several geometric curves in the mainline guideway alignment.

Figure 1: 2002 Project Programming Layout

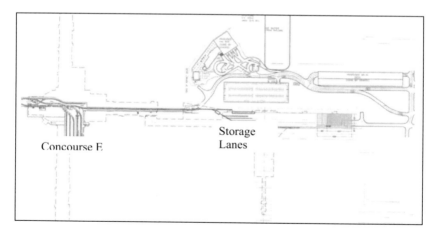

Figure 2: Layout from Prior Design (2005)

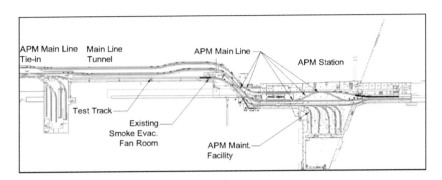

Program Criteria and Project's Design Development

Lea+ Elliott became a part of the project in its design development phase in 2004, and is an integral part of the project design team in the current design cycle. The programming criteria mandated that the extension of the APM System provide the following elements, integrated into the new terminal:

- Provide five light maintenance / storage tracks.
- Provide an added maintenance facility with offices for the AGTS System for its current and future larger fleet size.
- Extend existing test track by about 300 feet, to accommodate full dynamic test.
- Provide an automated car wash.
- Provide an extension to the mainline with a station at the new terminal for movement of passengers from/to the new terminal to the rest of the airport.

As a part of the facility's schematic design effort, several conceptual layouts were developed for the APM floor with different terminal configurations to analyze their interactions. Few of the options that were looked into during the workshops have been indentified in Figures 3A, 3B and 3C. The APM's seamless integration into the terminal was paramount in developing and selecting the preferred layout of the maintenance facility and location and layout of the mainline station. Figures 3 (A, B and C) illustrate some of the initial concepts that lead to the approved concept shown in Figure 4.

Figure 3: Concepts Developed for APM Integration in MHJIT Terminal

Figure 3A: Concept –A: Fully Developed Terminal

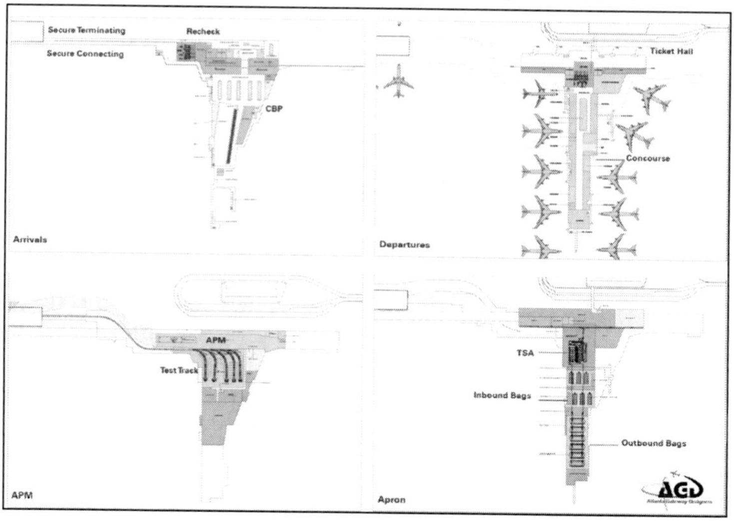

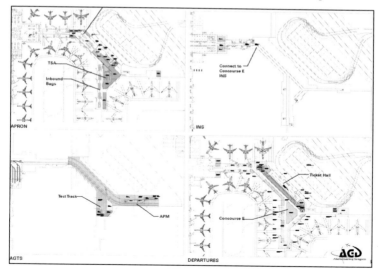

Figure 3B: Concept –B: Limited Size Terminal Development

Figure 3C: Concept –C: APM Floor Layout Plan and Terminal Integration

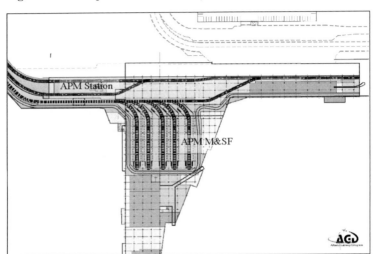

Figure 4: Approved Layout of the AGTS Level.

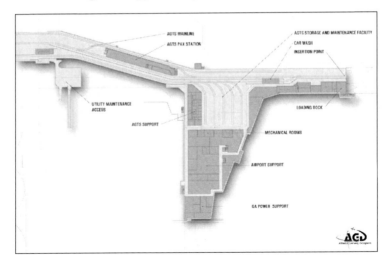

The primary advantage of the City selected layout is that it provides the optimum station location with most effective integration into the terminal. The location and orientation of the Station provided substantial benefits over the other alternative locations. They are as follows:

- The length of the APM guideway extension was shorter which provides a better round trip time and projects costs (for fleet and operations).
- The location permitted the inclusion of both front and back guideway crossovers, to provide increased operational flexibility.
- Use of larger guideway curve radii resulting in increased maximum train speed and an improved performance.
- Architectural effectiveness: the placement of the station outside, but adjacent to the main terminal building provided an efficient inter-relationship between the terminal space and AGTS space. This arrangement provided for simplicity of station design and space planning without its cumulative impact on multiple levels of terminal.

Fast-tracked Project Delivery Highlights and Challenges

The APM system configuration, guideway alignment development, design of the terminal building and finalization of the Bombardier Contract was developed around the overall fast-track delivery concept for the project. As described earlier, the primary factors that lead to an optimized interface between the facility development and APM contract execution was based on the following salient points:

- Location and layout of the MHJIT Station.
- Layout of the Maintenance Facility.
- Efficient "tie-in" and extension from existing east end of the track.
- Project phasing and construction sequencing, specifically for the tie-in area in existing Concourse E.
- Operational viability of existing system during construction.
- Integration of AGTS elements into the terminal.

Several of the aspects listed above were coordinated during planning and were validated by Bombardier early, during the negotiation for their contract. Comments and concerns were incorporated into the facility design drawings. This allowed an optimal level of design overlap between the terminal design and APM Operating System design.

Some of the key planning decisions and design level feasibility developments were vital to quickly establishing the final guideway alignment and its integration into the terminal design.

Two aspects/features that are site and project-specific will be discussed in this paper. These highlights a strong commitment of the project to provide comprehensive program, but also allow flexibility to review, plan and react to actual conditions encountered during design and procurement that can be incorporated with minimal impact to schedule and costs.

Project's Mainline Extension and Tunnel

The connection of the existing system to the new MHJIT requires the construction of a section of new tunnel of APM guideway that was proposed to traverse under the existing Concourse E. The project initially programmed a New Austrian Tunnel Method (NATM) type construction for the tunnel. NATM was selected to limit the impact of tunnel construction on the aircraft, gates and associated activities at the apron level. The NATM tunnel soil and structural requirements forced the tunnels to be about 40 feet apart. The alignment shown in Figure 5 was developed to support this tunneling methodology in coordination with the other site constraints described earlier such as the location of the station, existing smoke evacuation fan room and pile structure for Concourse E.

Figure 5: Design Alignment with NATM

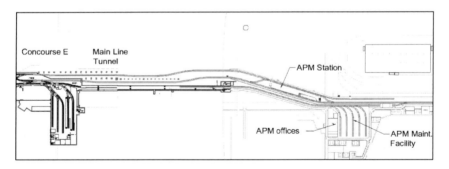

Based on the City's decision that the bid prices for the NATM tunnel were unacceptably high, the Project value engineered the tunnel construction. The airport decided to evaluate a "cut-and-cover" method of construction for the development of the tunnel. In order to optimize the impact of the "cut-and-cover," a large segment of the track alignment was brought closer to each other and maintained under a single bay of the piles. This resulted in the final alignment currently in design. This is shown in Figure 6.

Figure 6: Current Design Alignment

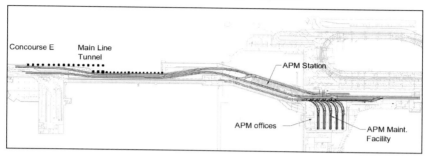

Critical Challenge: Operational Viability (Concourse E Tie-in)

One of the most critical aspects of the project is to ensure a viable tie-in between the existing system and the future extension with minimal impact on the existing service for the airport. A detailed review of existing layout, final layout and the possible intermediate and incremental phasing was undertaken during the planning phase of the project. This was developed and further validated by Bombardier and the project's Construction Managers. Construction level detailed breakdowns are being developed at the time of writing this paper. However, the incremental phasing, shown in Figures 7A, 7B, 7C and 7D identify the basic phasing framework of how the tie-in can be accomplished.

Figure 7A: Tie-in Phasing Sequence -1

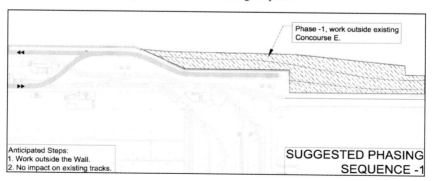

Figure 7B: Tie-in Phasing Sequence -2

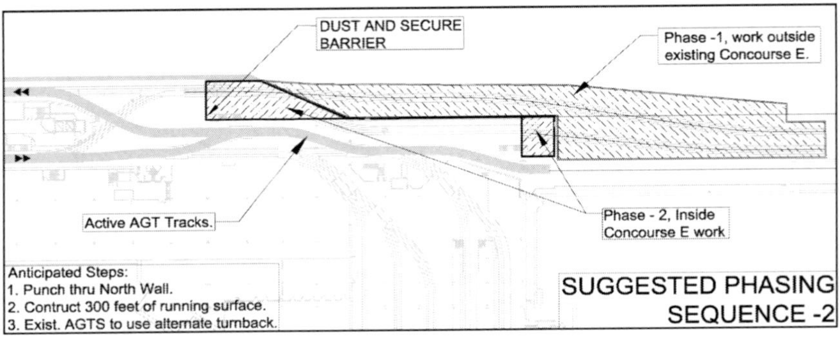

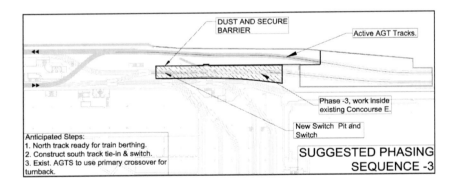

Figure 7C: Tie-in Phasing Sequence -3

Figure 7D: Tie-in Phasing Sequence -4

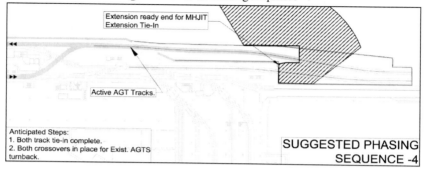

Additionally, a detailed breakdown of the activities within the sequence was developed as a flow chart (Figure 8) to illustrate the Bombardier and terminal infrastructure work. The flow chart was developed to facilitate the definition of activities for detailed schedule development. This is critical for the development of access dates and work interface between APM equipment installations and facility civil construction.

The above steps and processes have successfully moved the project through planning and procurement to the threshold of implementation and installation. The project is now under construction with the civil work underway. The APM extension and MHJIT is scheduled to open in 2012.

Figure 10: Concourse - E Tie-in Work Flow Chart

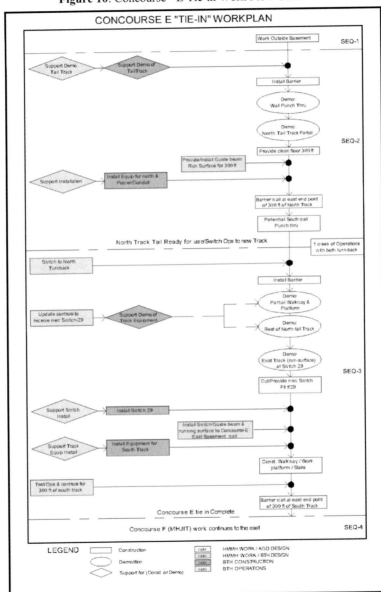

Expanding the APM System at Atlanta Airport

Gregory A. Adams[1], Project Manager, Atlanta APM Extension, **Thomas Sheakley**[2], Project Engineering Manager and **Frank Jeffers**[3], Automatic Train Control Manager Bombardier Transportation. Systems Division, 1501 Lebanon Church Road, Pittsburgh, PA, USA 15236; [1]PH (412) 655-6672; FAX (412) 650-6475 email: greg.adams@us.transport.bombardier.com
[2]PH (412)655-5165; FAX (412)655-5933; email: thomas.sheakley@us.transport.bombardier.com
[3]PH (412) 655-5206; FAX (412)650-3554; email: frank.jeffers@us.transport.bombardier.com

Abstract

The Automated People Mover (APM) system at Hartsfield-Jackson Atlanta International Airport is currently being extended to the new Maynard H. Jackson International Terminal (MHJIT). The MHJIT expansion project expands the current APM system east of Terminal E approximately 2400 feet to connect the new international terminal with the existing APM system. Bombardier's scope of supply includes ten (10) *BOMBARDIER*CX-100** vehicles and the system equipment necessary to extend the main line and the test track, the addition of five (5) new storage tracks, and a new light maintenance facility. Bombardier will upgrade the Automated Train Control (ATC) system for Concourse E with its *BOMBARDIER*CITYFLO** 550 driverless technology, a service–proven driverless, train-operating system.

The principal design considerations impacting the expansion are:

- Alignment extension
- Design for future expansion
- Fleet size expansion
- Addition of a light maintenance facility
- Addition of a new Power Distribution System (PDS) Substation
- Automatic train control design
- Closed Circuit Television (CCTV) and Operational Radio System (ORS) design
- Planning and executing civil and electrical demolition
- Planning and executing automatic train control cutover activities

This paper shows the complexity and challenges associated with the design, test, and placing into revenue service a system expansion while the existing APM system continues to operate at the world's busiest airport. This paper attempts to share the level of planning required to implement a "Brownfield" cutover, based on Bombardier Transportation experiences on the MHJIT expansion of the existing Atlanta Hartsfield-Jackson Atlanta International Airport APM.

Project General Scope

The MHJIT project expands the Atlanta Hartsfield-Jackson Atlanta International Airport APM that has been in successful operation since the early 80's. Design work began in January 2008. Electrical installation work begins in September 2009 followed by System Integration Testing in April 2011 which runs through Substantial Completion on 30 September 2011.

The engineering approach will maximize use of the existing service proven designs from previous projects and/or operating systems in instances of obsolescence or new design. The project encompasses both "Greenfield – new installation" and "Brownfield – modification to the existing installation" types of territory. The "Brownfield" installation requires another level of planning to assure that all of the logistical needs are maintained to assure a timely and efficient cutover process to avoid delays and to minimize interface or technical compatibility problems and to minimize risk.

Key success factors for the Atlanta MHJIT APM System Expansion consist of:
- On time project team performance: Design, manufacturing, quality, installation, and testing milestones within budget.
- Accurate final alignment, construction interfaces, and guideway construction.
- Detailed planning and synchronization of cutover activities.
- Timely site access.
- Minimal impact to current level of service.
- Until substantial completion, return the APM system back to service at the end of every test period for the resumption of safe, reliable passenger operations.

Alignment Extension

The APM System expansion (See Figure 1) includes ten (10) new *CX-100* vehicles, the associated wayside, guideway and station equipment, signaling, communications, power distribution, maintenance equipment and an expanded/upgraded Central Control facility.

The existing APM System alignment will be expanded approximately 2400 feet to connect with a new passenger station located in the new Maynard H. Jackson International Terminal (MHJIT) with Concourse E. The guideway for the expanded will be in tunnels similar to the existing system. Crossovers and switches are designed to function the same as the existing system.

The passenger station will be a center platform station with automatic station platform doors and barrier walls separating the passenger boarding/de-boarding area from the guideway. The station door/platform design will be for four (4) car trains and associated berthing positions on both the north and south platforms. Figure 2 is a schematic of the MHJIT expansion Automatic Train Control architecture.

Future Expansion

The MHJIT APM System Expansion is designed so that planned future expansion(s) of the System may be made with the same technology and of the same design as the MHJIT APM System Expansion installation and with minimal disruption to System operation. This applies to design of fixed facilities, vehicles, command, control and communications, system equipment.

Fixed facilities are designed with spare conduits and space capacity within conduits and wireways to permit future expansion/extension of the APM System. The guidance devices, guideway-mounted equipment, power distribution system, over-travel buffers, switches, wayside equipment will be designed to allow the guideway structure to be extended further east of the MHJIT terminal.

The vehicles are designed to operate interchangeably with existing vehicles and are capable of entrainment, in either orientation, in any train length from 1-car to 4-cars.

All command, control, and communication systems will be designed for expanding the system without replacing or destroying any of the initial System installation. This is accomplished by exchanging and/or adding equipment and modifying software in a modular fashion.

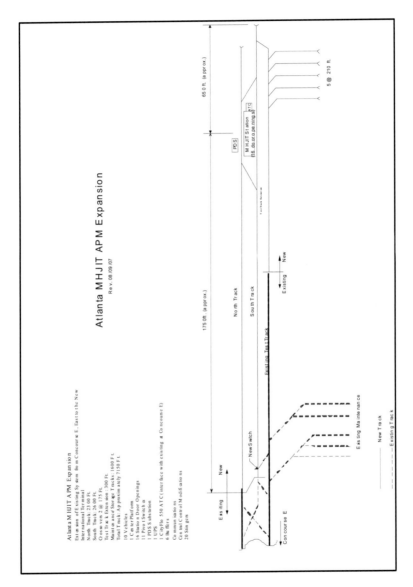

Figure 1 – MHJIT APM System Expansion Diagram

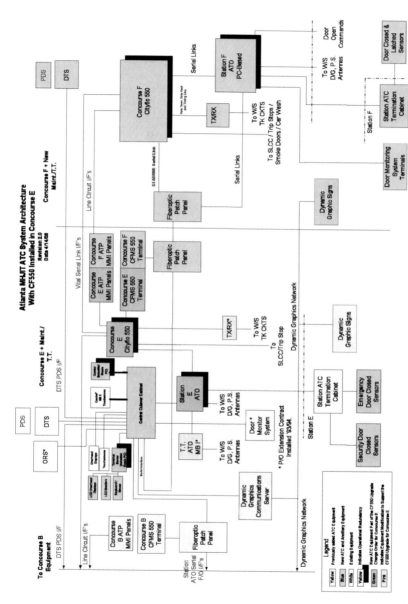

Figure 2 – MHJIT ATC System-level Architecture Diagram

Fleet Size Expansion

Bombardier will supply ten (10) new *CX-100* APM vehicles for the MHJIT APM System expansion. The fleet size increases from 49 to 59 vehicles.

The following are the key design parameters for Bombardier's *CX-100* vehicle. They represent the features from the previous 24-car procurement and are generally standard for this Bombardier product line.

Parameter	Requirement
Train Configuration	3- or 4-car trains
Maximum Speed	25 mph
Passenger Capacity	99

Addition of a Light Maintenance Facility

Due to the increase in *CX-100* vehicles from 49 to 59 vehicles, additional maintenance facilities and personnel resources were required to meet availability demands of the nation's busiest airport. Maintenance (including cleaning and housekeeping) is critically important to sustain APM operations with minimum disruption.

The Auxiliary Maintenance and Storage Facility (AM&SF) augments existing capability. The AM&SF will have five spurs (storage tracks) that can accommodate 4-car trains. It will be used in conjuction with the existing Maintenance and Storage Facility (M/SF). While the existing maintenance facility performs all types of maintenance at present, once the AM&SF is completed, heavy maintenance will continue to be performed in the existing M/SF while routine services/inspections (light maintenance) will be performed in the AM&SF. In addition, a train wash facility will be adjacent to the AM&SF.

The train wash facility is designed to automatically wash individual vehicles of a 4-car train. Maintenance personnel will drive the lead vehicle of a 4-car train into the car wash. Washing will take approximately 8 minutes per vehicle with a 4-car train completed in approximately 40 minutes.

The AM&SF will also provide additional space for workshops, test equipment, stationary equipment, administrative/ maintenance offices, personnel lockers, and toilet facilities. The system maintenance program will use a Site Information Management System that will be equipped to accommodate maintenance and storage of the fleet necessary to support APM operations with twelve (12) operating trains in the peak periods.

Automatic Train Control (ATC) System

The ATC System approach is based on the application of the *CITYFLO* 550, fixed block, solid-state ATP equipment. This will involve the replacement of the Vital Relay Interlocking plant located at Concourse E and the installation of a new plant of *CITYFLO* 550-based ATC equipment for the new MHJIT extension. Figure 2 shows the new equipment (Green Highlight) in a high-level System Block Diagram. The same figure shows the installation of the new *CITYFLO* 550 ATC equipment (Blue Highlight) in the MHJIT Station. The project will also involve the replacement of the existing RMX-based Central ATS Computer System with the new OPC-based equipment.

Identification of General Cutover Activities

In order to facilitate the construction sequence while maintaining uninterrupted system operation, the cutover plan consists of six phases.

All activities performed in the MHJIT extension and in any specific existing areas that would not cause system disruption are done at any time through the day. Activities in any area that may cause interruption to existing service will be performed during system non-revenue hours between 1:00 a.m. to 5:00 a.m., depending upon service demands of the day. All activities required to be performed during these hours will be noted within the activity section description. The following is a brief outline of the six cutover phases:

Phase 1: Relocate the Existing North Track Turn-back – Relocate the Existing North Track Turn-back. This phase of the cutover will involve straightening the exiting North Track so that it will run into the North Tunnel.

Phase 2: Modify Alternate Turn-back – This phase of the cutover involves the modification of the layout of the alternate turn-back to add a switch and connect it to the South Track. The switch will allow trains to traverse from the maintenance area to the North Track. The installation of the switch will require demolition of the existing guideway and constructing a new switch pit for the new switch.

Phase 3: Upgrade / Replace Concourse E ATC Equipment – This phase of the cutover will involve the upgrade / replacement of the existing Vital Relay Interlocking with the new *CITYFLO* 550 ATP equipment and the upgrade / replacement of the existing RMX-based Central Control with the new OPC-based Central Control.

Phase 4: Cutover Test Track – This phase of the cutover is an extension of the existing test track. It will include the new light maintenance and storage facility and an automated car wash. The longer test track area will be a significant improvement to existing facilities. It will allow speed testing of cars under maintenance.

Phase 5: Cutover South Track – This phase will place one lane of the dual-lane extension into service. This track will connect to the alternate turn-back through the new switch installed near Concourse E.

Phase 6: Cutover North Track – This phase will place one lane of the dual-lane extension into service. This track will connect to the North Track Turn-Back.

Planning and Executing Automatic Train Control Cutover Activities

In support of the key success factors stated earlier, significant effort will be spent in developing effective planning for the project implementation and coordination of activities between engineering, site startup team, O&M team, sub-suppliers at the site in accordance to the contract access, reliability and availability requirements. The goal of the planning will be to achieve the smooth flow of the work activities at the site by assuring the timely delivery of all materials, installation, cutover and test procedures, site staffing, training, security badging, spares provisions, tools by anticipating issues and opportunities early so that the proper actions can be realized.

An adequate level or rigor should be applied to all of the planning aspects to minimize and technical or scheduler risks at the site prior to initiating any installation or cutover activity.

The development of the detailed cutover planning steps and associated documents will be accomplished through the cross functional core project team members, comprised of representatives of system and design engineering, product assurance, project management, project scheduler, field engineering, test engineering, product introduction-startup team and key O&M team members, manufacturing, customer representatives and consultant.

The cutover planning for a complex project, such as the Atlanta MHJIT expansion project, occurred over a period of six to twelve months and is broken down into the following categories:

- Project Planning and Coordination
- Cutover Planning
- Testing, System Assurance and System Demo Planning
- Logistics Planning
- Design of Equipment Cutover Features / Capabilities into the Deliverable Systems
- Documentation & Analysis Requirements Associated with Major Cutover Phases
- Verification of Design Details

The Project Planning and Coordination:

- Develops and maintains an accurate Work Breakdown Structure and Scope Split Document to coordinate the roles and responsibilities of installation, cutover and testing activities for electrical installation, civil installation subcontractors, customers and suppliers to avoid delays and confusion at the site.
- Integrates each detailed cutover/test activity as a specific activity in the detailed project schedule with its own specific activity number.
    - Allows for adequate progress tracking at the site
- Clearly defines the role and responsibilities for all of the core project team members, along with site startup positions.
- Identifies key highly skilled and motivated O&M personnel who would be pulled over to the lead and to staff the site startup team charged with the installation, testing and commissioning the new equipment.
    - Supplement these experienced resources with additional resources as needed to perform the cutover work activities
    - Backfill and train additional resources to replace the experience resources pulled out of the O&M staff to form the site startup team
- Assures that coordination and adequate planning has been conducted with the site O&M and startup teams to plan dead time in the weekly cutover and testing activities to support the routine scheduled maintenance activities.
    - Identifies any unique access requirements or equipment needs (i.e. train quantity or make up requirements to support testing)
- Assures that the Detailed Project Schedule includes adequate schedule float (i.e. unused weekend days) as fall back to work additional hours to recover the schedule in the event of unplanned technical or interface problems, weather delays, late flight arrivals, among others.

The Cutover Planning:

- Defines scope of the Cutover Activities through the development of a detailed Cutover Plan, taking into consideration the System Architecture, products being installed, logical sequence of cutover steps, identifying the pre-requisite requirements, along with consideration to the operating constraints of the existing system.
    - Decomposes the cutover or activities with enough granularity in duration that it represents a complete work activity that can be accomplished in a day or a week
        - Reflects detailed cutover activities in the Detailed Project Schedule
        - Allows for adequate progress tracking at the site
- Develops work packages associated with each of the cutover steps and assigning them to specific phases and specific steps within each phase that can be completed within the system's operating constraints.

- o Ties the detailed cutover activity to the associated detailed test activities
  - Reflects detailed testing activities in the Detailed Project Schedule
  - Allows for adequate progress tracking at the site
- o Defines site access requirements for each work activity
- o Defines all of the associated tools, test equipment, test procedures, equipment, software any pre-requisites
- o Defines "backout procedures" and the re-validation test requirements that need to be fulfilled to reconfigure the equipment back to a previous configuration and to place it back into revenue operation, should a "mod" not be successful on a given night
- Allows adequate time at the end of the shift to reconfigure and retest the existing equipment installation so that it can be validated for revenue service or passenger carrying operation.
  - o Designs the cutover and testing activities with adequate forethought to permit the ease of backing out a cutover step in order to return the equipment to a configuration that is validated for revenue operation
  - o Defines clear "go/no go" criteria into the work plan for the shift to provide adequate time to restore the installation to a configuration that can be validated for revenue service or passenger carrying operation

The Testing, System Assurance and System Demo Planning:

- Includes a Site Acceptance Plan, which identifies all of the formal Site Acceptance Test Procedures, which will be used to validate the proper function of the equipment against the technical requirements of the contract in accordance to the System Compliance Verification Matrix.
- Defines a clear set of mutually agreed upon System Demo requirements at the end of each phase of the installation, cutover and testing and commissioning activity with the customer so that that phase can be accepted.
- Defines and adheres to an effective Site System Assurance process with the rigor of necessary "checks and balances" that does not impede nightly installation validation and readiness to return to revenue service operation.

The Logistics Planning:

- Defines the logistics of assuring that all of the adequate equipment and materials supporting the cutover are available at the site before they are needed.
- Defines any temporary barrier installation requirements to limit access, or to contain dust due to demolition activities (if applicable).
- Assures that adequate startup and warranty spares provisions are available at the site for the new equipment to assure that equipment failures can be addressed without impacting the contract reliability or availability performance.

- Defines staging area requirements or new vehicle delivery or insertion requirements (if applicable).
- Defines and aligns the procedure / process for scrap material disposition with regard to the customer process along with the associated customer approvals for any materials to be scrapped.

The Design of Equipment Cutover Features / Capabilities into the Deliverable Systems:

- Identifies and implements features in the system design and architecture to support the efficient switching between the existing equipment and systems and the new equipment systems being supplied on the contract (i.e., cutover cabinet (using keyswitch activated cutover relay interfaces and bussbar bypass jumpers), octopus cables, A/B switchover boxes to reconfigure serial links, reconfiguring parallel I/O system interfaces through interchanging, connecting disconnecting interconnect cables (where appropriate), or implementation of network interfaces).
    - To minimize wasted time to configure or reconfigure the equipment at the beginning or end of the shift
- Identifies equipment modifications to existing equipment that may be needed (early) to support the cutover installation and testing, and include these design activities into the Detailed Project Schedule to assure the design changes and associated materials, software, etc. are available to avoid delays at the site.
- Identifies and plans in advance, any intermediate software configurations needed to support interim equipment configurations and includes these activities into the Detailed Project Schedule so that they will be available to support not only the Factory Acceptance Testing but also site acceptance testing.
- Defines equipment space, access, and weight or clearance requirements, to fit the equipment into temporary and final locations in the equipment rooms including the need to repackage equipment to break multiple bay cabinet assemblies apart so that they can be transported down hallways.

The Documentation and Analysis Requirements Associated with the Major Cutover Phases:

- Defines the documentation and analysis deliverables that have to be provided as pre-requisites prior to beginning each phase such as:
    - O&M Manuals and Training Materials/classes for the O&M personnel to prepare them to maintain new equipment incrementally placed into revenue service
    - Safety Analysis Documentation are complete and submitted for customer review confirming the new equipment fulfills all safety requirements prior to placing it into revenue operation

- o Acceptance test procedures are complete, submitted for customer review & confirm all pass / fail requirements are verified prior to revenue service
- o Cutover Plans & procedures are complete, submitted for customer review & confirm all pass / fail requirements are verified prior to revenue service

The Verification of Design Details:

- Evaluates the incremental UPS power requirements for each of the incremental or interim configurations of the equipment for the existing equipment room, throughout the cutover process to assure that the UPS capacity is not exceeded.
- Evaluates incremental heat load for the incremental or interim configurations of the equipment in the existing equipment rooms throughout the cutover process to assure that the HVAC capacity is not exceeded.
- Evaluates incremental floor loads for the new cabinets to be installed in existing equipment rooms.
- Performs site inspections to visually inspect the fill factor for conduits and cable trays to determine if the existing installation can support the installation of additional cabling or if new conduits or cable trays need to be installed.
- Reviews "as-built" documents by the site personnel and engineering prior to the detailed design to mitigate any documentation errors.
- Assures the respective cutover steps or interfaces do not result in potential equipment damage to existing systems that have to be placed into revenue service at the end of the shift.
  - o If there is the potential of equipment damage to the existing equipment then adequate spares provisions need to be available at the site to mitigate any startup induced failures of that equipment

Potential Risks:

In spite of the effective planning of the phased installation cutover and commissioning, risks still exist when modifying a "brown field" installation. Following is a summary of general risks associated with the modification of a "brown field" installation:

- "As-built" drawings that do not accurately reflect the actual equipment installation, such as:
  - o Inaccurately documented design changes incorporated in the existing installation
  - o Field modifications incorporated by the O&M personnel, while performing corrective maintenance in the existing installation that have not been accurately documented or coordinated with the design engineers

- Field modifications incorporated by the customer or 3rd party suppliers in the existing installation that have not been properly documented
- Conflicting design requirements between former standards that original equipment was supplied versus new more restrictive standards relating to the new direct replacement equipment being provided.
- Latent defects, which may be discovered when modifying existing equipment that is being modified, within fixed budget requirements.
- Component obsolescence issues when having to supply direct replacement equipment for existing older systems, when required by contract.

Lessons Learned:

Based on the experiences gained on the MHJIT Expansion Project for the Hartsfield-Jackson Atlanta International Airport (APM) System, the following lessons can be shared:

- Accurate and effective communication of nightly configuration changes to the APM system equipment via passdown reports between O&M shifts is critical to maintaining reliable operation of the APM to assure the proper O&M response to system outages in light of ongoing cutover and commissioning of new equipment.
- Early initiation of a Failure Reporting Analysis and Corrective Action System (FRACAS) process to evaluate, assess and track equipment outages to accurately assess attributable outages to new equipment installation against un-attributable outages and to identify corrective actions, as needed.
- Obtaining customer approvals beforehand for any equipment modifications for the existing equipment being modified including review and approval of the retesting / validation procedures associated with the desired design change(s).
- Prior to performing an activity, conduct a thorough walk-through and site inspections during each step of the design, installation, cutover, or testing activity and associated procedures to identify / resolve problems before doing the activity.
- Thoroughly review the site installation, test procedures, and equipment interface documents; i.e. the track plan, schematics, equipment room layouts, Interface Control Descriptions (ICD's), and Site Acceptance Test Procedures.
- Site Startup Team conduct-detailed visual inspections of the equipment, rooms, and interfaces to identify potential documentation errors.
- Install software changes a couple days in advance of need for acceptance testing, to verify that the software is stable before beginning the actual formal acceptance testing and commissioning activity at the site.
- Successful cutover planning applies a high level of rigor to review the details of the cutover plan, procedures and equipment interfaces while ensuring there is sufficiently flexibility to not only minimize unexpected problems, but provide the ability to react to site problems and also recognize and exploit opportunities.

- Build time into each phase to leverage early setbacks and make corrections and adjustments so that a smooth flow of the work can then be achieved.
- Regularly perform a site "lessons learned" review process to identify and implement improvements and efficiencies to the cutover, testing and commissioning process to realize time savings.
- Assure that all "as-built" conditions at the site are accurately documented and the associated design documents and drawings are updated at the conclusion of the project and throughout the APM system's life cycle to provide an accurate basis for subsequent follow-on projects.

Conclusion:

The expansion of the APM system in Atlanta coupled with the upgrade of automatic train control for an existing concourse while continuing APM system operation at the world's busiest airport is one of the most complex undertakings in the APM business community.

Bombardier's disciplined approach of implementing complex "brown-field" APM System enhancement projects will leverage its knowledge and intimacy of customer needs to effectively plan APM System cutover activities as outlined above.

Based upon actions to date concerning the phased installation, cutover and testing employed on the MHJIT Expansion Project for the Hartsfield-Jackson Atlanta International Airport (APM) System, Bombardier expects to implement a complex expansion of one of the busiest airports in the world without delays or service disruptions in accordance with contract technical and schedule requirements.

* *BOMBARDIER*, *CX-100* and *CITYFLO* are trademarks of Bombardier Inc. or its subsidiaries

THE IMPACT OF APMS ON PROPERTY VALUE

David D. Little, AICP[1] and Margaret Picard[2]

[1]Principal, Lea+Elliott, Inc., 44965 Aviation Drive, Suite 290, Dulles, VA 22030; PH 703-537-7418; ddlittle@leaelliott.com

[2]Project Planner, Lea+Elliott, Inc., 44965 Aviation Drive, Suite 290, Dulles, VA 22030; PH 703-537-7450; mpicard@leaelliott.com

ABSTRACT

Landside airport APMs that go "off" airport property typically require multiple landowners and government agencies to agree on a multitude of elements—no easy feat. One of the most important of these elements is project finance and one of the most important components of project finance is the real estate value enhancement that a landside APM provides.

The positive impact of transit on property values has been well documented over the last four decades. A strong correlation between property value and proximity to fixed guideway transit (rail or APM) has been found for properties that are within walking distance of a transit station. That enhanced value of transit access is due to a number of factors, two of which are the relative cost and convenience of the private-auto transport option. Thus, increases in gasoline prices and in roadway congestion help to increase the value of transit access.

This paper helps define and quantify the positive impact that landside airport APMs can have on real estate property value and how that value enhancement can be "captured", to help improve a project's financial picture.

INTRODUCTION

There are currently forty-one APMs operating at airports worldwide. While the majority of these APMs are "airside" or beyond security serving aircraft gates, there are now a total of thirteen landside airport APMs. Landside APMs connect the main terminal with landside facilities such as parking garages, rental car centers, regional rail stations, etc. The frequency of landside APM implementations has been increasing, as shown in Table 1 below, with two in the 1980s, three in the 1990s, and eight to date in the first decade of the new century. A number of these APMs go off airport property connecting to other, non-airport facilities. An airport landside system under construction at Atlanta will join the list later this year serving both a rental car facility and hotel/convention center. Future landside systems in design and construction respectively at Miami and Phoenix will also serve off-airport facilities.

Table 1. Landside Airport APMs

Airport	Opening Year	Vertical Nature	Guideway Length (Dual-lane)	LANDSIDE FACILITIES SERVED			
				Parking	Rental Cars	Region. Rail	Off-AP Facility
Houston [1]	1981	Underground	3.2 km single-lane				No[2]
London Gatwick	1987	Elevated	1.2 km			X	No
Tampa [1]	1990	In garage	1.0 km single-lane	X	X	-	No
Chicago	1993	Mostly elevated	4.3 km	X		X	No
Newark	1996	Elevated	5.1 km	X	X	X	Yes
Minneapolis/ St. Paul [1]	2001	Underground	0.4 km	X	X	-	No
Düsseldorf	2002	Elevated	2.5 km	X		X	Yes
New York—JFK	2003	Elevated	13.0 km	X	X	X	Yes
Birmingham	2003	Elevated	0.6 km	X		X	Yes
San Francisco	2003	Elevated	4.5 km	X	X	X	No
Singapore Changi [1]	2006	Elevated	1.0 km				No
Toronto	2006	Elevated	1.5 km	X		future	No
Paris- CDG [1]	2007	N.A.	3.3 km	X			No

Notes:
1. Airport has both the landside APM listed here and a separate airside APM.
2. Houston has station at an on-airport hotel.

The longer distance (and cost) and the multi-stakeholder nature of landside APMs add to the implementation challenges of each individual project. Such challenges include the feasibility of the project in terms of affordability. Benefits must exceed costs for each stakeholder of the project; stakeholders that may have very different goals and objectives. The system must be affordable to the stakeholders in relative and absolute terms. These are among the many elements that must align for such a project to be built. Many more systems have been planned than have been built.

The affordability or finance star is among the most important. Project costs, funding and revenues are all critical components of the finance equation. Landside APM costs typically vary in system length and "vertical nature" (elevated, at-grade, or tunnel. In terms of funding, landside APMs going off airport property are often eligible for a wider range of funding sources such as regional and/or federal transportation funds.

System revenue sources may also increase with off-airport APMs, and are evolving in ways similar to urban rail systems. While airport APM systems do not typically charge a fare or ridership fee, systems can generate revenues through real estate value enhancement. This is the additional value to a property created by the increased local and potentially regional access of the APM system.

AIRPORT INDUSTRY TRENDS INFLUENCING REAL ESTATE VALUE

Two current trends in the airport industry that are helping to increase the level of this real estate value enhancement: Regional rail access to major airports and the Airport City concept. The following sections explore these trends and their relationship to real estate value and APMs.

Regional Rail Access to Airport Trend

Rail access to major airports has been steadily increasing in recent years. The range of rail technologies providing metropolitan access includes light rail, rapid (heavy) rail, and commuter rail. Rail technologies serving airports providing inter-city access include commuter rail and high-speed rail.

Regional rail access to a major airport presents challenges in terms of optimally locating the rail station(s) at the main terminal(s). The spatial requirements of the rail line connection to the station or stations within the congested airport landside with its roadways and parking facilities is a geometric challenge. The geometric constraints (minimum horizontal curves and maximum vertical grades) of the regional rail lend itself to level, straight alignments. Unfortunately, the landside environment at major airports can rarely accommodate such an alignment and consequently compromises are made that balance level of service with cost.

In some cases the rail system is brought in below grade (a cost compromise) so as to avoid other structures. In other cases the rail system and its airport station are located away from the main terminal (level-of-service compromise). Such locations require high-capacity connections to the main terminal such as landside APMs.

Major European airports have had rail access for many decades with some airports having multiple rail lines accessing their facility. Oslo, Zurich, Frankfurt, Amsterdam, Paris-CDG and London's multiple airports have routinely achieved rail mode shares of between 20 and 35 percent.

Rail access to major U.S. airports was slower to take hold but has grown significantly in the past 20 years. Rail mode share for the largest U.S. airports (passenger levels and land area) remains low due to their longer distance to the regional central business district (CBD). The U.S. airports with high rail mode share all have rail stations located near the airport's main terminal(s) entrance. A list of some of the major U.S. airports with rail access, sorted by rail mode share is provided in Table 2. More recently, airports in Asia and the Middle East have implemented rail access, or are in the process of doing so.

Table 2. U.S. Airports with Direct Connection to Rail

AIRPORT	MAP 2007	COMMUT. RAIL	RAPID RAIL	LIGHT RAIL	APM LINK	ACCESS YEAR	MODE SHARE	STA. LOCAT
Washington National	19		●			1977	14%	Near
Atlanta	90		●			1988	10%	Near
New York – JFK	46	●			+	2003	8%	Near
San Francisco	36		●		+	2003	7%	Varies
Chicago-Midway	19		●			1993	6%	NA
Portland	14			●		2001	6%	Near
Chicago O'Hare	76	●	●		+	1984 & 1996	5%	Varies
Newark	36	●			+	2001	5%	Varies
Baltimore-Washington	22	●		●		1980 & 1997	3%	Far
Philadelphia	32	●				1985	3%	Near
St. Louis	15			●		1994	3%	NA
Cleveland	11		●			1968	2%	Near
Minneapolis-St. Paul	35			●	+	2004	NA	Near

In addition, many major U.S. airports are currently planning to implement rail access to their facilities. Table 3 below provides a list of some of these airports and the type of rail service planned to access the airport.

Table 3. Major U.S. Airports Planning Rail Access

Airport	Access Type
Dallas /Ft. Worth	Direct LRT and Commuter Rail
Denver	Commuter Rail
Ft. Lauderdale	TBD
Honolulu	Automated Rail
Las Vegas	APM (monorail)
Miami	Rapid Rail + APM link
Oakland	Rapid Rail + APM link
Orlando	TBD
Phoenix	LRT + APM link
Salt Lake City	LRT
Seattle Tacoma	LRT
Washington Dulles	Rapid Rail

Airport Cities Trend

The Airport City, or high-density commercial (offices, hotels, retail) development within proximity of the airport's main terminal is another growing trend in the industry. As with regional rail, these developments compete for valuable space on (or near) airport property. The developments are typically located some distance away from an airport's main terminal. The roadway requirements and traffic generated by such developments in a sense "compete" with those of the airport users (airline passengers and airport/airline employees) and therefore the provision for mass transit access to/from the Airport City directly benefits the Airport by reducing non-airport traffic on the airport's and regions roadways. In some cases an APM can provide such mass transit access.

Airport City development, also called Aerotropolis or Sky City at some airports, has occurred at a number of major airports and is planned for at many more. The Airport City concept is that the airport is more than just aviation infrastructure; it is a multimodal, multifunctional enterprise generating commercial development within and beyond its borders. The multiple functions include commercial office, retail, hotels, convention/exhibition centers, free trade zones, cargo processing and distribution. The Airport City attracts commercial tenants with its accessibility and connectivity to customer and enterprise partners.

The Airport City can also develop a brand image that further enhances property value. Airports that pioneered the Airport City concept include:

- Frankfurt
- Amsterdam
- Paris CDG
- Hong Kong
- Seoul
- Kuala Lumpur

Frankfurt Airport is a good example of how a major international hub airport (home to Lufthansa Airlines) is expanding well beyond its initial air transportation function to become a bustling commercial center in itself. In addition to the transport of airline passengers and cargo, Frankfurt Airport City includes shipping, restaurants, hotels and offices. Similarly, Amsterdam's Schiphol Airport, the commercial development helps the Airport diversify its revenue sources. Its CEO, Gorlach Cerfontain, said in 2007 "Once again, our unique Airport City proved its worth, with the consumers and real estate business areas contributing 75% of the company's operating result." Non-aviation revenues accounted for 43% of total revenues for the airport that year.

Kuala Lumpur International Airport had recently earmarked approximately 2,700 acres, or just over ten percent of its total area, for the development of an airport city. The airport city development's further commercial center would house a retail district, trade center, exhibition center, hotels, and a transportation hub.

Many airports have major Airport City developments in the planning, design and/or expansion phases including Dallas/Ft.Worth, Dublin, Abu Dhabi, Dubai, Detroit, Washington Dulles and numerous airports in China. In general, major airports are increasingly developing commercial office parks, hotel/conference centers, and even educational and cultural centers for the traveling customer and the local population.

Dallas/Ft. Worth has land available for commercial development over an 8,000 acre area. The airport focused on logistically oriented facilities in its 400-acre International Commerce Park which opened in 2000. Meanwhile, at the opposite end of the airport is the planned Southgate Park which will consist of hotels, restaurants and commercial office buildings.

The economic slowdown of the late 2008 and 2009 will certainly slow the implementation schedules at some of these airports, but the relative strength of a well integrated commercial development at a major international airport will continue in the future. The high density nature of Airport City commercial developments and the need for airports and regional governments to keep their surrounding roadway network from becoming overly congested both lend themselves to the high level of service (capacity, frequency, image, non-emission) of APMs.

For the thirteen current airports with landside APMs, connecting into the APM either at current station(s) or via a system extension is a relatively simple way to improve access to/from and within an Airport City development. In 2009, Atlanta will open the fourteenth such landside APM, a 2.3 km system with three stations including one at a major convention center. The APM station at the airport main terminal is located adjacent to the airport's regional rail (MARTA) station, connecting the convention center to the region's 36 other MARTA stations. Clayton County is currently pursuing a feasibility study of Airport City development.

Even without a current Airport City development, Atlanta International Airport illustrates the economic impact and operating revenue potential of a major international airport with approximately 55,000 people working at the airport, their 400,000 regional jobs tied to the airport, and a 5 billion dollar direct and indirect impact to the regional economy. Rental income help contribute approximately 20 percent of the airport's operating revenue.

In general, Airport City developments help to both expand and stabilize airport revenues whose traditional dependence on landing and parking fees has seen high variability over the years. Airport City development is considered to continue as a strong long-term trend at major airports.

Airport Rail Access + Airport Cities = APM

In the right combination, the two industry trends of rail access and Airport City development help improve the revenue side of the financial equation for a landside APM through enhanced real estate value. Table 4 below provides a qualitative assessment of the applicability of a landside airport APM given the potential range of airport rail access and of Airport City development density.

		Table 4. Rail Station Proximity and Commercial Density relating to Landside APM Applicability		
		Proximity of Rail Station to Main Terminal		
		None	Near	Distant
Airport City Commercial Density	None	○	○	◐
	Limited	◔	◐	◑
	High	◐	●	●

Key: ○ = Poor ◑ = Good ● = Excellent
 ◔ = Poor/Moderate ◐ = Moderate

The focus of the remainder of this paper is on the scenario in which the airport has regional rail access with a station located at (close proximity) the main terminal and a high-density commercial development (Airport City) on or near airport property.

REAL ESTATE VALUE ENHANCEMENT

In the landside airport scenario described above, a regional rail station at the main terminal and high-density commercial development near the airport, a landside APM becomes an extension of the regional rail system. In this example the commercial development is not served by the regional rail. The APM is assumed to serve airport-related facilities, such as parking and rental car, as well as the commercial (office buildings, hotel and retail) development. The ridership on the APM therefore consists of office workers, airline passengers and airport employees. The peaking characteristics of these different rider groups do not tend to coincide, resulting in a more even distribution of bi-directional ridership throughout the day. In the morning there is an office worker ridership surge in the terminal-to-office direction, but an opposite ridership flow of airport employees and airline passengers.

In this scenario, the commercial development is now connected to the regional rail system providing commuter access to the office, hotel, and retail workers from around the region. Given the typical location of major airports on the periphery of the urban core, these rail commuters would be traveling in the counter-commute direction: outbound from the core in the morning and inbound towards the core in the evening. Such non-peak directional ridership tends to "fill empty seats" for the rail transit authority, adding ridership but not requiring additional capacity (trains).

The impact of rail transit on real estate values has been documented for many years. Many of these studies have been site (city) specific but a synthesis of this research can be summarized as follows:

- Proximity to rail positively impacts property values.
- The increase in regional accessibility is the most important factor in the value enhancement.
- Density bonuses tied to rail access can have a very positive effect on value.
- Value enhancement is greatest at the station and declines as the distance from the station increases.
- The correlation between value enhancement and distance depends on local ambient conditions.
- Value enhancement varies by type of regional rail (rapid rail vs. light rail).
- Local traffic congestion, and the availability and cost of parking influence the real estate value enhancement.

QUANTIFYING REAL ESTATE VALUE ENHANCEMENT

The estimation of the value of real estate value enhancement due to a landside APM takes into account site specific as well as regional data. Site-specific estimates for commercial (office and retail) value are obtained in terms of rental charges assuming no transit access. In cases where there is no comparable office space near the subject airport, historical data comparing other airport to downtown value ratios can be used.

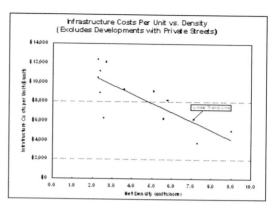

The extent of the rail system's coverage within the greater urban region is an important factor. Correlations between daily regional rail system-wide ridership and the airport's rail modal share have been determined in prior studies. These correlations can then provide guidance on whether a specific airport real estate value enhancement would be in the low, medium, or high range of values.

The specific Airport City development size (total floor area) and the distance of specific buildings to the future APM station are then calculated. A recent such estimation at a major international airport found value enhancements measured in rental premium of rail transit access to over 300,000 square meters of office space were within walking distance of two of the landside APM stations would cover approximately 60 percent of the O&M costs of APMs. Looked at another way, if the annual rental premium was capitalized, it would "cover" between 15 and 20 percent of the APM's capital costs. These metrics, such as the percent of capital costs covered or of annual O&M costs, are a function not just of the amount of commercial space accessed but of the APM system's capital and O&M costs. In the above example, the planned system was of low to moderate length compared to the existing landside APMs (Table 1).

For landside APMs of shorter length serving a larger commercial development, the real estate value enhancement can significantly improve the project's financial picture. Longer APMs have the opportunity, with multiple stations, to compensate for their higher costs by serving commercial developments at multiple stations. A critical factor is the development's density as measured in floor area ratios which is a function of building heights and property setbacks. A cluster of 100-floor buildings might pay for the entire APM but not be practical (or allowable) in an airport environment. They key is to explore the density ranges in terms of the cost/revenue impact of the APM system. This needs to occur in the early planning stages of the airport landside.

CAPTURING REAL ESTATE VALUE ENHANCEMENT

Capturing the real estate value enhancement has traditionally been a real challenge for rail transit authorities, as the value typically went directly to the property owners adjacent to, and near the rail station. For an airport landside APM the capture of value should be more straight-forward: the value is captured by the airport for property on the airport and captured by project stakeholders for off-airport property. If the transit authority or another government authority is one of the stakeholders then value capture could be achieved via mechanisms such as property tax increment revenues, special assessments, nonlocal public match and/or user fees.

CONCLUSION

Every major airport has its own unique characteristics and surrounding urban environment. The landside mobility needs of major airports are met with either buses, APMs or a combination of the two technologies. These two technologies provide very different levels of service in meeting an airport's landside mobility needs.

APMs can provide the benefits of high-capacity, convenient, fast transport that meets the mobility goals and objectives of the airport, surrounding commercial developments, and the greater regional government. APMs provide a higher level of service in their grade-separated, exclusive right-of-way compared with that of a bus in mixed traffic on the airport's road network. The APM's higher level of service typically comes at a price: a higher capital cost--though operating costs tend to be lower, and life cycle costs are often very similar over 25 years. But the higher initial costs can stop a project even if its longer term life cycle costs are competitive (or even lower). Therefore, it is a critical part of an APM feasibility study or multimodal study to consider the additional airport revenues that rail access to commercial development provides.

When a landside APM serves a commercial development in addition to the usual airport facilities (parking, rental car, etc.) and can connect that commercial development into a regional rail system, then the real estate value enhancement can help offset the higher capital costs. The degree of this offset is a function of the size of the commercial development served. The property value enhancement, property tax revenue increase, reduced congestion/emissions all combine to help meet the wide range of needs of their diverse group of stakeholders.

The growing trends of rail access to airports and high-density commercial development on or near airports can combine to make APMs a center piece of landside airport planning. The resulting landside can help to diversify airport revenues and help the airport achieve both its own and the regions sustainability goals.

If the elevated nature of an APM provides access to an otherwise inaccessible property than a much greater amount of value enhancement is attributable to the APM.

London Heathrow Terminal 5 APM Project

Jon Brackpool
Portfolio Technical Leader – Heathrow, APM & PSE
BAA Capitol & Solutions
44 (0) 1293 507603
jonbrackpool@t5.co.uk

Glenn Morgan
Project Director, Heathrow and Gatwick APM Projects
Bombardier Transportation, Systems Division
1501 Lebanon Church Road
Pittsburgh, PA   15236-1491 USA
412-655-5244
glenn.morgan@us.transport.bombardier.com

Abstract

Demand for air travel in the southeast of the United Kingdom is expected to double in the next 20 years. London Heathrow Airport, one of the world's leading international destinations is at the center of a network of London-based airports. Heathrow previously consisted of four terminals. A fifth terminal, known as "T5" was developed and constructed by BAA (formerly the British Airport Authority).

To execute the T5 project, BAA selected an innovative approach by implementing a true partnering philosophy. Design firms and major system suppliers were pre-selected based on qualification submittals.

The true uniqueness of the T5 project was in not utilizing a typical and traditional contract as would normally be used in the construction industry. It was a very different arrangement for all parties.

This paper elaborates on the details of this unique project approach: a success story that uses an unconventional partnering agreement and a design philosophy that is unprecedented in the industry, especially when considering a project of this complexity, under considerable constraints, and of such magnitude and duration.

Introduction

A fifth terminal at Heathrow, known as "T5", was developed and constructed by BAA. (See Figure 1) The T5 Programme consisted of 16 major projects and an Automated People Mover (APM) System that was included as part of the infrastructure to allow passengers to move conveniently and efficiently between the main concourse (T5A) and remote concourse (T5B) delivered under phase 1 (March 2008) (see Figure 2). A second remote concourse (T5C) is under phase 2 construction (Dec 2010).

Figure 1 - Terminal 5: Views of the terminal and concourses

The T5 "toast rack" layout is extendable to meet the future strategic development plans of Heathrow and similar to the configuration of the US.-based Atlanta and Denver airports. In addition to the *BOMBARDIER* INNOVIA** APM, the Heathrow Express and the London Underground Piccadilly transit lines were also extended from Heathrow's central terminal area to the new T5 main terminal to provide direct links to central London. With the creation of a major bus hub and the direct link from one of the world's busiest city orbital motorways (M25), this led to a truly inter-modal transportation center capable of increasing Heathrow's passenger handling capability up to and above 90 million passengers a year with only two runways. All the guided transport systems (the APM, Heathrow Express and Piccadilly line) have also safeguarded expansion and extension capability beyond T5 in their designs.

The BAA realized at an early stage that considering the history of large construction programmes around the world, there would be a likelihood that:

- Costs could be 10-25% over budget
- Completion could be 3 – 6 months late
- There could be costly and time consuming claim disputes over change orders, retribution and liquidated damages
- Environmental damage and impacts could occur
- And most importantly, according to statistics during construction of major infrastructure projects, there could be loss of life

Figure 2 - T5 Concourse: A transit system departures platform

As the funding for the project was inextricably linked to finance obtained from the City of London and with considerable losses to BAA and Shareholders for late delivery, to execute the T5 programme, BAA decided to implement a true partnering philosophy to attempt to avoid the pitfalls of so many other large construction programmes. In 1998, BAA selected the former Adtranz, which was acquired by Bombardier in 2001, as the APM supplier of choice and subsequently signed what was called a "People Contract", as with all the other major suppliers, also known as 1st tier supplier, on the programme. Work under this arrangement was similar to what a transit consultant would traditionally perform in defining system requirements to the customer's brief and ultimately the APM system specification. BAA recognized that by working with suppliers early on to develop requirements, innovation and designing for manufacturing benefits could be realized. This initial work included feasibility studies, design optimization, system definition and schedule integration with Bombardier ultimately contracted as the APM supplier for T5 under a cost plus fixed fee (fixed profit) arrangement: the T5 Agreement.

The T5 Agreement was no traditional contract typical to the industry. All the major suppliers involved were pre-selected as "best in class" and co-located in a true "under one roof" team/working arrangement from the very beginning. In addition, BAA assumed *"all the risk"*; there were no liquidated or consequential damages, and exchange rate and inflation risks were held at the customer level. It was a very different arrangement for all parties involved.

Additionally, the *INNOVIA* APM system (see Figure 3), which is critical to the efficient operation of T5, is driverless and operates in a tunnel environment within the airside security restricted zone, and was required to meet the constraints of this challenging operating environment. These challenges included meeting the highest standards set by the new UK fire regulations (BS6853) and for the underground rail vehicles being fully accessible to all through compliance with UK Disability Discrimination Act regulations [similar to ADA] and maintaining security segregation between international arriving and departing passengers to UK and EU regulations. Since it was also one of the first applications of a moving block signaling system in the UK, deploying the *BOMBARDIER*CITYFLO** 650 train control technology, all of the relevant statutory bodies also had to be engaged and partnered with.

Figure 3 - An *INNOVIA* vehicle sees daylight for the last time before being lowered into the transit system tunnel

The T5 Agreement

On most projects, customers, consultants and suppliers would typically form independent teams, each delivering its own work scope. It was decided very early on that for a project of the size, complexity and longevity of T5 it would be almost impossible to manage T5 like a "normal" project. This was especially so for risk management.

BAA found a way of addressing this situation with a ground-breaking and legally binding document: the T5 Agreement. The decision to introduce the T5 Agreement rather than a contract was a bold move for BAA. The foundation of the Agreement was trust, working together, avoidance of conflict, openness and the integrity of all First Tier Suppliers. (See Figure 4) This was evident in the insurance policy arrangements. BAA paid a single premium for the multibillion-pound project to the benefit of all the main suppliers, providing one insurance plan for coverage of the main risks. Risks resulting from acts of God, industrial action, legislative or government change or failure to deliver were all covered by BAA. Supplier's exposure was limited to loss of profit or insurance excess payments. This allowed for the removal or risk monies and contingency from suppliers budgeted costs.

Figure 4 - The Transit System teams celebrate a major milestone by installing the *INNOVIA* vehicles into the tunnels at T5

The T5 agreement also integrated and co-located teams and implemented a "best supplier for the job" approach. It meant having people work in the spirit of cooperation and working to deliver at a minimum, best practice, and continually striving to achieve and maintain exceptional performance; a culture of problem solving versus finger pointing.

A team-based incentive plan was also put in place providing additional bonuses if exceptional performance was achieved across various stakeholder teams.

Ultimately, the T5 agreement helped suppliers make the right choices and deliver the program without worrying about most typical commercial implications. It encouraged suppliers and teams to work in a collaborative way to solve problems together and get the job done.

The Challenge

Some of the key challenges with the T5 APM system involved its integration into a constrained site located between two operational runways, (see Figure 5) working to a very tight schedule under constant review by local authorities and resident groups to contain noise, road traffic, dirt and dust and while under constant review from various statutory parties and stakeholders, including health and safety (the T5 site twice achieved a duration of two million hours between reportable accidents). BAA was also the first customer to order Bombardier's *INNOVIA* technology (See Figures 7 and 9) and was the second to put it into service (Dallas Fort Worth International Airport's was the first to commence operations). All of the above, coupled with the very unique T5 project specific partnering arrangement, offered a situation which called for an innovative approach, as well as a completely different type of project mind set.

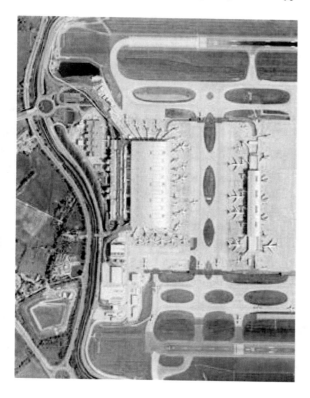

Figure 5 –
T5 - The constrained nature of the site, between two fully operational runways and the busiest motorway in Britain

T5 Statistics

- Longest public inquiry in UK history
- 4.3bn GPB
- 30-35 million additional passenger capacity
- 60,000 individuals having worked on the site
- Up to 8,000 workers on site at any one time
- New 87 metre (285 ft.) tall traffic control tower
- 13.5 km (44,300 ft.) of new bored tunnel
- New spur road linking T5 to M25 motorway
- Sole occupancy of T5 by British Airways
- 2 of the longest free standing escalators in Europe (see Figure 6)
- 650-bed 5 star hotel
- New 4,000 vehicle car park
- New baggage system
- New PRT linking T5 to preferred parking
- New 6 platform underground rail station for Piccadilly line and Heathrow Express extensions (See Figure 8)
- 2 rivers diverted
- 90% of all waste recycled

Figure 6 - The escalators from departures level to the transit departures platform – two of the longest free-standing escalators in Europe

T5 APM Facts

- Moves 5800 to 6800 passengers per hour at peak (more than Piccadilly and Heathrow Express rail lines combined)
- 90 second headways
- 95% of passengers able to board 1st train
- Minimum system availability of 99.75%
- 50 km/h (31 mph) maximum speed
- *CITYFLO* 650 moving block
- Two 3-car trains initially operating as a shuttle
- Expansion capabilities to pinched loop with 4-car trains
- Switches: 5 left hand pivot, 4 right hand pivot and 2 turntable (see Figure 7)
- Highest compliance for fire performance of any system in the U.K.; meets cat 1a of BS6853
- Statutory approvals obtained at every stage
- Achieved every interim milestone on time
- Zero reportable accidents throughout the life of the project

Figure 7 - T5 concourse A crossover: Transit system guidebeam and switch installation

The Experience

BAA leadership developed and implemented the concept of "Technical Leadership" (TL) on T5, which focused on delivering the right product, right the first time:

- "One team" devoid of any company branding
- Trust and open approach to challenges
- Single model environment
- Best team/best person for the role
- New solutions/rapid responses
- Openness/communication
- Clear requirements
- Empowerment to deliver

Figure 8 - The new London Underground, Piccadilly Line terminal platform beneath Terminal 5

The Foundation of the Experience

To mitigate customer-held risk, BAA implemented programme-wide project controls through mandatory procedures and reporting covering, but not limited to: safety; cost; programme and quality. Each was considered of equal importance and in the words of the T5 Programme Construction Director "not an OR but an AND conversation". Risk identification was undertaken jointly with all parties contributing and the most appropriate people owning. Responsibility for management of risks was held by each respective supplier within their respective project area.

Technical collaboration between interfacing suppliers and their contractors and clear direction on technical matters was essential to ensuring quality and 'right first time' design and construction. To this end, BAA established "Technical Leadership" a small customer team of no more than 25 people with expertise in different engineering and design systems to functionally support across all project areas. Technical Leaders' accountability comprised engineering/design integrity and performance of their systems, including the APM. The role was to provide support by leading the technical community as one team, devoid of company alliance, enable delivery, facilitate successful integration, be assured and be the receiving customer.

The Technical Lead for APM also focused on achieving high reliability. A reliability programme was established with, and run by Bombardier using the principles of Failure Reporting, Analysis and Corrective Action [FRACAS]. The Dallas Fort Worth [DFW] implementation of *INNOVIA* (see Figure 9) and *CITYFLO* 650 technologies had preceded Heathrow T5. This was identified as an opportunity for learning, therefore relationships with DFW APM operation were established to share data. Together, with findings from factory and site acceptance testing, improvements were implemented to the benefit of T5, DFW and the *INNOVIA* product development. The APM System demonstration target of 99.75% availability was achieved within 30 days of the scheduled 60 days and achieved 100% on multiple occasions.

The APM system was also part of the Track Transit System (TTS) Project comprising suppliers delivering civil structure, tunnel ventilation, station fitout, vertical circulation and other specialist systems including CCTV, Access Control, Public Address, Fire Alarm, etc. The APM project was recognized within the T5 Programme for exemplary team work and repeated delivery to milestones by winning a number of T5 "high five" awards. The TTS team was also awarded the prestigious UK Construction News - Quality in Construction Award for Collaborative Engineering Design.

The Results

There is a history of large capital infrastructure projects that suffered delays and cost overruns.

Overall, the T5 project is viewed as being completed on time, to budget, with extremely high "ground breaking" levels of Health and Safety, to the highest quality and with care and consideration of the environment. The achievements were also true for the APM portion of the project:

- The T5 APM was the first wholly underground driverless passenger train in the U.K. to be delivered on time and to budget.
- The vehicle reached the highest possible standards of fire compliance; cat 1a of BS6853.
- The T5 APM achieved the required reliability in the absolute minimum time period of 30 days (System demonstration) with the required performance of 99.75% and with seven days at 100% and only 20 minutes of cumulative downtime.
- The APM programme met every interim milestone during construction and Testing and Commissioning for both T5 and the Statutory Approval process.

In addition, and on a broader scale, the BAA, in partnership with Bombardier and Balfour Beatty (the electrical installation contractor and infrastructure fitout team) was awarded the United Kingdom Quality in Construction award for "Collaborative Engineered Design".

Cited by the awarder as "a clear continuity of design thinking" and "a truly superb piece of engineering teamwork," the Track Transit System is described as a crucial part of the terminal infrastructure that is used 20 hours per day, 365 days a year. The collaboration between the BAA and Bombardier is portrayed as a complete team involvement and seamless design strategy that raised this project above the other award finalists in this category. The partnership is also lauded for its technical leadership (TL) that focused on delivering the right product, right first time, rather than concentrating on time and cost. The article continues: "At all times, the emphasis was on best product and even with Bombardier's design office located 5,000 miles away in Pittsburgh, the team operated very much as a single entity . . . the system worked first time, achieving its required performance level and making the 60-day post commissioning period an irrelevance."

Conclusion

The APM solution for Terminal 5 took maximum advantage of the T5 Agreement, enabling the entire effort. It was also clear at an early stage that a new approach (TL) was necessary in order to provide excellence in engineering and delivery to challenge the norm and seek innovation in design.

The T5 Agreement has in many ways set a benchmark for future major project procurements in terms of efficiency, safety and innovation

Figure 9 - An *INNOVIA* vehicle viewed from the servicing pit in the maintenance base at Terminal 5 Concourse C

BOMBARDIER, INNOVIA and *CITYFLO* are trademarks of Bombardier Inc. or its subsidiaries.

MIA Mover APM:
A Fixed Facilities Design-Build Perspective

B. M. Schroeder, P.E.[1]

[1] Parsons Corporation, 7600 Corporate Center Drive, Suite 500, Miami, FL 33126; PH (786) 464-1000; FAX (786) 845-7119; email: barbara.schroeder@parsons.com

ABSTRACT

As part of the Miami Intermodal Center (MIC) development program, the MIA Mover Automated People Mover (APM) System is Miami-Dade County's contribution to the MIC Program. The MIA Mover is under development by the Miami-Dade Aviation Department (MDAD). MDAD is currently implementing the MIA Mover development under a DBOM (design-build-operate-maintain) delivery method, with the Phase I design-build components scheduled for completion in September 2011. The MIA Mover will operate on a 1.27-mile elevated guideway between the airport terminal (MIA Station) and the intermodal center (MIC Station).

Under design-build delivery, fixed facility design and construction activities are interconnected. This presents unique challenges for the fixed facility design-build contractor and the owner. A relatively new procurement technique for APM systems in the United States, the design work is phased so construction work can be initiated prior to completion of the design; a foundation to roof/super-structure (bottoms-up) approach is deployed to accelerate the work. Engineers and contractors coordinate closely to address potential field conditions during the early design development. The design-build contractor and the owner meet frequently to address opportunities and constraints to expedite reviews and approvals. As with every transportation project, cost is a dominant factor. With a budget fixed in 1996, the design-build contractor developed and negotiated multiple technical and administrative alternatives to design and construct the 2008-2011 MIA Mover APM System within the owner's budget constraints.

Incorporating a new transportation facility into an existing built-out environment presents distinguishing design and construction opportunities and constraints. The MIA Mover eastern terminus (MIC Station) is currently under construction by the Florida Department of Transportation (FDOT) as part of the MIC program. MIC Station construction will be complete before MIA Mover fit-out work begins in 2010. The western MIA Mover terminus (the MIA Station), located between two existing parking garages, uniquely incorporates the APM maintenance and storage facility below the station to generate operating efficiencies and cost savings.

BACKGROUND

Miami-Dade County Aviation Department (MDAD) selected the Parsons-Odebrecht Joint Venture (POJV), as the design-build contractor to design, build, operate and maintain (DBOM) the MIA Mover Automated People Mover System at Miami International Airport (MIA). This $250 million, 1.27-mile transportation facility features an automated, rubber-tired people mover connecting the Miami Intermodal Center (MIC), a consolidated rental car, long-term parking facility, and local and regional transit services to the MIA main terminal. The MIA Mover will shuttle airport users and workers between the stations at the MIC and the MIA terminals.

The MIA Mover will serve the growing number of MIA passengers who rent vehicles or use public transportation to reach their local destinations. The MIA Mover will also provide transportation for the daily work commute of airport and airline workers to the airport. MIA accounts for approximately one out of every four jobs in Miami-Dade County.

The MIA Mover's elevated dual guideways enable trains to travel in both directions simultaneously. Its approximate path runs east from the MIA Station to Central Boulevard, over Le Jeune Road, and along NW 21st Street, curving north into the MIC Station directly above the MIC terminal access roadways (***Figure 1. MIA Mover Site Map*** and ***Figure 2. MIA Mover Guideway Traversing Central Toll Collection Facility***).

The guideway is elevated approximately 40 feet above grade and is supported on concrete piers typically spaced at 120-foot centers. Foundations consist of either augered cast-in-place piles or drilled shafts, depending on service and access constraints. Conventional reinforced concrete hammerhead piers support standard bulb T beams. As part of the MIC and connecting road works, FDOT has constructed several of the guideway foundations.

The MIA Station is located at the airport terminal on the third level between the Flamingo and Dolphin parking garages (see Figure 1. MIA Mover Site Map). Passengers access the terminal via existing walkways. Vehicular traffic will pass under the MIA Station/Maintenance and Storage Facility (M&SF) at the ground/street level. The M&SF is located on the first level, allowing 16'-6" clearance above the roadways. The station platform rests above the M&SF on the second level (***Figure 3. MIA Station Cross-Section***).

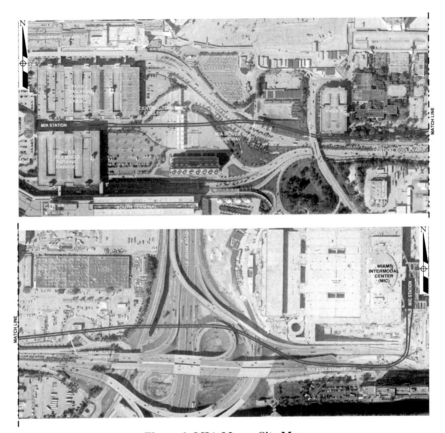

Figure 1. MIA Mover Site Map

Figure 2. MIA Mover Guideway Traversing Central Toll Collection Facility

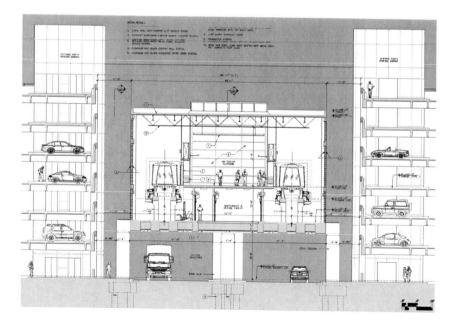

Figure 3. MIA Station Cross Section

The overall MIA Station footprint is approximately 32,500 square feet. Concrete columns are supported on deep foundations. The first and second levels bear on concrete slab floors and are enclosed by concrete masonry unit (CMU) walls. The steel framed roof and most of the CMU walls are encased by cladding and standing seam.

The MIC provides multi-modal connections to MIA, making the MIA Mover APM System a vital component for transferring airport customers to the MIA terminal. The MIA Mover connects the airport terminals to: Miami-Dade County's local and regional bus service; Amtrak; Tri-Rail (South Florida commuter rail service); and Metrorail (Miami-Dade County's heavy rail service), upon completion of the MIC-Earlington Heights Metrorail Extension project.

As part of the MIC program's decongestion strategy on roadways in and around the airport, rental car shuttles will be eliminated and replaced by the MIA Mover. Current estimates project that once the MIA Mover is operational, 30 percent of current vehicular traffic will be removed from the roadways, greatly reducing passenger/vehicular traffic conflicts at the terminal curbs.

DESIGN/BUILD STRATEGIES

The MIA Mover APM System DBOM contract comprises two phases. Under the design-build contractor responsibility, Phase I includes: fixed facilities design (MIA Station, guideway, east substation, MIC fit-out, and ancillary roadway, drainage, and utility systems, design management, and fixed facility design services during construction); operating system design (automated, rubber-tired people mover); and fixed facilities construction. Phase I requires design-build contractor responsibility for analysis, manufacture, supply, fabrication, assembly, factory testing, shipping and installation of the initial operating system (provided under subcontract for the operating system supplier).

The design-build contractor will provide on-site inspection and testing of the fixed facilities during construction and is responsible for the on-site integration and verification testing and other preparations for start-up of the initial people mover system by the operating system supplier. The design-build contractor provides related project management, control, and administration, including independent quality oversight, a comprehensive safety program, business services, project controls, construction management, field engineering, and oversight of construction operations by trade subcontractors.

Phase II covers operations and maintenance (O&M) of the initial system during the five (5) year period after system start-up, with two five-year extension options.

Fixed Facilities Work Plan

Schedule and cost drive the successful delivery of the MIA Mover project under the design-build delivery mechanism.

Schedule. An accelerated design and construction schedule is a primary advantage of the design-build project delivery mechanism. To develop schedule efficiencies, the MIA Mover design-build contractor organized design activities to facilitate construction sequencing. Organizing the design work into realistic and feasible components was an initial priority for meeting schedule requirements. The Owner defined thirteen (13) Design Units (DUs) to complete the fixed facilities. The design-build contractor consolidated the work into six (6) design units to achieve additional time savings:

- Design Unit I: MIA Station/M&SF
- Design Unit II: Guideway from MIA Station to Pier WB2
- Design Unit III: Guideway from WB2 to MIC Station
- Design Unit IV: East Substation
- Design Unit V: MIC Station Fit-Out
- Design Unit VI: Park 8

The development of the fixed facility design generally follows the four contract-stipulated submittal stages (preliminary design, in-progress design, final design and readiness for construction). To expedite the design development to meet schedule and cost constraints in a design-build environment, DUs I, II, and III are divided into component phases to advance the designs to readiness for construction before the full fixed facility design unit is complete. These phases represent a bottoms-up construction approach, beginning with the foundations and utilities, then station shell and guideway substructure, and concluding with the station finishes and guideway superstructure.

This organization of the work enables the design-build contractor to initiate construction on the foundation works as the design achieves readiness for construction. The above-ground works progress through design following the Owner-stipulated review process. For the MIA Station/M&SF (DU I) and guideway (DUs II and III) facilities, design submittals constitute the progress to build the works.

The foundation and utility designs are fast-tracked through preliminary design, permit set, and readiness for construction submittals. Design development for the permit set combines the in-progress and final design level work. Station shell and finishes and guideway substructure and superstructure design development follow the conventional submittal approach (preliminary, in-progress, final design, readiness for construction).

Preliminary Design Submittal. The preliminary design submittal (PDS) advances the concept level design developed by the Owner and further detailed as part of the design-build contractor's proposal (during the procurement period from December 2004 through June 2008). During preliminary (approximately 35 percent level of design), the design-build contractor identifies additional opportunities to expedite the schedule and gain cost efficiencies. The design for each component phase (foundations and utilities, station shell and finishes, and guideway substructure and superstructure) is addressed in this submittal.

Foundation and Utilities Permit Set Submittal. During the foundation/utility permit (FUP) submittal, the design-build contractor focuses on utility identification and location, subsurface investigation and geotechnical evaluation, and augered cast-in-place pile, drilled shaft, pier cap, and grade beam structural designs. The permit set is signed and sealed for submittal to the permitting agencies, while a complimentary submittal is issued to the Owner for review and acceptance.

In-Progress Design Submittal. Development at an approximate 60 to 75 percent design level constitutes an in-progress design submittal (IPDS). The Owner, Authorities Having Jurisdiction, and permitting agency comments are incorporated into the in-progress design. The designer and the constructor coordinate frequently to address potential design parameters that could generate problems during construction. During this phase of development, issues are resolved and further cost and schedule saving opportunities are incorporated into the design, generally freezing the design for final development.

Final Design Submittal. With overall design accepted by the Owner, AHJs, and permitting agencies, the design-build contractor focuses its resources on completing the final design submittal (FDS). This phase of design development achieves the 100 percent design level.

Readiness for Construction Submittal. The readiness for construction submittal is equivalent to the bid drawings and specifications package used for a conventional design-bid-build project. Once accepted by the Owner, these documents are used by the design-build contractor to initiate the construction works.

The progress of design unit development and construction sequencing drive the project schedule and the critical path. The critical path is dictated by contractual site access constraints and by the construction duration established for the major work components.

Construction sequencing plays a major role in schedule acceleration. After field surveys and utility location and identification are complete, the design-build contractor establishes construction sequencing to build the project in logical order and to leverage resources, site access, and early start activities.

The guideway construction is sequenced as follows:
- Maintenance of traffic (MOT)
- Augered cast-in-place piles

- Pile cap construction
- Piers, pier caps, and bearings
- Setting beams
- Diaphragm
- Cast-in-place deck
- Parapets, running surface (plinth), and switches

Once the civil works are finished, the guide-beam and power rail will be installed by a specialty subcontractor. Installation of operating system electrical and communication equipment, video surveillance, and signage will follow.

The MIA Station will be constructed concurrently with the guideway sections west of Le Jeune Road. After field surveys and utility location and identification are complete, the station construction will be sequenced as follows:

- MOT implementation and fencing of the working area between the parking garages
- Relocation of existing utilities that may be in conflict with the foundations
- Installation of the auger cast piles and elevator drilled shaft
- Construction of the pile caps
- Construction of the columns
- Construction of beams and elevated slab
- Structural steel and concrete deck for the interior platform
- Concrete pilaster and beams for the concrete guideway
- Exterior wall and roof construction

After completion of the civil works, the operating system subcontractor will be granted access to the station to complete the installation command, control and communication (CCC) infrastructure and the M&SF equipment.

In parallel the guideway and station construction, the design-build contractor will build the substation to distribute power to the operating system. The civil work includes construction of the duct banks and the installation of the power feeder cables.

Verification and acceptance testing of the fixed facilities and the operating system will take place upon completion of the fixed facility construction and operating system installation. The final six months of Phase I is dedicated to operation with minimum requirement testing. Completion of the tests will establish Phase I substantial completion.

Cost. Preliminary engineering and construction bidding by competing proposers during the procurement phase is an Owner-attractive feature of the design-build delivery methodology. The Owner receives design and construction innovations at the proposer's expense. During the four-year procurement phase, the design-build contractor was invited to develop cost-saving alternatives to deliver the MIA Mover APM System at the 1996 engineer estimate value.

The design-build contractor submitted 26 innovative technical and administrative cost-saving ideas to help the Owner procure the project within the budgetary constraints. During contract negotiations, the Owner accepted twelve of the innovations to achieve approximately $60 million in cost savings. Those related to the fixed facility design-build work include:

Superstructure Modifications. This innovation modifies guideway superstructure to reduce construction costs. The changes reduce the width of the guideway structure by eliminating a raised emergency walkway; modifying the vehicle design (end-door emergency egress instead of side door egress); replacing the solid parapet wall with metal handrails; and replacing the trapezoidal box superstructure (***Figure 4a. Typical Section Trapezoidal Concrete Box Girder***) with bulb T girders (***Figure 4b. Typical Section Bulb T (BT72) Precast Girders***) and a cast-in-place deck.

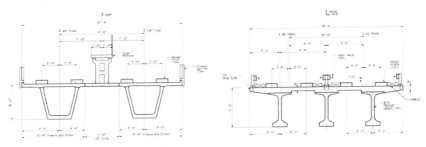

Figure 4a. Typical Section Trapezoidal Concrete Box Girder

Figure 4b. Typical Section Bulb T (BT72) Precast Girders

The proposed end door emergency egress arrangement for the MIA Mover vehicle offers a safe and economical egress solution. By eliminating the center raised walkway, the distance between the dual tracks is reduced, allowing a narrower superstructure. Metal handrails are lighter than concrete parapet walls, reducing the dead loads carried by the superstructure. The bulb T superstructure section is more economical to construct.

Modified Foundation System. The design-build contractor recommended augered cast-in-place pile foundations for the guideway structure instead of drilled shafts. The augered cast-in-place piles meet the intent of the specifications and are used extensively and successfully in the Miami area. Because a segment of the guideway over-crosses Florida Department of Transportation (FDOT) rights-of-way (Le Jeune Road), drilled shaft foundations are required in this area.

MIA Station/M&SF Consolidation and Modified Finishes and Roof Structure. An initial innovation by the design-build contractor is the integration of the MIA Station with the operating system maintenance and storage facility (M&SF) as a multi-level, multi-function facility. By combining the station and the M&SF, a significant property area is reapportioned for other airport purposes and construction mobilization and work forces are consolidated to one location.

The originally proposed MIA Station roof design and station finishes (*Figure 5a. MIA Station Original Design Concept*) were simplified to produce additional cost savings (*Figure 5b. MIA Station Current Design*). While maintaining the original station footprint and associated functionality and aesthetics, structural and architectural components were substituted to provide a more economical solution.

Figure 5a. MIA Station Original Design Concept

Figure 5b. MIA Station Current Design

North and South Corridor Pedestrian Bridges/Moving Walkways. The most significant cost-savings are achieved by eliminating the direct pedestrian connection between the MIA station entrance and the existing parking lot walkways to the north and south. The design-build contractor conducted numerous analyses and determined that elimination of these pedestrian facilities will not substantially affect the MIA Mover system performance. Adjacent facilities are not disturbed during construction of the walkways.

Vehicular Bridge. The demolition and reconstruction of an out-of-service vehicular bridge connecting Park 2 (Dolphin parking garage) and Park 3 (Flamingo parking garage) was included in the original MIA Mover project requirements (*Figure 6. Park2/Park 3 Vehicular Connection*). Because this facility has no functional relationship with the APM, the design-build contractor recommended removal of this work scope from the contract to realize significant cost savings for the Owner. The Owner can exercise an option to design and construct a replacement facility when the future southeast parking garage is built.

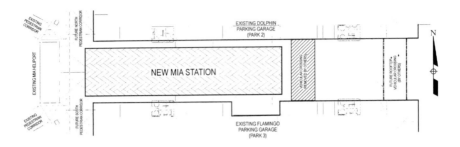

Figure 6. Park 2/Park 3 Vehicular Connection

Track Crossovers. Four of the six original track crossovers are eliminated to achieve additional cost savings. This simplifies the guideway alignment, contributing to cost reduction for guideway construction and guideway switch equipment (***Figure 7. Initial vs. Final Guideway Switch Configuration***). The operating system supplier thoroughly analyzed APM operations to ensure that elimination of the track crossovers would meet contract specifications. By placing four-car trains in service for interim and ultimate operations, no change in normal mode management capacity will occur. The elimination of the crossovers simplifies the configuration for the automatic train control system and the operation pattern at failure mode.

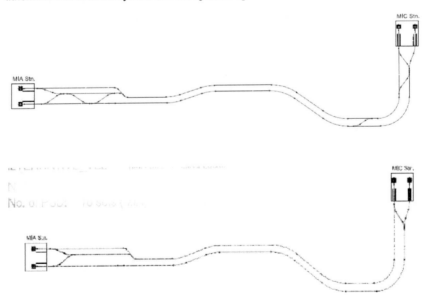

Figure 7. Initial vs. Final Guideway Switch/Crossover Configuration

West Substation. The contract technical provisions require two redundant substations. The equipment provided in each substation is further redundant as double ended switchgear. In effect, this provides quadruple redundancy, instead of the intended double redundancy. The west substation was eliminated and the size of the east substation increased to achieve the required double redundancy. The substation equipment is sized to provide sufficient power to operate the APM system as stipulated in the technical provisions.

Direct Material Purchase. The design-build contractor recommended the County implement a direct material purchase program to gain substantial savings through a county sales tax exemption for materials and equipment purchased for MIA Mover construction. Miami-Dade County is permitted to directly purchase equipment and material without paying sales tax. A direct material purchase program is an effective way to reduce the equipment and material purchasing costs and has proven effective in previous Miami-Dade County capital construction projects.

Owner-Provided Office Space. The original contract requirements allow 5,000 square feet of Owner-provided office space to house Owner and design-build contractor management staff. The design-build contractor, under this scenario, as in conventional design-bid-build procurements, is responsible for providing design and construction field offices, the cost of which is included in the contractor bid. Because an Owner-owned office building, situated within one-quarter mile of the project site is available, the design build contractor proposed that this facility be used as the MIA Mover project command center for Owner, design-build contractor, and operating system supplier work forces. An additional 10,000 square feet of office space is needed. The design-build contractor is responsible for remodel and renovation of office interiors to accommodate all parties. The Owner provides upgrade of other building systems, as required. The implementation of this alternative generates substantial cost savings for the Owner.

CONCLUSIONS

During design development, the design-build contractor continues to investigate strategies to save time and money to enhance design and construction strategies. Continued coordination among all parties (the design-build contractor, the Owner and its representatives, and the Authorities Having Jurisdiction) exposes new opportunities and constraints where innovative thinking generates additional project benefits.

The fixed facility design-build delivery method enables design and construction efficiencies that a transportation facility provider may not realize under a conventional design-bid-build project delivery procurement. The integration of the fixed facility designer and contractor forces facilitates design and construction cost and schedule savings. Incorporating a new transportation facility into an existing built-out environment is advanced when design and construction forces jointly identify opportunities and resolve issues early during project development.

REFERENCES

Miami-Dade Aviation Department (2008). *Final Contract Documents, MIA Mover APM System, MDAD Project No. J104 A.* Volumes I-V and Supplemental. Jun. 3, 2008.

FDOT MIC Major Project Overview (2008). *Miami Intermodal Center Project Overview*, http://www.micdot.com/mic_program/11_10_08_MIC_Major Project Overview.pdf(Nov. 6, 2008).

Parsons-Odebrecht, Joint Venture (2006). MIA Mover APM System Technical Proposal, Parcel A, Part 2, Volume I, Operating System and Volume 2, Fixed Facilities and Aesthetics, Feb. 22, 2006.

Parsons-Odebrecht, Joint Venture (2008). MIA Mover APM System Supplemental Technical Proposal, Parcel A, Envelope A, Part 2, Volume I, Operating System and Volume 2, Fixed Facilities and Aesthetics, Jan. 9, 2008.

Parsons-Odebrecht, Joint Venture (2008). Design Units I, II, and III Preliminary Design Submittals.

MIA MOVER PROCUREMENT

Sanjeev N. Shah [1], P.E., M.ASCE (Primary Contact), Larry Coleman, [2] Margaret Hawkins Moss [3], and Franklin Stirrup [4], P.E.

(1) Principal, Lea+Elliott, Inc., 5200 Blue Lagoon Drive, Suite 250, Miami, FL 33126; PH 305 500 9390; Fax 305 500 9391; snshah@leaelliott.com
(2) Manager of Planning Projects, Lea+Elliott, Inc., 5200 Blue Lagoon Drive, Suite 250, Miami, FL 33126; PH 305 500 9390; Fax 305 500 9391; lcoleman@leaelliott.com
(3) Sr. Procurement Officer, Miami-Dade Aviation Department, P.O. Box 592075, Miami FL 33159; PH 305 869 1421; Fax 305 876 8067; MMoss@miami-airport.com
(4) Project Manager, Miami Dade Aviation Department, P. O. Box 592075, Miami, FL 33159; PH 305 876 7922; Fax 305 876 8067; FStirrup@miami-airport.com

Abstract

The MIA Mover is an elevated Landside Automated People Mover (APM) system that is being implemented at Miami International Airport (MIA) by the Miami-Dade Aviation Department (MDAD). The MIA Mover will provide a convenient and reliable means for transporting passengers between a centrally located MIA station and the Miami Intermodal Center (MIC). One of the first elements of the MIC currently being constructed under the supervision of the Florida Department of Transportation is a Consolidated Rental Car Facility which is scheduled for completion in mid-2010. Other elements of the MIC programmed to come on line in the near future include links to the regional transit and commuter rail systems.

MDAD commitments for the operational readiness date for the MIA Mover necessitated that the entire project (infrastructure and operating system) be procured under a single Design-Build-Operate-Maintain contract (inclusive of infrastructure and operating system). Draft documents were issued for an Industry Review in the second quarter of 2004 to solicit industry input, and the Request for Proposals was advertised in December 2004. The DBOM Contractor has been selected and the Contract was awarded in July 2008. This paper examines the various factors that resulted in the extended procurement duration including the competitive environment that resulted proposals for different classes of technologies, the effects on the construction market in the post-Katrina and the overall global demand that resulted in hyper-inflationary pressures, the economic reality of budgets and related policy implications.

Background

The MIA Mover (previously known as the MIC/MIA Connector) is an elevated Landside Automated People Mover (APM) system that is being implemented at Miami International Airport (MIA) by the Miami-Dade Aviation Department (MDAD). The MIA Mover will provide a convenient and reliable means for transporting passengers between a centrally located MIA station and the Miami Intermodal Center (MIC). The MIC includes the Consolidated Rental Car Facility (CRCF), which will be one of the first elements of the MIC to become operational.

After 9/11 and its impacts on the Aviation Industry and MIA, efforts were undertaken to re-evaluate the overall MIA Capital Improvements Program in conjunction with the uncertainty presented by business and economic conditions. This re-evaluation resulted in several significant changes to the project. First, the MIA Mover was modified to operate in a "straight configuration," with approximately 1.25 miles of dual lane guideway with a single Station within and compared to the previous plan with approximately 1.75 miles of dual lane guideway and 3 stations within MIA(1). Second, the capacity requirements for the system were modified to support ridership commensurate with 39 million annual air passengers (MAP) in the 2015 time frame compared to 48/55 MAP previously projected for 2015 and reaching approximately 48 million annual air passengers by 2033. These modifications resulted in reduced project budgets without compromising the primary goal of the system - provide a connection between the Miami Intermodal Center and Miami International Airport. However, under the modified plan, passenger distribution within the MIA Terminal area will be from a single station resulting in longer walk distances.

The DBOM contract procurement for the project with the adopted straight configuration was authorized for advertisement in December of 2004. The procurement documents were structured as performance specifications to permit system suppliers to propose their proprietary technologies to fit within the established site specific constraints. Since the adopted configuration had two stations, instead of multiple station stops, cable-propelled technologies could be applied and the procurement strategy and documents were structured to accommodate multiple classes of technologies in an effort to encourage competition.

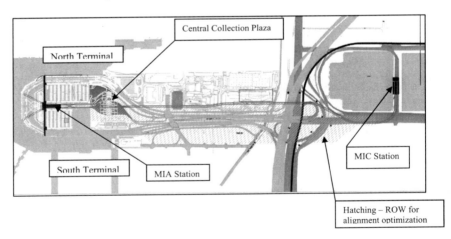

Figure 1 – MIA Mover APM System – Straight Configuration Procurement Strategy

---

1.) The process followed in reexamining the various project parameters such as ridership, alignment, need for compatibility with future airport needs and other operational considerations was previously described in a paper entitled *MIA Mover – Post 9/11 Strategy* that was presented at the 2005 ASCE APM Conference

MDAD and Miami-Dade County commitments for the operational readiness date for the MIA Mover necessitated that the entire project (infrastructure and operating system) be procured under a single Design-Build-Operate-Maintain contract (inclusive of infrastructure and operating system). Draft documents were issued for an Industry Review in the second quarter of 2004 to solicit industry input.

A full turnkey (Design/Build/Operate/Maintain or DBOM) approach was evaluated wherein the Infrastructure and the Operating System are procured through a single contract. This approach was designed to permit each potential supplier to define its infrastructure requirements specific to its technology (providing competitive benefits) and was selected for the procurement of the MIA Mover project for the following reasons:

- Lowers capital costs due to increased competition among available APM Systems (different classes of technologies)
- Provides for faster completion of the capital project (a time savings of approximately one year over the other approach, where infrastructure would be procured separately from the operating system)
- Reduces Owner risks since the Contractor has a single source of responsibility for infrastructure and operating system procurement
- Reduces potential for delays due to integrated and centralized project management and a cost effective infrastructure design/compatibility with the operating system.

The DBOM approach was fully endorsed by an airport Peer Review Group that included representatives from Dallas Fort-Worth International Airport, the Port Authority of New York & New Jersey, Los Angeles International Airport, Greater Toronto Airport Authority, Metropolitan Washington Airports Authority and San Francisco International Airport. Further, the Miami Airlines Affairs Committee also supported the DBOM method of delivery for the MIA Mover.

The MIA Mover procurement approach was made open to a range of viable APM Operating System technologies that could meet the specified (minimal) project performance parameters. The range of viable technologies, that could be applied to the MIA Mover project, fell into two main categories; (a) full performance self-propelled technologies (that have been largely service-proven) that due to their inherent capabilities will exceed, in some cases, the minimum contract requirements; and (b) cable-propelled technologies, (that due to the minimal project parameters may require some degree of innovation) and will likely meet the minimum contract requirements.

Competitive procurement was a primary goal for the MIA Mover procurement process. Due to their inherent features, the full performance technologies in some cases exceed the minimum contract requirements and are generally more expensive than cable-propelled technologies. Since ranking of proposals was based on a combination of technical merit and price, it was important that the ranking methodology (representing a combination of technical merit and price) be structured in way that would encourage the range of technology suppliers to view their technologies as viable with a good chance of success, if they propose competitively. For example, if the price weight factor is too high, then it is likely that the full performance technology suppliers will view the process as unfavorable and may not participate – resulting in a loss of competitive environment. Conversely, if the price weight factor is too low, then it is likely that the proposal prices will be higher.

The technical scoring methodology was closely examined to provide a fair environment for self-propelled and cable-propelled technologies. Various potential proposal scenarios were simulated to develop technical score ranges that could be anticipated. Cost estimates were developed for each of these scenarios, combined with the range of technical scores to come up with a total score formula that could maintain a fair, yet competitive environment for the different classes of technologies. The appropriateness of the scoring/ranking methodology was underscored by the fact that two self-propelled technology proposals and one cable propelled technology proposal were received.

Procurement Process

The Request for Proposals, as advertised in December 2004, initially established the proposal due date as March 23, 2005. To assure that a competitive procurement environment was fostered with the goal of obtaining responsive and responsible proposals the proposal due date was extended on several occasions until February 22, 2006 at which time the proposals were received. Key reasons for the proposal due date extensions were:

- Teaming structure change for one potential respondent.
- Requests by potential respondents (and sureties) to address the insurance, risk management and surety requirements included in the RFP documents. Industry concerns were generated due to surety, risk and insurance market conditions that were further exaggerated by the global construction demands and the impacts of Hurricanes Katrina and Wilma
- Adjustment to the MIA Mover baseline alignment to accommodate a possible future westward extension of the Miami-Dade Transit Metrorail system line (MIC-Earlington Heights Connector and the East-West Corridor) from the MIC.
- Modifications to the project participation goals to reflect appropriate opportunities for Community Business Enterprise and Community Small Business Enterprise firms to participate in the procurement as part of proposer teams.

Three (3) proposals were received on the proposal due date of February 22, 2006 – from Bombardier-PCL, LLC (utilizing Bombardier's self-propelled Innovia technology), Parsons Odebrecht Joint Venture (utilizing Mitsubishi Heavy Industry's self-propelled Crystal Mover technology) and Slattery Skanska, Inc. (utilizing Doppelmayr's cable-propelled technology).

The proposal submittal consisted of:

- CSBE Envelope containing the information required to demonstrate compliance with the Community Small Business Enterprise goals established by Miami-Dade County for the construction aspects of the project.
- CBE Envelope containing the required information to demonstrate compliance with the Community Business Enterprise goals established by Miami-Dade County for the professional design services aspect of the project.
- Parcel A containing the Technical Proposal addressing proposed designs, operating system technology, management, qualifications and the operations and maintenance approach, to comply with the Contract requirements, including future expansion opportunities.
- Parcel B containing the Lump Sum Pricing Proposal commensurate with the Technical Proposal in Parcel A, the maximum anticipated 15 years of Operations and Maintenance and the potential future expansion of the System (as an Owner option).

Parcel B, the price proposal, from each proposers was to be opened only after the CSBE, CBE and Parcel A information had been evaluated, the proposals were determined to be responsive to the requirements in the RFP and the technical scoring completed. After the evaluation of the CSBE and CBE Envelopes the Slattery Skanska, Inc proposal (with the DCC technology) was found non-responsive by the County Attorney's Office and eliminated from further consideration. Parcels A, the technical proposals for the remaining two proposers were evaluated and clarifications were sought. Prior to the technical scoring by the County's appointed Selection Negotiation Committee (SNC), the County Attorney's Office found the Bombardier-PCL, LLC proposal non-responsive resulting in its elimination from further consideration.

Parsons-Odebrecht Joint Venture (POJV) was the sole remaining proposer, whose price proposal was opened after the technical scoring was completed. The price proposal exceeded the project budget, namely due to the hyperinflationary pressures in the construction industry driven by the global demand and the impacts of the Hurricanes Katrina and Wilma. Negotiations commenced shortly thereafter, during which stage, the County Attorney's Office found POJV non-responsive, bringing the process to a halt.

Structured Negotiation Process

The County evaluated various options including re-advertisement of the solicitation to re-initiate the procurement process. Since all the proposals were for various reasons deemed non-responsive, it was determined that a structured negotiation process with all the three (3) original proposers would be fair to the proposers (who had already made investments in preparing the original proposals) and in the best interests of the County in terms of schedule, the County's commitment to service the Rental Car Facility with an APM within two years after opening and the potential to avoid a new process wherein the risk of being found non-responsive would still remain. As such, the County Manager recommended a structured negotiation process that was authorized by the Board of County Commissioners.

At the start of the structured negotiations process, the County requested and received confirmation from the three (3) original proposers of their interest and desire to continue to participate in the process.

The structured negotiation process involved four basic steps:

Step 1 - Level the playing field between the proposers by (1) evaluating the technical proposal of Slattery Skanska, Inc. to the same level of detail as of the other two proposers; (2) publicly opening Price Proposals from Bombardier-PCL, LLC and Slattery Skanska, Inc., as was previously done for the Price Proposal from Parsons-Odebrecht Joint Venture.

At the conclusion of this step, the previously unopened price proposals from Slattery Skanska, Inc. and Bombardier-PCL, LLC were publicly opened.

Step 2 – Supplemental Instructions to Proposer were issued with information on the process and instructions on the permissible and required **updates to their previously submitted proposals.** The requested updates permitted the proposers to cure/remedy previously identified responsiveness/conformance issues. Recognizing the project budget limitations, the County requested proposers to update their proposals to address identified value-engineering opportunities that were acceptable to the County that may permit the proposers opportunities to provide a better price and possibly meet the project budget; proposers had the flexibility to take advantage of each or any of the value-engineering opportunities identified by the County. The updated proposal, addressing the County identified value-engineering opportunities, became the Baseline proposal for the evaluation purposes.

Additionally, proposers were requested to identify other value engineering alternatives specific to their proposed operating system and/or infrastructure, referred to as Proposer Initiated Alternatives or PIAs, for the County's consideration. These Proposer Initiated Alternatives would be evaluated by the County for the highest ranked Baseline proposal as a further opportunity to improve the Best Value to the County.

Further, the updated proposals were to address "Must Haves" such as a new Good Faith Proposal Guaranty; updates licenses/certifications/authorizations and any team modifications; updated Project Participation forms to comply with the County's project participation provisions; and updates demonstrating compliance with the project's minimum requirements (technical as well as insurability, ability to provide surety bonds, etc.). Price Proposal updates to reflect modifications to the technical proposals or other conditions were also required to be submitted in a separate sealed package.

The Supplemental Instructions to Proposer were issued in October 2007 with updated proposals due on or before January 9, 2008.

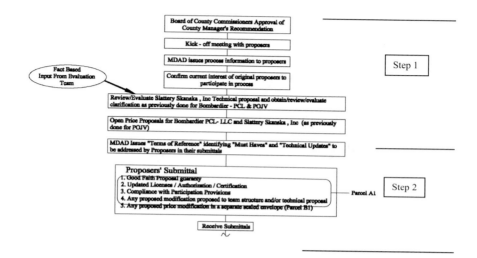

Figure 2 – Step 1 and 2 Flowchart

Step 3 – Evaluations and Negotiations following a structured process wherein the County's Selection/Negotiation Committee (SNC), supported by a fact finding technical team, was to evaluate, obtain clarifications, and negotiate contemporaneously with all proposers in the best interests of the County. The process primarily focused on the Updated Baseline technical proposal as outlined in Figure 3 below.

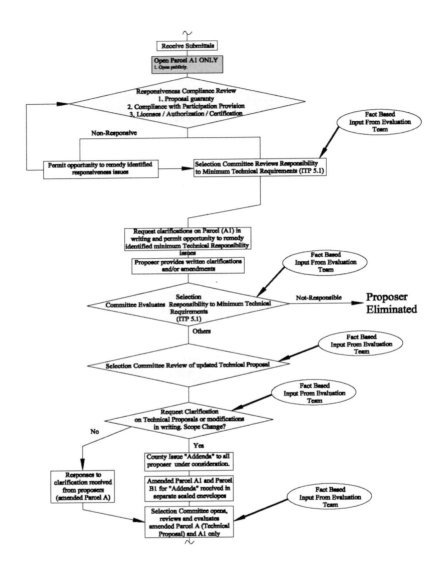

Figure 3 – Outline of Step 3 Evaluations

Step 4 – Technical Scoring, Price Proposals Evaluation and Ranking was then to be conducted by the County's SNC based on best value to the County with due consideration of Technical merit, price, budget and local preference.

For the highest ranked proposer, the SNC was to then determine the necessity to consider Proposer Initiated Alternatives. If deemed necessary, the Technical Proposal(s) for Proposer Initiated Alternative(s) from the highest ranked proposer was to be opened and the corresponding Price Proposal(s) for the Proposer Initiated Alternative(s) could be opened on the same date or at a later date, as determined by the SNC. SNC would then conduct negotiations with the highest ranked proposer including on any of their Proposer Initiated Alternative(s). If negotiations were not concluded to the SNC's satisfaction within 10 business days of commencement, the SNC could consider the next ranked proposer and initiate the process including necessity of considering their Proposer Initiated Alternative(s) and conduct negotiations. This process could continue until negotiations are concluded as determined by the SNC.

Upon completion of these steps, the Selection/Negotiation Committee would provide their recommendation to the County Manager, who would make an appropriate recommendation to the Board of County Commissioners for their action.

Updated Proposal Submittal

On January 9, 2008, an Updated Proposal was received from Parsons Odebrecht Joint Venture. Bombardier-PCL, LLC submitted a letter "regretfully withdrawing from this procurement." Legal counsel for Doppelmayr (supplier to the Slattery Skanska, Inc. team) submitted correspondence expressing regret for not submitting a proposal and noting that the team collectively could not resolve various commercial issues in time to present an acceptable proposal.

The POJV proposal was evaluated in accordance with the outlined process. The price proposal was opened for the Baseline proposal and was found to be substantially in excess of the established project budget. As such, the Selection Negotiation Committee elected to consider POJV's Proposer Initiated Alternatives. Several Proposer Initiated Alternatives such as advance payments were rejected as not acceptable under MIA's Trust Agreement. Other alternatives were found acceptable and negotiations on the final price were commenced. Among the value engineering alternatives accepted were:

- Simplify guideway superstructure aesthetic design
- Reduce the number of crossover switches
- Use a flat versus curved roof design for station
- Eliminate several pedestrian bridge connections between station and passenger terminal
- Use auger cast versus drilled shaft foundations
- Arrange for direct purchase of materials by the Owner to realize sales tax savings
- Owner to provide contractor with additional temporary office space
- Reduce the allowance account from 10% to 5%

Negotiations concluded on March 27, 2008 and the initial and final agreed upon prices are summarized in the table below:

Description	Base Proposal Price	As Negotiated
Fixed Facilities*	$ 220,059,258	$ 152,396,640
Operating System*	$ 99,066,445	$ 94,103,360
Phase 1 System	**$ 319,125,705**	**$ 246,500,000**
Allowance Account	$ 31,912,571	$ 12,325,000
sub-total	**$ 351,038,276**	**$ 258,825,000**
Parking Allowance	$ 3,250,000	$ -
IG Audit Account	$ 797,814	$ 616,250
Total Phase 1	**$ 355,086,090**	**$ 259,441,250**

Final contract documents were then prepared and the contract was awarded by the County on July 1, 2008. Notice to Proceed was then issued on September 8, 2008 and the project is scheduled for completion in three years or September 2011.

Conclusion

In constrained economic conditions, it is critical to bridge the relationship between what the Owner one can afford to pay for a project, their goals and objectives for the project and what a project competitively costs. Value engineering in this case was driven by consideration of the airport's needs and service objectives, potential cost savings and the benefits/consequences associated with each option. The MIA Mover structured negotiations process provided a mechanism by which a competitive environment could be maintained while considering value engineering opportunities at the same time. In the end, when it came down to a choice between compromising on certain features of the project that were not affordable and not having a system, the Owner's options became quite clear. It is crucial to note that the process permitted the proposers to offer the advantages of their inherent design/system capabilities while still providing a level playing field that would retain the interest among multiple teams. A fair and competitive procurement process is to the advantage of the Owner as well as the APM supplier industry and construction industry. Thinking out of the box can provide innovative solutions that will be even more applicable in the constrained economic conditions that the global market is currently facing.

For the procurement process to be successful, it is very important that consultants (engineers, planners, architects) act fully and solely in their role as the technical experts providing necessary technical input and information in assisting the Owner in framing the appropriate policy issues and make the appropriate policy decisions. This approach was found to be beneficial to MIA in general and the MIA Mover project in particular.

APM FEASIBILITY STUDY FOR THE VIENNA CENTRAL STATION DEVELOPMENT AREA

Heimo Krappinger*

* Associate, AXIS Ingenieurleistungen, Rainergasse 4, 1040 Vienna, PH +43 (1) 50670-230, krappinger@axis.at

Abstract

In 2013 the new Vienna Main Railway Station will be constructed on the location of the old Vienna Southern and Eastern Railway Stations which will be demolished. A large section of the area of the old stations, north of the new Railway Station, will be available for development (500,000 square-meter / 5.4 mill. square feet of gross floor area).

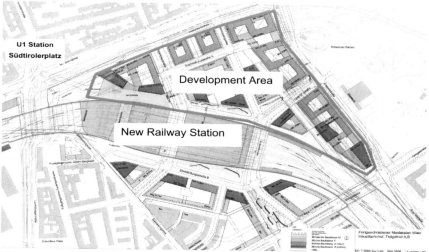

This paper focuses on the feasibility of different public transport systems (PT systems) as supplementary systems to the existing and already planned PT systems (tram, bus and S-Bahn) for this area. The main target was a better connection between the development area, the new Main Railway Station and the Underground Station of Line U1 (Südtirolerplatz). Four different PT systems (conveyor belt / moving walkway, APM, tram bus) were compared and finally the cable driven pinched-loop system (multiple vehicles with a short headway and a detachable grip system) was chosen as the preferred alternative. In a second stage this system was further investigated and additional routes worked out. Currently the stakeholders (City of Vienna, Austrian Railway Company ÖBB and the developers) negotiate the financing of the project.

Introduction

The development of high density sub-centers within the urban areas of our cities require efficient PT services within such centers and high quality connections to the primary PT systems that already exist there. Very often development areas like the said railway station area are already connected via PT lines and have existing lines within their boundaries. Such existing PT systems are sufficient to meet the former requirements of these development areas but do not cope with the planned new and mostly much more intense uses.

In case of the Vienna Southern and Eastern Railway Stations, both railroad terminals, they have been used mostly for passenger transport, as office space as well as cargo facilities.

The new planned Vienna Main Railway Station will be a through station with much higher passenger frequencies compared with today's two railroad terminals. Furthermore there will be intense retail, office as well as residential uses. Additionally, hotel and culture infrastructures will be developed.

All these uses generate up to 75,000 trips per day taking place from and to the new development. 56% of these trips are expected to be PT trips, 21% will be trips covered by the non-motorized transport (walking and cycling) and only 23% will be car trips. This shows the significance of offering a high quality PT connection to the primary PT network as well as a very good distribution within the area together with a pedestrian and bicycle friendly roadside development.

The existing and already planned PT systems as shown in the masterplan for the Vienna Main Railway Station area offer already a quite good PT coverage, using the existing S-Bahn, tramway and bus lines. However, it lacks the direct connection to the main PT system – the Underground Station of Line U1 (Südtirolerplatz). Further it does not offer a PT system running through the whole development area.

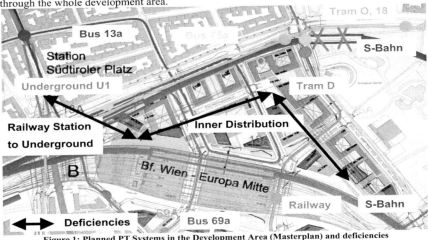

Figure 1: Planned PT Systems in the Development Area (Masterplan) and deficiencies

Therefore a feasibility study was commissioned. The contents are to address the deficiencies and to find suitable routes for different supplementary PT systems offering increased comfort for the travelers and a better (traffic) development quality for the office and residential buildings to be constructed on the area north of the new railway station.

The main targets to be achieved are the improvement of the quality of the PT supply (route and stations) and its quantity. In order to reach a high quality of supply the following criteria were defined:

- Local availability
- Adequate frequency of service of the system
- Short travel times
- High comfort
- Easy and barrier-free access to the stations

The quantity of PT supply is described through the capacity and the hours of operation of the new system.

The targets to be reached with the new supplementary PT systems are shown below:

Preservation	- of the accessibility
Improvement	- of the development core areas
	- of the local availability (stations) and frequency (intervals)
	- of the operating time (adjustment to the Underground Line 1 intervals)
	- of the accessibility of the primary PT systems, especially Underground Line 1
	- of the PT supply
	- of the traffic behavior
	- of the reliability
	- of the travel times from and to the U1
	- of the travel comfort
	- of changing PT systems

Avoidance	- of noise and exhaust gas
	- of accidents
	- of business decline
	- of stranded investments
	- of undesirable developments
	- of high operating and personnel expenses
Gain new benefits	- raise the attractiveness of the development area
	- high quality for the users
	- attractive PT system (feeder system) as incentive to use PT
	- trend-setting alternative PT system as pilot-project for Vienna
	- more diversity in PT

Characteristics of the compared systems

The systems compared were:
- Conveyor belt / moving walkway
    Bundling of passenger streams on short routes (approx. 200m/655 feet), speed up to 0.75m/s (2.5 feet/s) and a theoretical capacity of up to 13,000 pphpd (passenger per hour per direction)

- APM Shuttle system (cable driven with fixed grip on the drive cable, 2 trains)
    Automatic driverless PT systems for short distances (4-5 km/2.5-3.1 miles) running on a reserved track; with train sizes of about 30-500 persons; speed up to 14 m/s (46 feet/s) and a theoretical capacity of 3,000-13,000 pphpd.
    As reference and base for technical details the systems of Doppelmayr Cable Car GmbH (Cable Liner Shuttle) and Leitner Ropeways (Minimetro) were used.

Image 1: Doppelmayr Cable Liner Shuttle; © DCC

- APM Pinched-Loop System (multiple vehicles with a short headway and a detachable grip system for attaching/detaching to the drive cable)

    Automatic driverless PT systems for short distances (4-5 km/2.5-3.1 miles) running on a reserved track in short intervals (up to 50s); with vehicle sizes of about 30-50 persons; speed up to 8.5 m/s (28 feet/s) and a theoretical capacity of up to 5,000 pphpd.

    As reference and base for technical details the systems of Doppelmayr Cable Car GmbH (Cable Liner) and Leitner Ropeways (Minimetro) were used.

Image 2: Doppelmayr Cable Liner; © DCC ;   Leitner MiniMetro; © Leitner

- Tram

    Traditional urban PT system for short to medium distances with vehicle sizes of about 135 – 205 persons; speed up to 70 km/h (44 miles/h) and a theoretical capacity of about 2,500 pphpd depending on the headway.

- Bus (gas driven, low-floor)

    Traditional urban PT system for short to medium distances with vehicle sizes of about 90 – 150 persons (12m/40 feet – 18m/60 feet); speed up to 70 km/h (44 miles/h) in urban areas and a theoretical capacity of about 2,500 pphpd depending on the headway

Routes for the different systems

Based on the masterplan for the development area, for all systems different routes were proposed.

Conveyor belt / moving walkway:
For this system two principal solutions were found (Level +1 above ground, Level -1 underground). The capacity of the specific system is about 3,600 pphpd.

Level +1:

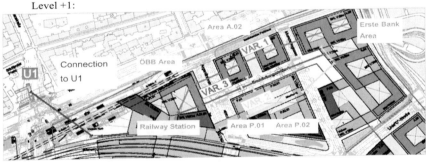

These solutions connect the U1-Station with the railway station and furthermore most of the other planned buildings within the development area. The crossing of the roads takes place elevated with headroom of 4.70m (15.5 feet). The bridges for the conveyor belts / moving walkways carry one conveyor belt / moving walkway per direction and a pedestrian way.

Level -1:

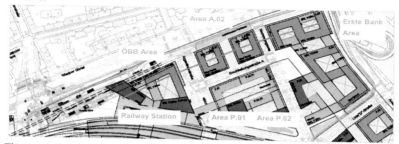

The connection to the underground station can be realized as shown as for the level +1 solution. The connection of the buildings are planned to take place on basement level (-1).

APM:

The APM can either use a single track (permanently attached to the drive cable, shuttle system) with a turnout track switch or a dual track system (for both APM systems). All solutions run elevated in the second floor.

Single track:

When using a single track there are two principal solutions possible – either with 3 or 5 stations (symmetric station distribution). The eastern end-stations can additionally have two different positions (green or red). There are two trains running against each other and using the turnout track switch to pass-by. The capacity is about 2,500-2,700 pphpd.

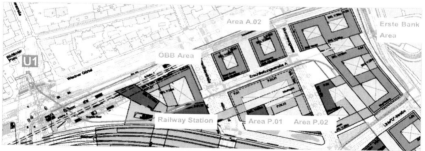

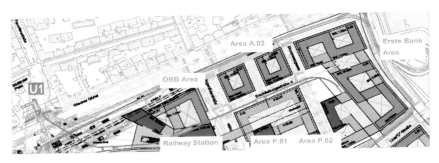

Dual track:
When using a dual track system, one optimized route was developed. Both APM systems (permanently attached/shuttle system and automatic attaching and detaching/pinched-loop system) can be installed and the number and location of the stations can freely be chosen. For this solution 4 stations were found sufficient, whereas the eastern end-station can additionally have two different positions (green/red). Extensions however are possible for the red solution only. The capacity for the pinched-loop system is about 3,400 pphpd; for the shuttle system it is about 2,700 pphpd.

Extensions:
Additionally the following extension and special solutions were considered.

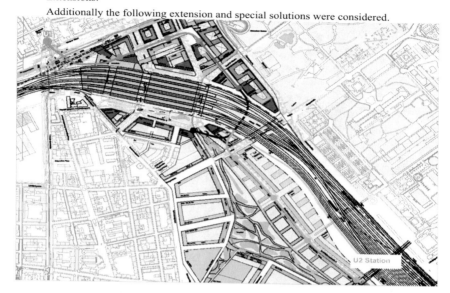

Extensions are possible towards the Area B (black) or towards the Underground Line 2 Station Gudrunstraße when using the pinched-loop system.

Tram:
For the tram one base solution was worked out with an additional extension scenario.

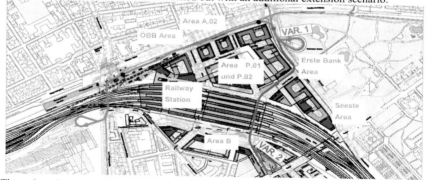

The red track runs partly on existing tracks and does not offer a direct connection to the U1 station. The purple extension connects the Area B with the Railway Station main entrance. The capacity for these variants is about 2,700 pphpd.

Bus:
The bus solution shows a ring-line connecting all the areas surrounding the railway station. Capacity 1,800-3,000 pphpd, depending on bus-size.

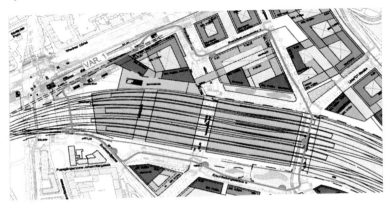

Operational specifications

The new systems are supposed to fulfill a feeder-function for a connection to the U1 station Südtirolerplatz. Due to this target the intervals and the operation time need to be adapted to the intervals and operation time of the underground line U1.
Based on that and other data, the operation time for the new systems was set with 20 hours per day, starting from 5:00 o'clock in the morning until 1:00 o'clock in the morning, 7 days a week and 365 days per year on an average peak-hour interval of 3 minutes.

Comparison of the different systems

Based on the main criteria of short travel time, high local availability (stations) and short headway, capacity and waiting time, the pinched-loop APM system turned out to be the best of the compared systems.

Headway:	1:00 minute compared to 7:30 minutes (tram) and 3:00 minutes (bus)
Travel time:	4:30 minutes compared to 9:35 minutes (tram), 9:25 minutes (bus) and 14:00 minutes (conveyor belt / moving walkway)
Capacity:	3.360 pphpd compared to 1.650 pphpd (tram), 3.000 pphpd (bus) and 3,600 pphpd (conveyor belt / moving walkway)

Besides the criteria shown above, another 60 criteria were used to compare the different systems. A conclusion is shown below.
- The tram and bus systems cannot connect directly to the underground station, leading to a long walk and generally to long travel times.
- All systems show low emissions and their capacities are sufficient to cope with the demand.
- The pinched-loop APM adapts best to the changes in demand during the day, since its capacity can easily change due to adding or taking out of vehicles.
- The influence of the tracks (elevated structure) of the APM and the conveyor belts / moving walkways within the roadside environment are quite large compared to the other systems.
- All road-bound systems have conflicts with the other means of transport.
- The conveyor belts / moving walkways cannot easily be used by handicapped people. Therefore the criterion of use for everybody is not fulfilled. Additionally they are quite slow in comparison.
- The automatic running systems like conveyor belt / moving walkway and APM need the least personnel and show high safety standards and do not interfere with other means of transportation.
- The tram does not run through the development area and therefore does not offer the same quality of service as the APM

In conclusion, the automatically attaching and detaching pinched-loop APM system turned out to be the "best" system which can be expected to have the highest acceptance by the users.

Cost

After taking into consideration all the boundary conditions of the different systems and finding comparable solutions for all the systems the rough investment and operation cost were calculated.

System	Investment Cost, in €, approx.	Operation Cost / year, in €, approx.
APM autom.	21.3-25.5 million	1.2 million
APM fixed	14.7-19.8 million	1.0 million
Conveyor Belt / Moving Walkway	12.7 million	0.7 million
Tram (on existing tracks)	11.0 million	2.0 million
Bus	1.9 million	1.3 million

The APM was the most expensive system when considering the investment cost, but the operating cost were considerably less than the tram operation cost. Based on the APM's advantages over the other systems (60s headway, automatic operation, etc.) it was chosen as preferred alternative for further investigation.

APM AS SUPPLEMENTARY PT SYSTEM

Based on the findings of this study the preferred alternative – a cable driven pinched-loop APM system– was further worked and scenarios within the development area as well as extensions into the surrounding areas were developed. The route of the APM was changed due to the necessities of the developers and owners of the area.

The length of the finally proposed system (first phase) was about 820 meters, crossing a main traffic corridor (the "Gürtel" with a 4-lane road, two tramway lines and the S-Bahn) and running through the planned main railway station. This route led to specific considerations of the numerous underground installations in the Gürtel area and structural and fire protection issues when running through buildings. A major issue was the location of the APM-stations within the development area and the connection of the planned buildings and uses to the new PT system.

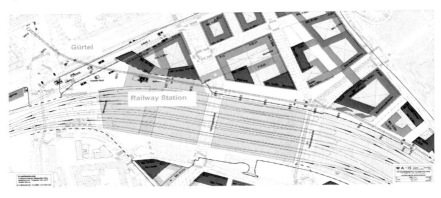

Eventually, different extension-scenarios were developed, connecting the existing underground line U1 with the planned extension of the underground line U2, targeting on a better PT infrastructure for development areas in the surrounding of the new railway station.

Summary and outlook

Currently the City of Vienna, the Austrian National Railway Company (ÖBB) and the developers investing in the area are discussing the shared funding of the investment and operation cost. If these negotiations are successful, the APM system may be constructed by 2014-2015 and then running integrated within the PT network VOR (**V**erkehrsverbund **O**st-**R**egion; the public transport cooperation covering the east of Austria), requiring no extra ticket in addition to the regular Vienna PT ticket which is valid for the whole city of Vienna. The old Southern Railway Station will then look like the picture below.

Las Vegas People Mover Integration Potential
Wayne D. Cottrell, Ph.D., P.E.[1]

[1] Advanced Transit Association; 1853 Santa Rita Drive, Pittsburg, CA 94565-7653, PH (909) 204-0260, email: waynecottrell@advancedtransit.net

ABSTRACT

As of 2008, the city of Las Vegas, Nevada featured five operating or under-renovation automated people movers (APMs), excluding those serving McCarran International Airport and the Primm resorts located some 40 miles south of Las Vegas. This paper investigates the feasibility of linking four of the five APMs into an integrated, seamless system. The vision would be of an interconnected, driverless transit loop that would supplement traditional ground transportation modes (private auto, taxi, shuttle bus, walking). One challenge is the varying technologies of the APMs: three are cable-propelled, while one, the Las Vegas Monorail, runs on electricity. Also, none of the track infrastructures are similar, making it infeasible to simply connect them. Differing lengths, speeds, acceleration rates, train sizes, and other aspects are also barriers to integration. Further, with the exception of the monorail, all of the APMs were funded by Las Vegas resort owners for the purpose of serving travel within their properties. The benefits and additional costs of a public, fully-connected system would need to be rectified with the owners. Three integration alternatives are proposed: extending the monorail to form a continuous loop, linking the APMs and the monorail, and connecting the APMs with moving walkways. The latter alternative is the most infeasible, and potentially the least attractive. The former alternative has been discussed as part of the monorail's Phase 3 plans, although the author estimates a $1 billion price tag. Coordination of trains under the second alternative is theoretically feasible, with a 40-min round trip and sustained, 10-min headways. The challenges of full connection are not insurmountable, but would require extensive technological modifications, alterations, and reconstruction. Compromises would need to be reached among the owners regarding the technology and operations adjustments. Another challenge would be identifying sites for and funding four new stations, at which passengers would be able to make cross-platform transfers.

INTRODUCTION

Las Vegas, Nevada has one of the highest concentrations of automated people movers (APMs) in the world, with five systems either operating or under renovation along the Las Vegas Strip as of 2008. The five APMs are:

- Bellagio-Monte Carlo line (scheduled to reopen in Nov. 2009)
- Circus Circus Shuttle
- Las Vegas Monorail
- Mandalay Bay Express
- Mirage-Treasure Island line

Additional APMs are located at McCarran International Airport, as well as in the small resort community of Primm, which is located 40 miles south of Las Vegas. The five Strip APMs are shown in Figure 1. It is apparent that the lines serve the same geographical area, yet are disconnected. The objective of this research was to investigate the prospects for linking the lines to create a continuous system.

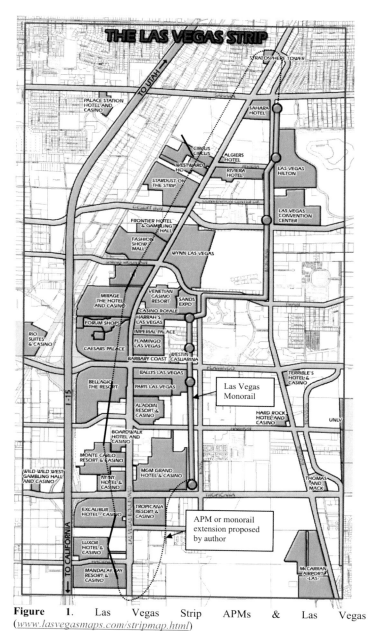

Figure 1. Las Vegas Strip APMs & Las Vegas Monorail (www.lasvegasmaps.com/stripmap.html)

(*Note: The Boardwalk, Frontier, Stardust and Westward Ho are all closed. CityCenter is under construction on a site between the Bellagio and Monte Carlo.*)

The longest APM in Las Vegas is, by far, the Las Vegas Monorail, at 6.5 km one way. As shown in Figure 1, the monorail runs generally north-south on an alignment located east of the Strip. The monorail extends from the Sahara on the north to the MGM Grand on the south, stopping at the Las Vegas Hilton, Las Vegas Convention Center, Harrah's Las Vegas and Imperial Palace, Flamingo Las Vegas and Caesar's Palace, and Bally's Las Vegas. The monorail was originally planned for extension northward to downtown Las Vegas, with a stop at the Stratosphere Tower, but the federal government denied the project's funding (Illia 2005). (Several technological mishaps during startup influenced the decision, including parts falling from the elevated track structure, and a power outage). The monorail *is* slated for extension to McCarran International Airport as part of a Phase 2 plan, however (the northern portion of the airport is squeezed into the bottom right corner of Figure 1). Funding for the extension is to come from private sources, and construction has been projected to be completed in 2012.

Along the west side of the Strip, a series of APMs connect pairs or trios of resorts, with no APM links between them. Starting from the south end of the west side of the Strip, the Mandalay Bay Express links that resort with the Luxor and with the Excalibur near Tropicana Road. Just to the north, a line links the Monte Carlo and the Bellagio. This line is currently being rebuilt to connect to the CityCenter development, which is being constructed *between* the Monte Carlo and Bellagio, adjacent the former Boardwalk Hotel & Casino. Farther to the north, an APM connects the Mirage with Treasure Island. Finally, the Circus Circus Shuttle connects that resort's two main buildings.

As suggested in Figure 1, it is possible to draw imaginary lines connecting the northern and southern ends of the Las Vegas Monorail with the "west side" APMs, thereby creating a closed-loop system. Connection of the west side APMs was suggested as early as 2006 by the Las Vegas Monorail Company, as part of a broad-based Phase 3 plan. The west side resorts and developments that are *not* served by an APM include the Tropicana, New York-New York, Caesar's Palace, and Fashion Show. (Note that the Frontier, Stardust, and Westward Ho had all been closed as of this writing). Circus Circus is served by an APM (its own), but it is not connected to any other resorts by APMs. Gottdiener et al. (2000) wrote that the Mirage proposed to construct an APM connector to the Bellagio, bypassing Caesar's Palace. Caesar's objected to the use of its property's "air rights," particularly for a private APM, however. Ultimately, the APM was not built, and Caesar's Palace remains one of the Las Vegas resorts without an APM.

HISTORY

The notion of a "Las Vegas People Mover" was proposed during the 1980s, primarily as a potential showcase for magnetic levitation (maglev) technology. A short-distance maglev line was intended to be a forerunner to a long-distance line connecting Las Vegas and Los Angeles. Two offerors, in fact, pursued separate alignments. HSST, a subsidiary of Japan Airlines, proposed a line extending from the Tropicana to Circus Circus, along Industrial Road. The north-south alignment would have been on the west side of the Strip. Transrapid, a German consortium, planned a 7.1-km line from McCarran International Airport to downtown Las Vegas. A "high-technology transportation systems study" was sponsored by Clark County (in which Las Vegas is located) in 1988 (Gertler and Donaldson 1990). A groundbreaking ceremony, attended by Nevada's lieutenant governor, was held in 1989 to mark the beginning of construction of the Transrapid line. A 2.1-km line in downtown Las Vegas was to be built. Despite the exciting prospects, and the completion of a number of technological studies, neither maglev line was built. Construction of the Transrapid ceased in

1990 amid financial and political issues. Although a dozen support columns had been built, all of them were "deconstructed" (i.e., torn down). A partially finished maglev station was converted into library administrative space (Schumacher 2005). The maglev proposal eventually "morphed" into the Las Vegas Monorail, the latter of which began as a link between the MGM Grand and Bally's. A major step in the development of the monorail was taken in 1997, when the Nevada state legislature approved a bill that enabled a private entity to own, operate, and charge a fare for use as a public transport system. The legislation enabled the developers of the monorail to generate revenue, enhancing the feasibility of the investment, as well as its expansion. The Las Vegas Monorail Company acquired the MGM Grand-Bally monorail in 2000, and proceeded to expand it into a multi-station, public transport operation. The Las Vegas Monorail is hailed as the first privately-owned public transportation system in the U.S. It is also the first driverless transit system with fully-automated controls in the U.S. (Roembke 2004).

TECHNOLOGY OVERVIEW

Circus Circus Shuttle

This 209-m long, cable-propelled line connects the main Circus Circus hotel and casino with Skyrise Manor, a secondary Circus Circus hotel building. The line, completed in 1985, uses only one vehicle that shuttles back and forth. The line's capacity is 1,250 persons per hour direction. The vehicle can carry 54 persons, at a speed of 6.1 m/sec. An I-beam in the center of the track provides horizontal guidance, and a single cable tows the vehicle back and forth. The shuttle supplemented a cable-propelled, two-vehicle shuttle that began operation in 1981. This shuttle was discontinued and dismantled in 1996, however, to make way for new construction. The 1985 shuttle is still in operation. The *discontinued* shuttle is shown in Figure 2.

Figure 2. Circus Circus Shuttle (discontinued version) (www.subways.net/usa/vegas1.JPG)

Mirage-Treasure Island People Mover

This line, 309 m in length, connects the Mirage and Treasure Island (TI) resorts. The line was completed in 1993, and carries one two-vehicle train. The train has a capacity of 120 passengers (60 per vehicle), and the line capacity is 1,800 persons/hr. Propulsion, similar to the other "west of Strip" APMs, is by cable. The operating speed is 8 m/sec. A photo of the line is shown in Figure 3.

Bellagio-Monte Carlo

This APM was in the process of being upgraded as of this writing. The former line operated as a shuttle between the Bellagio and Monte Carlo, from 1998 (Jakes and Ang 1999) until 2006. An upgrade to the former line was made during the interim. The new APM will have a midpoint station at the new CityCenter development. The new APM is being built by Doppelmayr, is cable-propelled, and will be 650 m in length with a capacity of 3,000 persons/hr/direction. The opening of the line is scheduled for November 2009.

Figure 3. Mirage-Treasure Island (TI) People Mover (http://world.nycsubway.org/perl/show?17738)

Mandalay Bay Express

This APM, also built by Doppelmayr, is cable-propelled and 838 m in length. The line opened in April 1999. Unlike the other "west side" APMs, the Mandalay Bay Express operates on dual, parallel tracks. One track is for "local" service, making all station stops, while the other is used for "express" services, stopping at the endpoints only. The Mandalay Bay Express also has a much larger consist (i.e., train passenger capacity) than the other APMs. Trains consist of five vehicles, seating 32 passengers each. Local trains have a capacity of 1,300 passengers/hr and express trains have a capacity of 1,900 passengers/hr. The running speed is 8 m/sec. The system was built with no intention for expansion (i.e., it is "self-contained"). Its maximum potential length is 4,000 m, given the limitations of the cable propulsion mechanisms. Two independent haul ropes provide propulsion to the local and express trains. The track infrastructure features a steel truss, trough-shaped guideway with dual lanes. The one-way express travel time is 151 sec, including terminal dwell times of 25 sec each. The APM was honored by *Elevator World* magazine in 2000-2001 as the project of the year (see Figures 4 and 5).

Figure 4. Mandalay Bay Express at the Luxor
(*www.railway-technology.com/projects/mandalay/mandalay1.html*)

Figure 5. Mandalay Bay Express at the Excalibur
(www.dcc.at/doppelmayr/references/en/tmp_1_435423795/Las_Vegas_Mandalay_Bay_detail.aspx)

Las Vegas Monorail

The initial phase of the Las Vegas Monorail was a 1.1-km, dual-lane "casino connector" between the MGM Grand and Bally's which opened in 1995. The system integrator was Bombardier, who provided M-Series Monorail vehicles, while VSL designed the guideway structure. The train capacity is 240 passengers (72 seated). The technology was selected specifically for its potential expandability. Monorail expansion was indeed pursued starting in 1997, when proposals were submitted. In 2000, the Las Vegas Monorail Company was established as a non-profit organization that subsequently acquired the MGM Grand-Bally's system. With the involvement of a number of stakeholders, and working under the authority of the Regional Transportation Commission of Southern Nevada, "Phase 1" of the Las Vegas Monorail was completed in 2004. The Phase 1 system, which was the existing system as of this writing, is 6.5-km in length, as briefly described in the "Introduction." "Phase 2," originally intended to be an extension to downtown Las Vegas, is to be an extension to McCarran International Airport. "Phase 3" *may* feature an extension to the west side of the Strip.

The Las Vegas Monorail does not operate on a set schedule, but headways range from 4 to 12 min. The directional line capacity is 3,200 persons/hr, and the maximum running speed is 80 km/h. Between-station travel times range from 1 min (Harrah's-Imperial Palace to Flamingo-Caesar's Palace) to 4.5 min (Harrah's-Imperial Palace to Las Vegas Convention Center). A single-ride ticket is $5, and a one-day pass is $12. Revenue is generated from fares and from advertisements. The photo in Figure 6 displays one form of advertisement.

Figure 6. Las Vegas Monorail
(www.reviewjournal.com/lrvj_home/2004/Jun-30-Wed-2004/photos/mono.jpg)

APM INTEGRATION

It is a straightforward exercise to draw lines between the existing Las Vegas APM lines, forming an imaginary loop around the east and west sides of the Strip. Although some argued that a "Strip monorail" would not solve all of the area's transportation needs (Caporale 1997), about 34 resorts, shopping centers, and other Strip developments would be served by a monorail loop. A similar loop was proposed as "Concept 9" in the aforementioned "Potential High-Technology Transportation Systems Study" (Gertler and Donaldson 1990). In Concept 9, the Strip loop was actually concentric with a larger, outer loop that linked McCarran International Airport, Thomas and Mack Center (arena), and the downtown. Construction – which was ultimately abandoned – began in 1989 on a much shorter, 2.1-km maglev line between the Cashman Field Center and downtown (Huss and Sulkin 1987), coming nowhere near the Strip.

It is a much greater challenge, however, to physically connect the APMs, making a seamless system. One issue is the differing means of propulsion: the "east side" monorail runs on electricity delivered through powered rails that are mounted to the guide beams, while the "west side" APMs are all cable-propelled. Another issue is the differing guideways: the Las Vegas Monorail, along with the Mandalay Bay Express, feature dual-lane "twin" tracks, while the other APMs operate on single-lane, out-and-back "shuttle" tracks. Also, there are differences in the guideway structures: The Las Vegas Monorail and Bellagio-Monte Carlo line are single-beam monorails, while the other APMs operate on dual-rail (albeit narrow gauge) guideways. Interestingly, the Mandalay Bay Express train *resembles* a monorail, but the steel truss track structure was specially designed to accommodate cable propulsion. It is also interesting to note that the Mirage-Treasure Island and Circus Circus APM guideways are similar.

The evolution of public transportation in the U.S. featured a brief period during which cable cars were converted to electric railways. Most of these conversions occurred during the first half of the twentieth century, with a few even occurring during the last decade of the 19th century. In all cases, the conversion was to an electric railway with an overhead wire, as opposed to a powered-rail setup. Also, in all cases, the cable car lines were effectively dismantled, except for the basic components of track infrastructure, to make way for electrical power. Hence, there is some precedence for converting cable to electric propulsion, although not vice versa.

Given the constraints, it is evident that there are three potentially feasible alternatives for creating a closed APM loop along the Las Vegas Strip:

1. Extend the Las Vegas Monorail along the Strip's west side; supplement existing west side APM service, or discontinue the west side APMs, replacing them with the monorail.
2. Extend the Las Vegas Monorail southward to the Mandalay Bay, establishing a transfer point to the Mandalay Bay Express. Extend the Mandalay Bay Express northward to the Monte Carlo, establishing a transfer point to the Bellagio-CityCenter-Monte Carlo line. Alternatively, extend the latter southward to the Excalibur, establishing a transfer point to the Mandalay Bay Express. Extend the Bellagio-CityCenter-Monte Carlo line northward to the Mirage; or, extend the Mirage-Treasure Island southward to the Bellagio. In either case, establish a transfer point. Finally, extend the Las Vegas Monorail southward from the Sahara to Treasure Island, possibly incorporating the Stratosphere Tower by aiming northward, before heading south.
3. Leave all APMs as is; establish linkages between all APMs with moving walkway systems, preferably at similar grade levels, to enable convenient transfers.

Although all of the west side APMs are cable-propelled, their technologies differ enough to preclude seamless connections. With the first alternative, one of the main concerns would be the cost of extending the Las Vegas Monorail. Another concern would be the effect of replacing the free west side APMs with a fare-based monorail. With the second alternative, one of the key issues would be the space requirements of transfer stations, along with the coordination of trains for passenger transfers. There would be four transfer points. It may be possible to establish new transfer stations in areas that currently do not have stations, rather than convert existing stations.

ALTERNATIVE 1: EXTEND THE LAS VEGAS MONORAIL

The first alternative would be to extend the Las Vegas Monorail southward to the Mandalay Bay, then northward along the west side of the Strip to the Stratosphere, and then southward again on the east side of the Strip. It would be a 9.6-km extension, increasing the length of the entire loop to 16.1 km (from 6.5 km, a 148% increase). Although a fully-developed cost estimate is beyond the scope of this paper, it has been widely reported that the existing monorail required a $650 million investment (Roembke 2004). The investment translates to $100 million per km. Applying the rate to the extension suggests a $960 million cost. In truth, the cost might be less than this, presuming that certain infrastructure components (e.g., maintenance facility, power-related interfaces) already exist, and would not need to be introduced. New monorail stations may offset some of these assets, however. New stations might be located at the Tropicana, Mandalay Bay, Luxor, Excalibur, New York-New York and Mirage, Bellagio, Caesar's Palace and Forum Shops, Mirage, Treasure Island, Fashion Show Mall and Wynn Las Vegas, Circus Circus and Echelon Place, and Stratosphere Tower. If the format of the existing monorail alignment is followed, then the west side alignment would probably parallel Frank Sinatra Drive and Industrial Road, traveling along the *back* sides of the west side resorts. A more centrally located Strip alignment would be preferable, but would probably be very difficult and expensive to retrofit. Integration with the west side resorts could be ensured with grade-separated pedestrian connections and moving walkways. The reader is referred to Kitagawa (1990) and Moore and Opthof (1993) for discussions of monorail extension projects in Tokyo and at Newark International Airport, respectively. It is recognized that the existing monorail alignment has been criticized for not being proximate to the Strip; some have attributed a loss of potential patronage to the alignment choice.

ALTERNATIVE 2: EXTEND SEVERAL APMS AND ESTABLISH TRANSFER STATIONS

The vision of a linked APM network would be that of extended APM and monorail lines, meeting at new stations, with passenger interchanges at each station facilitated by walking across a platform. One Las Vegas Monorail extension would be southward from the MGM Grand to the Four Seasons Hotel, just south of Mandalay Bay. An en route station would be placed at the Tropicana, and a new, transfer station would be built at the Four Seasons. From here, the Mandalay Bay Express, extended southward from the Mandalay Bay, would take passengers northward to a new transfer station at New York-New York. The Mandalay Bay Express would, thus, also be extended northward from the Excalibur to New York-New York. From here, the Bellagio-Monte Carlo line, extended southward from the Monte Carlo, would continue northward to a new transfer station at Caesar's Palace. The Bellagio-Monte Carlo would, therefore, be extended northward from the Bellagio to Caesar's Palace. From here, the Mirage-Treasure Island line would continue northward to a new transfer station at Fashion Show Mall, Echelon Place, Circus Circus, or Stratosphere Tower, depending on how far the Las Vegas Monorail is extended from the Sahara. The revenue-generating potential of the monorail may make it a better candidate for extension to Fashion Show Mall, however, as opposed to a lengthy northern extension of the Mirage-Treasure Island line. From this fourth and final, new transfer station, passengers would continue on the monorail. To summarize, new transfer stations would be established at:

- Four Seasons Hotel
- New York-New York
- Caesar's Palace
- Fashion Show Mall

Estimated monorail and people mover extension lengths would be as follows:

- Las Vegas Monorail from MGM Grand to Four Seasons: 2,100 m
- Mandalay Bay Express from Mandalay Bay to Four Seasons: 700 m
- Mandalay Bay Express from Excalibur to New York-New York: 300 m
- Bellagio-Monte Carlo line from Monte Carlo to New York-New York: 300 m
- Bellagio-Monte Carlo line from Bellagio to Caesar's Palace: 500 m
- Mirage-Treasure Island (TI) line from Mirage to Caesar's Palace: 500 m

- Mirage-Treasure Island (TI) line from Treasure Island (TI) to Fashion Show Mall: 400 m
- Las Vegas Monorail from Sahara to Fashion Show Mall (via Stratosphere): 3,800 m

The cost of extending the monorail a total of 5.9 km would be about $590 million, signficantly less than the $960 million cost of building a continuous monorail loop. The costs of extending the three west side APMs are unknown, however; the author was unable to readily find such data for the existing APMs. Special stations built with passenger transfer platforms would be an additional expense. Cross-platform transfers between two different rail modes or types are not common, but do exist. One example is the Millbrae BART station in Millbrae, California, where passengers can walk across a platform to transfer from BART trains (rapid rail) to Caltrain trains (commuter rail) (see Figure 7). The author is not aware of any monorail-APM or APM-APM cross-platform transfer stations, however (although there is a *monorail-monorail* transfer station at the Magic Kingdom in Walt Disney World).

Figure 7. Millbrae BART Station: View from Caltrain (*http://world.nycsubway.org/perl/show?57714*)

The round trip travel time on a linked APM-monorail system would be longer than that of a round trip on an exclusive, all-monorail system. Although this would be a disadvantage of the linked lines, it is likely that the system would be used primarily for short trips. A study that examines passenger trip length activity on the Las Vegas Monorail could be used to plan for trip lengths on the linked west side APMs. Another issue is that the Las Vegas Monorail operates on twin rails, such that reverse direction travel takes place on different guideways. The Mandalay Bay Express has twin rails that are each used for bidirectional travel. Perhaps the tracks could be converted to unidirectional travel, similar to the Las Vegas Monorail. The other APMs are single-track, bidirectional lines. Coordination between the separate lines would facilitate short transfers.

Coordination of the APM Lines and the Monorail

It is theoretically possible to coordinate the three extended west side APMs, along with an extended Las Vegas Monorail, using a methodology described by Vuchic (2005). The following operating parameters are assumed:

Table 1. Las Vegas APM Operating Parameters

Extended Line	Original Length (m)	Original + Extended (m)	Running Speed	Stations
Las Vegas Monorail	6,500	12,400	Variable*	*
Mandalay Bay Express	838	1,840	8 m/sec	5
Bellagio-CityCenter-Monte Carlo	650	1,450	11 m/sec	5
Mirage-Treasure Island (TI)	309	1,210	8 m/sec	4

* Monorail speed is based on current Sahara to MGM Grand (6.5 km) travel time of 13 min.

To determine APM operating speeds from running speeds, station dwell times of 25 sec are assumed. Vuchic (2005) noted that a timed transfer system is usually introduced when lines have headways of between 10 and 30 min, implying that shorter headways do not necessarily require precisely timed transfers. It may be useful to employ timed transfers between the Las Vegas APMs, however, to heighten their attractiveness, demonstrate efficiency and seamlessness, and minimize APM travel times. Round trip travel times of 660 sec for the extended Mandalay Bay Express, 504 sec for the extended Bellagio-Monte Carlo line, and 453 sec for the extended Mirage-TI line were estimated. The Mandalay Bay Express would be the critical link, with an 11-min round trip. The round trip would be too long to meet the Las Vegas Monorail more than once every 11 min. It would be prudent, therefore, to transform the Mandalay Bay's second track into a reverse-direction rail. Construction of a Mandalay Bay Express extension would, therefore, need to include crossovers to enable trains to transfer from one track to another. The crossovers would allow more than one train to use each track at the same time, enabling substantial reductions in Express' headways.

If the Mandalay Bay Express is transformed into a "pinched loop," then the Bellagio-Monte Carlo line would become the critical link. The 504-sec round trip would meet the Mandalay Bay Express (and Mirage-TI line) every 8.4 min, rounded to 10 min to make the headway divisible into 60. The round-trip travel time on the entire linked monorail-APM system would be 38.3 min (24.8 min on the monorail, 5.5 min on the Mandalay Bay Express, 4.2 min on the Bellagio-Monte Carlo line, and 3.8 min on the Mirage-TI line). Rounding this up to 40 min, it should be possible to sustain a 10-min headway around the entire loop, in both travel directions. It would not be possible to coordinate shorter headways; longer headways could be accommodated, but would not be advised.

ALTERNATIVE 3: LINK THE APMS WITH MOVING WALKWAYS

The third alternative would be to leave the APMs and monorail as they are, and link the systems with moving walkways. This alternative was not investigated in great detail in this study, since it would not truly represent an "integration" of the APMs. Moving walkways are considered suitable for trips up to 300 m in length, although longer-distance walkways were being studied (Kusumaningtyas and Lodewijks 2008). The author found that the gaps between all of the APMs were excessive. The only moving walkways connection that might work would be between the Excalibur and the Monte Carlo, at about 600 m. Despite this potential, Alternative 3 was not investigated further.

DISCUSSION

Of the three alternatives proposed for establishing a continuous, automated transit loop around the west and east sides of the Las Vegas Strip, the second one has the greatest potential for making use of existing APM infrastructure. The first alternative, extending the Las Vegas Monorail around the west side of the Strip to supplement – but not necessarily replace – the west side APMs, has been discussed by the Las Vegas Monorail Company as part of its "Phase 3." The advantage of this alternative would be the continuity of the technology, with no need for transfers between lines. The disadvantage may be the alignment, which is likely to be along the "back sides" of the west side resorts, similarly to the "east side" monorail. The cost, estimated to be nearly $1 billion, is also a disadvantage. The third alternative, linking the existing APMs with moving walkways, is not appealing because of the long distances between the endpoint stations. The *shortest* distance between APM endpoints is 600 m, which is twice the recommended maximum length of a moving walkway. It is possible to simply install successive walkways to cover long distances, but the author is not aware of the attractiveness of such an arrangement.

Although it is theoretically feasible to coordinate the APMs and monorails as part of the second alternative, several issues would need to be addressed. First, this study did not examine the level of interest in extending the existing west side APMs to new transfer stations. Gottdiener et al. (2000) discussed one case in which adjacent resort owners did not agree on APM alignments and connections. Second, and similarly, this study did not examine the level of interest in building transfer stations at resorts that currently do not have APMs (e.g., Four Seasons, New York-New York). Third, it is not clear how the free west side APMs would be coordinated with the fare-based monorail. Would it be awkward for users to pay to ride the monorail, but not the APMs? Would the transfer from an APM to the monorail be delayed by the purchase of a ticket? Or, would a special mechanism for transfers be established? Fourth, the coordination between the APMs and the monorail would work well with 10-min headways provided that there are no service disruptions. Four separate transit entities would need to communicate consistently and efficiently to facilitate service continuity. Fifth, the costs of the various APM extensions, along with the investment needed for constructing transfer stations, are not well known. Finally, the differing capacities of the monorails and APMs may create difficult transfer situations under certain scenarios. For example, if all of the passengers of a full monorail train (~240 passengers) wish to transfer to the smaller Mandalay Bay Express or Mirage-TI line, then some spillover riders may not be accommodated. The number of waiting passengers could accumulate if successive, full monorail trains arrive at an APM transfer station. The potential for this scenario, as well as a mitigating strategy, would warrant further study. Despite these issues, a coordinated and linked people mover system has the potential to enhance passenger movement along the Strip. The benefits of APM integration, which include a lower cost than Las Vegas Monorail expansion, a better west side alignment than an extended monorail, and the use of existing APM infrastructure, deserve additional investigation.

REFERENCES

Bivens, John A., "The Las Vegas People Mover Experience," *Proceedings, 2nd International Conference on Automated People Movers*, Miami, FL, 1989, pp. 765-772.

Caporale, Robert S., "Las Vegas – Expanding Oasis," *Elevator World*, Vol. 45, No. 5, May 1997, pp. 56-57.

Castaneda, Steven and Dean Hurst, "A Face Lift for the Bellagio-Monte Carlo People Mover," *Proceedings, 10th International Conference on Automated People Movers*, Orlando, FL, May 1-4, 2005, pp. 295-306.

Gertler, Peter B. and Vicki Donaldson, "High-Technology Transportation System for Las Vegas, Nevada," *Compendium of Technical Papers, Institute of Transportation Engineers 60th Annual Meeting*, Orlando, FL, Aug. 5-8, 1990, pp. 368-372.

Glorioso, Sharon, "Las Vegas Monorail Inaugurated Service in July," *Public Works*, Vol. 135, No. 9, Aug. 2004, p. 90.

Gottdiener, Mark, Claudia C. Collins, and David R. Dickens, *Las Vegas: The Social Production of an All-American City*, Blackwell Publishing, Hoboken, NJ, 2000.

Huss, Hans-Werner and Maurice A. Sulkin, "Las Vegas/M-Bahn Magnetically Levitated People Mover System," *International Conference on Maglev and Linear Drives*, Las Vegas, NV, May 19-21, 1987, pp. 125-128.

Illia, Tony, "Odds Go High Against Plan to Extend Las Vegas Monorail," *Engineering News Record*, Vol. 254, No. 5, Feb. 7, 2005, p. 16.

Jakes, Andrew S., "Economic Analysis of a Monorail Link Between the Stratosphere Tower and Downtown Las Vegas," *Proceedings, 6th International Conference on Automated People Movers*, Las Vegas, NV, Apr. 9-12, 1997, pp. 213-223.

Jakes, Andrew S., "The State of People Movers circa 2000," *Mass Transit*, Vol. 26, No. 4, Jun. 2000, pp. 36-46.

Jakes, Andrew S. and Michael A. Ang, "From a Prototype to Revenue Service in Two Years: Yantrak System Opens at Bellagio in Las Vegas, Nevada," *Elevator World*, Vol. 47, No. 1, Jan. 1999, pp. 74-75.

Kitagawa, Takashi, "Extension Project on the Haneda Line of Tokyo Monorail," *Japanese Railway Engineering*, No. 113, Mar. 1990, pp. 9-13.

Kusumaningtyas, Indraswari and Gabriel Lodewijks, "Accelerating Moving Walkway: A Review of the Characteristics and Potential Application," *Transportation Research A: Policy and Practice*, Vol. 42, No. 4, May 2008, pp. 591-609.

Lechner, Edward H. and Andrew S. Jakes, "Procurement Management and Design Review for Automated Guideway Transit Systems for Mirage Resorts in Las Vegas," *Proceedings, 6th International Conference on Automated People Movers*, Las Vegas, NV, Apr. 9-12, 1997, pp. 731-740.

Moore, Harley L, III and Raymond A. Opthof, "Expansion of the Newark International Airport Monorail," *Proceedings, 4th International Conference on Automated People Movers*, Las Vegas, NV, Mar. 18-20, 1993, pp. 425-434.

Mori, David J. and Alfred Fruhwirth, "Mandalay Bay Express: In Eight Months from Concept to Revenue Service," *Proceedings, 8th International Conference on Automated People Movers*, San Francisco, CA, Jul. 8-11, 2001, pp. 165-172.

Neumann, Edward S., "Las Vegas – A Significant APM Market," *Proceedings, 4th International Conference on Automated People Movers*, Irving, TX, Mar. 18-20, 1993, pp. 206-219.

Neumann, Edward S., "Small Scale Systems in Las Vegas – Appropriate Technology for the Setting," *Proceedings, 6th International Conference on Automated People Movers*, Las Vegas, NV, Apr. 9-12, 1997, pp. 272-283.

Neumann, Edward S., "Las Vegas – A Showcase for Automated People Mover Technology," *ITE Journal*, Vol. 69, No. 3, Mar. 1999, pp. 36-40.

Neumann, Edward S., "Past, Present, and Future of Urban Cable Propelled People Movers," *Journal of Advanced Transportation*, Vol. 33, No. 1, Spring 1999, pp. 51-82.

Roembke, Jackie, "The Las Vegas Monorail: First in its Class," *Mass Transit*, Vol. 30, No. 6, Sep./Oct. 2004, pp. 36-43.

Schumacher, Geoff, *Sun, Sin & Suburbia: An Essential History of Modern Las Vegas*, Stephens Press, LLC, Las Vegas, NV, 2005.

Snyder, Tedd L., "Las Vegas Monorail Innovations," *Proceedings, 10th International Conference on Automated People Movers*, Orlando, FL, May 1-4, 2005, pp. 797-807.

Stone, Thomas J., Jeff Kimmel, and Carlos Banchik, "Las Vegas MGM Grand to Bally's Monorail System," *Proceedings, 6th International Conference on Automated People Movers*, Las Vegas, NV, Apr. 9-12, 1997, pp. 284-296.

Stone, Thomas J., Carlos A. Banchik, and Jeffery B. Kimmel, "The Las Vegas Monorail: A Unique Rapid Transit Project for a Unique City," *Proceedings, 8th International Conference on Automated People Movers*, San Francisco, CA, Jul. 8-11, 2001, pp. 99-126.

Vuchic, Vukan R., *Urban Transit Operations, Planning, and Economics*, John Wiley & Sons, Inc., Hoboken, NJ, 2005.

"Bellagio-Monte Carlo Tram Repair," Schwager-Davis, Inc., San Jose, CA, www.schwagerdavis.com/pdf/13_JR13BellagioJR.pdf. <accessed on Feb. 3, 2009>

http://en.wikipedia.org/wiki/Las_Vegas_Monorail. Las Vegas Monorail information. <accessed on Feb. 15, 2009>

www.lvmonorail.com. Las Vegas Monorail website. <accessed during Jan. and Feb. 2009>

www.railway-technology.com/projects/mandalay. "Mandalay Bay Cable Liner Cable-Drawn Rapid Transit System, USA." <accessed on Jan. 31, 2009>

www.ropeways.net/index.htm?aktuell/doppelmayr2/index.htm. "New CABLE Liner Shuttle for Las Vegas." <accessed on Jan. 31, 2009>

Planning and Procurement of the Doha, West-Bay APM

Kamel-Eddine Mokhtech, PhD, PE. *
Sanjeev Shah, P.E. **
Hassan Eisa M. Al-Fadala, PhD, ***
Ghanim Hassan Al-Ibrahim****
Hassan Qaddoura, P.E. *****

*Senior Associate, Lea+Elliott, Inc., 5200 Blue Lagoon Drive, Suite 250, Miami, Florida 33126; PH305-929-8118; Kmokhtech@leaelliott.com
**Principal, Lea+Elliott, Inc., 5200 Blue Lagoon Drive, Suite 250, Miami, Florida; 33126; PH305-929-8122; Snshah@leaelliott.com
*** DCEO Operation, Qatari Diar Real Estate Investment Company, P.O. Box 23175, Doha, Qatar; alfadala@qataridiar.com
****Group Director, Engineering and Ventures, Qatari Diar Real Estate Investment Company, P.O. Box 23175, Doha, Qatar; galibrahim@qataridiar.com
***** Project Manager, Qatari Diar Real Estate Investment Company, P.O. Box 23175, Doha, Qatar; hqaddoura@qataridiar.com

Abstract

Qatar is a country that occupies about 11,500 square kilometers on a peninsula that extends north into the Persian or Arabian Gulf from the Arabian Peninsula. Doha is the capital of the country and a major administrative, commercial, financial, and population center.

The West Bay area of Doha is experiencing phenomenal growth, spurred by a governmental mandate to transform it into an international business magnet and the principal commercial and financial center of Doha. The area currently includes a vibrant commercial center including hotels, office buildings, residential towers, and government offices. Future development plans include a convention center, a high rise tower (over 100 stories tall), multiple new mid-rise and high-rise towers, as well as a 2,500-space underground parking facility.

The Lea+Elliott team is performing preliminary engineering and preparing tender documents for the Doha West Bay Automated People Mover (APM) system. This includes design of the APM operating system and establishing the infrastructure interfaces and requirements.

The first phase of the APM will include a five-kilometer dual-track underground tunnel, 10 stations and a fleet of approximately 42 cars, configured into two and three-car consists. The first car in each consist will be reserved for VIPs and their families. This paper provides an overview of the APM system and discusses its key planning, design and procurement elements.

Introduction

The state of Qatar is experiencing a phenomenal growth spurred by natural resources wealth as well as judicious investment worldwide. Qatar is the world's biggest exporter of liquefied natural gas. With Gulf countries witnessing a steady annual population growth, cities are increasingly facing congestion, compelling governments to spend on improving infrastructure. A key element to increased and sustainable economic growth is a reliable and efficient transportation system. Several railway and Metro projects both inside Qatar and connecting to the neighboring emirate of Bahrain via one of the world's longest bridges are in the planning stages. The common point of these transportation modes is in the heart of the Capital City of Doha, i.e.: the West bay District. In response to the continued growth of West Bay, an APM is planned to relieve congestion and increase mobility and accessibility to the main attraction poles such as the Convention Center, the Barwa Financial Center, hotels, embassies and the multiple residential and business towers in West Bay and serve as feeder to the future metro and rail lines. Qatari Diar, one of the region's most influential and innovative real estate investment companies leading many major projects domestically and internationally, was mandated by the Qatari government to oversee the planning and construction of the APM system.

Lea+Elliott Team are performing preliminary engineering and preparing tender documents for the Doha West Bay APM Operating system. This includes design of the APM operating system and establishing the infrastructure interfaces and requirements. The APM will share a common underground, multi-level station with a future metro system that is planned between the City of Lusail and Doha International Airport, a regional rail line and a high speed rail connection to Bahrain. The APM will connect the convention center (and the regional metro) to the major activity hubs within the West Bay area.

The first phase of the APM will include a five-kilometer dual-track underground tunnel, 10 stations and a fleet of approximately 42 cars, configured into two and three-car consists. The first car in each consist will be reserved for VIPs and their families. An extension to the Barwa Financial Center is under consideration.

West Bay

Figures 1, 2 and 3 illustrate the geographic location of the state of Qatar, the West Bay district, including an identification of the major commercial and entertainment sites and transportation arteries, as well as the main shopping mall in the West Bay District.

Figure 1: Qatar

Figure 2: West Bay

Figure 3: City Center- Doha

The Need for an APM System

Previous studies commissioned by Qatari Diar highlighted the need for an automated people mover. Ridership projections indicated that the peak link load or passenger per hour per direction could be close to 6000 pphpd. Subsequent analyses performed by Colin Buchanan and Partners resulted in a slightly lower peak link load. Considering that the more recent analysis did not include some of the latest planning elements such as the railway and high speed lines that would originate or transit through West Bay, as well as other initiatives such as a parking policy that would provide a few thousand parking spaces at the outskirts of West Bay, it was determined that a number close to 6000 pphpd would be the design target for year 2026. Ridership projections were performed using three out of the nine alignment alternatives that had been initially identified. The base and the preferred alignments, identified as option 1 and 9 respectively, are shown on Figures 4 and 5. Option 9 achieves the highest usage and provides an improved quality of service to the passengers.

Figure 4: Option 1 (Base Alignment)

Figure 5: Option 9 (Preferred Alignment)

Stations

The preferred alignment includes 10 stations located, on average, about 450 m apart. Station 1 located underneath Al-Wahda street will be an intermodal station where a future high speed rail line to Bahrain as well as a Metro line connecting Lusail, a future big development north of Doha, and Doha International airport will converge. The details of the station layout, designed by MIL Architects, are being refined in close coordination with Qatari Diar as well as other transportation stakeholders.

Stations are configured with side platform to accommodate a single bore tunnel selected due to right-of-way limitation precluding the use of the originally planned twin-bore tunnels. Stations have typically three or four access points depending upon their location. The concourse is at the upper level, while track and platform are at the lower level. The concourse includes retail areas as well as fare gates to allow access to paid passengers.

Technology Assessment

Prior to the selection of the APM system for Doha West Bay, different transportation alternatives to facilitate passenger movement between the traffic generators were evaluated. An initial screening considered Personal rapid Transit, Monorail, Guided Buses, Maglev, Cable propelled as well as self-propelled Automated People Mover. Automated People Movers were identified as the applicable technology. The following includes a description of the screening process.

Considering the projected ridership, as well as the existing and future site conditions, at-grade technologies were eliminated; indeed an at-grade LRT capacity of 185 passengers (using 4 passengers per square meter) yields a throughput of 32 LRT trains per hours operating at less than two minutes headways. It is difficult to see how this can be achieved while competing with the existing road traffic. The same conclusion can be reached for BRT where a throughput of more than 40 buses an hour would be required, with operation at less than ninety seconds headway.

PRT remains an unproven technology and as such is not considered any further.

Cable technologies are not applicable to the Doha West Bay potential alignments which include several stations with variable spacing, switching of trains for failure management and routing of trains in the Maintenance and storage facility. Cable propelled systems are more appropriate for a shuttle type service application.

Maglev (High Speed Surface Transport) is not applicable to the Doha West Bay APM. The tight curves in Doha West Bay (approximately 55 m curve radius) are too small for a proven maglev technology such as the one in operation in Chubu (Nagoya, Japan).

Self-Propelled Automated People Mover technologies are applicable for the Doha West Bay APM system. A multicriteria analysis, using the criteria listed below, was subsequently performed to assess elevated vs. underground APMs.

1. Traffic Impacts
2. Duration of Maintenance of Traffic
3. Environmental Considerations such as aesthetics, impact on adjacent facilities and business, Maintenance and Storage Facility (M&SF) location, impact of environment,
4. Operational flexibility,
5. Reliability/ Safety, and
6. Costs.

Operationally, both options are comparable. However, the underground/tunnel option provides for fewer disturbances in terms of maintenance of traffic, business impacts and aesthetics, but at a higher cost. Furthermore, the lack of availability of land in a prime real estate district close to the blue waters of the gulf constitutes a significant drawback to the elevated option.

As a result of the analysis, underground APM was selected as the technology of choice.

Geotechnical

A geotechnical survey performed by the Dynamic Management Group (DMG) of Qatar along the preferred alignment indicated that the top six meters of soil under the road surface are mainly sand and dirt, while the Simsima rock layer starts about 10 meters below ground level. This information, along with the utility mapping, was used as an input to determine the tunnel invert.

Alignment

Alignment was analyzed, in conjunction with tunnel configuration for the various alternatives. The initial configuration consisted in twin-bore tunnels and center platform stations. However, right-of-way constraints south of the convention center precluded the use of twin-bore tunnels. The decision was then made to select a single bore tunnel and a side platform arrangement. Given the nature of the soil (Simsima rock) at the elevations being considered, a Tunnel Boring Machine (TBM) was deemed adequate as a construction tool. Discussion with tunnel builders led to the target of 120 m radius of curvature in areas that would be excavated using a TBM. Stations will be constructed using the cut-and-cover technique. Traffic disruption will be minimized using deck plates to cover the area under construction.

Maintenance and Storage facility

The scarcity of available land in West Bay led to the original selection of the Sheraton Park Site, where a two storey underground parking structure sized for about 2000 parking spaces will be located. The Maintenance and Storage facility will be located underneath the parking structure. The Park site construction is consistent with the construction of the Convention Center as well as a 110-storey tower who are scheduled for completion prior to the start of construction of the Automated People Mover. This imposed the constraint that a detailed construction interface between the Park Site and the M&SF could not wait until the selection of a contractor for the Automated People Mover. In order to support the design of the Park site, a detailed construction interface was then initiated. Some of the issues that were considered are described below:

Insertion of cars in the maintenance facility will be performed using an insertion shaft sized to accommodate the largest car of the applicable technologies being considered. Given that the surface level of the Park Site includes landscaping, water and other architectural features, the location of the insertion shaft required a detailed coordination effort including a structural review of the crane requirements, loading zone dimension and road access. A review of the originally planned Park site column grid spacing (9m x 9m) was performed by laying out the Maintenance and Storage facility track work using the switching characteristics of the applicable technologies (crossover layout, linear distance between crossovers, track centerline distance, heavy and light maintenance lanes locations, forklift movement and access, overhead crane location, offices and shops footprint etc.). It was concluded that an 18 m x 18 m column grid was required in the area of special trackwork (crossovers), a 9 x 18 m column grid in the linear portions of the maintenance and flow through lanes, and a 9 x 9 m column grid in the shops and offices areas. Figure 6 shows a layout of the maintenance and storage facility under the Park Site.

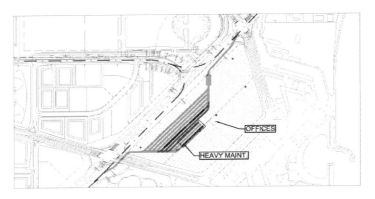

Figure 6: Maintenance and Storage Facility

System Operation

Figure 7: Operating Modes

System operation is configured in three loops, an outer loop where trains run counterclockwise and two inner loops where trains run clockwise. This configuration provides a comprehensive coverage of the area and allows a reduction of the overall trip time. Another advantage is that the offered capacity and train consist size can be tailored to the demand for each loop and not based on the highest link load, as would be the case for a pinched loop alignment. Given the expected ridership in the three loops, it became apparent that the size of the train consist required flexibility to tailor the offered capacity to the demand. A review of the expected ridership led to need to specify trains consists with two to three car length. Discussions with local authorities led to the determination that one car should be dedicated to VIPs and families where the other cars will be dedicated to other riders. This requirement dictates that the smallest operational train consists of two cars.

Fare Collection, Communication, and Traction Power

The Doha West Bay APM will be one of the first transit systems in operation in Doha and the transportation regulatory bodies are in the process of being established. Given that an integrated fare collection policy is under development, and in order to be able to preserve flexibility an open APM fare collection system based upon smart cart has been specified. Space has been allocated for fare gates and full-feature ticket vending machines at station concourse levels. Integration with the rest of the transit systems fare collection will therefore be preserved.

The communication system will be state of the art and will include the latest technologies and features. Extensive video surveillance, in the stations and on-board the trains, will be implemented.

The traction power distribution system will be Direct current, 750 V DC. A preliminary power simulation has been performed and resulted in three substations on the mainline and one substation in the maintenance and storage facility. If required, Assured Automatic Receptivity Units (AARU) will be used on the mainline to absorb excess regenerative power. The AARU's will be located outdoors and integrated within the station entrances architectural features. Kahramaa, the local electricity and water provider, indicated that they would be able to bring the high voltage feeders (11 KV) to two mainline substations and to the distribution substation (66 KV/11 KV) planned at the Park Site. The APM Contractor will then distribute the traction power to the other substations along the tunnel.

Procurement

The procurement approach will follow a conventional design-build structure, where the facilities are tendered to a facilities Contractor, while the Operating System Contract will be awarded to an Operating System supplier at the end of a competitive procurement process that will require the submittal of a technical and price proposal. The operating systems procurement documents include the Conditions of the Contract based on the FIDIC (Federation Internationale Des Ingenieurs Conseil) Yellow Book, Instruction to Tenderers, Special Provisions, Technical Provisions, Reference

Drawings and Operations and Maintenance Provisions. The initial Operations and Maintenance Contract will have five year duration, with two optional five year options for a total of fifteen years.

System Expansion

Several system extensions, that are consistent with the planned urban development, are being considered. The extension that seems to more likely to occur first will extend the APM towards the Barwa Financial Center, one of future financial poles of the region. Figure 8 shows a suggested alignment that will serve, in addition to the Financial Center, a stadium as well as several housing and commercial developments. In conjunction with the system extension, several alternative M&SF sites are being considered.

Figure 8: Potential System Extension

Conclusion

The Doha West Bay APM is expected to relieve the ever growing congestion in West

Bay. It will connect all major traffic generators such as hotels, convention center, City Center and residential as well as office towers in the West Bay district. It is being planned, and will be built, as an integral part of the other regional and local transportation systems in the city of Doha and in the state of Qatar. The preliminary engineering activities are proceeding and have accounted for the latest urban development plans in West Bay.

PRT CASE STUDY AT THE VILLAGE WEST DEVELOPMENT IN KANSAS CITY, KANSAS

Stanley E. Young[1], Peter Muller[2], Moni El-Aasar[3], Dean Landman[4] and Steven Schrock[5]

[1] Research Engineer, University of Maryland, Center for Advanced Transportation Technology, 5000 College Avenue, Bldg #806 #3103 College Park, MD 20742 PH 301-403-4593; seyoung@umd.edu
[2] President, PRT Consulting, 1340 Deerpath Trail, Suite 200, Franktown, CO 80116, PH 303.532.1855; PMuller@prtconsulting.com
[3] Principal, BG Consultants, 4806 Vue De Lac Place, Manhattan, KS 66503 PH 785.537.7448; moni@bgcons.com
[4] Principal, LTR, P.A., Transportation Research & Planning PH 785.266.4467; dlandman1@cox.net
[5] Assistant Professor, The University of Kansas CEAE Department, 2159B Learned Hall 1530 W. 15th Street, Lawrence, Kansas 66045-7609 PH: 785.864.3418; schrock@ku.edu

Abstract

In June of 2008 a Personal Rapid Transit (PRT) application study commenced at a popular development on the western edge of the Kansas City metropolitan area. The purpose of the project is to investigate the potential of PRT to solve transportation and mobility issues in a popular commercial development that encompasses retail and entertainment commonly known as the Village West development. In addition to retail, restaurants, and entertainment, the Village West area borders on a major NASCAR race track, and a former greyhound racing facility available for redevelopment. Village West also contains a minor league ball park, and a regional medical facility is located immediately to the east. Planned additional development includes a casino and water resort. As with most modern development, the dominant uses of land are parking lots and access roads to serve the attractions. This study investigates the ability of a PRT system to improve the inter-accessibility of the existing and planned facilities, to serve as a feeder to any existing and planned transit systems, and to reduce the quantity of land dedicated to roads and parking lots, freeing up land for additional development.

Introduction

Kansas State University in cooperation with PRT Consulting investigated the potential of PRT to solve transportation and mobility issues in a popular retail and entertainment district commonly known as the Village West development on the western edge of the Kansas City, Kansas metropolitan area. This area is located north of the intersection of Interstates 70 and 435. Existing development in the northwest quadrant includes the Kansas Speedway, the Legends at Village West (and upscale shopping district), mega retail stores of Nebraska Furniture Mart and Cabela's Outfitters, and many other smaller retail shops and restaurants. Also in the vicinity is the Woodlands Race Track and the Providence Medical Center. Planned development includes a casino and major water resort. See Figure 1 for a contextual map of the study area.

[Figure 1]

Despite the number of attractions, the dominant use of developed land however, is parking lots and roads to serve these attractions. This study investigated the ability of a PRT system to:
- Improve the inter-accessibility of the existing and planned facilities
- Serve as a feeder to any existing and planned transit systems
- Reduce the quantity of land dedicated to roads and parking lots freeing up land for additional development or open areas
- Reduce green-house gas emissions, run-off, and mitigate other environmental consequences related to development
- Estimate cost of implementing and operating a PRT system

The analysis is planned to include not only the impact of a PRT system to serve the existing development, but also how development may have been designed differently had it incorporated PRT from the beginning. The work commenced in June of 2008 and is currently in progress at the time of this writing. This paper reports on the issues, and work completed as of Feb 2008. [Presentation at the Automated People Movers conference will be updated to be current at the time of the conference.]

Stakeholders and Facility Description

The initial task was to identify stakeholders and invite their participation. Representatives from various organizations attended the project kickoff meeting, and subsequent one-on-one briefings. These organizations include:
- The Kansas Speedway
- Wyandotte County Unified Government
- Federal Transit Administration
- Federal Highway Administration
- Kansas Department of Administration
- Kansas Department of Transportation
- Mid-America Regional Council
- Department of Aging
- RED Development Corporation

On June 16, 2008 a stakeholders meeting was held to initiate the project and solicit input. Highlights of input provided by the stakeholders are given following:

"... the Casino proposals for the area ... included a stipulation that the casino provide transit service."

(a representative from the department of Aging) noted that the aging population would like and benefit from the convenience and practical nature of such a system."

" ADA parking at the track is filled by about 8:30 am on race days and others are turned away. Track operators would like to see a remedy so that folks needing special transportation accommodations can be more readily served."
"... a remote parking lot might not be needed if existing parking could be used more efficiently."

"... the Racetrack wants to move people around the racetrack and the nearby businesses as a benefit to all on race days and other days."

" Wyandotte County is implementing a Bus Rapid Transit system that will link Kansas City, Kansas with the Legends development. A circulator system at the development would greatly improve the efficiency of the transit link."

A detailed map of the development area is shown in Figure 2.

[FIGURE 2]

The initial stakeholders meeting led to many follow up meetings with suggested contacts. As a result of the contacts, the project team met with the developers of the Schlitterbahn water resort at their headquarters in Texas. Schlitterbahn had previously proposed a monorail system for the development area, including detailed cost estimates and proposed methods of financing (see section on finance). The latter contained analysis that was still relevant to the PRT study. As an additional benefit, the prior proposal to investigate monorail concepts had initiated discussions at the county government of methods to better circulate patrons in the area using methods other than traditional vehicles and buses.

Structural aspects of PRT Options at the Legends West

A structural study was preformed to determine the governing codes for building the guide-way for a proposed Personal Rapid Transit System (PRT) in the area. The study focused on the compliance of the Cardiff Network guide-way for the ULTra PRT system as designed by Advanced Transportation Systems Ltd. (ATS Ltd.). Code governance is determined by jurisdictional authority. The project being located in Kansas City, Wyandotte County, Kansas falls under the Wyandotte County public works. Wyandotte County uses KDOT and AASHTO standards for bridges. Additionally regulations require any bridge structure that exceeds $200,000 be reviewed and approved by KDOT.

CARDIFF NETWORK GUIDE-WAY DESIGN REVIEW

The Cardiff Network guide-way structure consists of longitudinal spanning side beams and cross members all in standard rolled sections. The surface for the rubber tired vehicles is constructed of pre-cast concrete planks approximately 4 in. (95mm) thick with nominal reinforcement (1). The spans for this track are 18m (approx. 59') long. This results in a very elegant and trim design.

The KDOT bridge unit performed a review of the Ultra design for conformance to KDOT and AASHTO standards. The review cited two main concerns with this design related to non-redundancy and the presence of fracture critical members.

> Structural redundancy allows the bridge to continue to carry loads after the damage or the failure of one of its members. *AASHTO LRFD Bridge Design Specifications, Interim 2005,* states "Multiple-load-path and continuous structures should be used unless there are compelling reasons not to use them"(4).

> A fracture critical member (FCM) is defined as a tension member or a tension component of a member whose failure would be expected to result in collapse of the bridge (5). In 1978 guidelines for design went to effect placing more stringent set of criteria for design, manufacture and inspection of Fracture Critical Members (FCMs).

As stated in the *NCHRP Synthesis 354* "International scanning tours for bridge management and fabrication have noted that Europe does not have special policies for FCMs. A risk based approach, coupled with more rigorous three-dimensional analysis techniques, is used to ensure that a sufficient level of structural reliability is provided. Consequently, steel bridge designs that would be considered fracture-critical in the United States are still commonly built without prejudice in Europe. At this time, the governing codes used in the United States and Kansas put stringent (and costly) requirements on FCM based designs.

In considering the Cardiff Network guide-way design as presented, both redundancy and fracture critical conditions are noted:

> The two longitudinal spanning side beams are made from standard rolled steel sections. These beams would be considered both non-redundant and fracture critical. When a two member bridge system is compromised by loss of one load bearing member there is no load path redundancy.

> Because the Cardiff Network system would be considered non-redundant and fracture critical it would require fracture critical inspection. The rolled steel design would be difficult if not impossible to inspect for internal corrosion.

Due to the current design philosophy in the United States and in Kansas, the Cardiff Network guide-way design as presented would not be allowed as a PRT guideway in the Village West development without significant justification and/or mitigation of the above concerns.

ALTERNATIVE INVERTED "T" GUIDEWAY

A viable alternative to the Cardiff Network guide-way, would be a pre-stressed concrete inverted T-beam structure. The inverted T-beams are placed side by side across the width of the structure. The concrete deck is formed directly on the web of the beams with falsework that is left permanently in place. Concrete curbs/parapet would provide the required guide-way safety. Once the deck is cured the beams and deck become a composite structure with continuous beams. The underside of the inverted T-beam structure is completely enclosed, leaving a smooth, finished appearance. Between the webs of the beams is a void area which can house any mechanical or electrical elements needed to run the PRT system. Openings for manholes and junction boxes can be provided for easy access to any systems within the T-beam voids. A simple diagram of such a structure is shown in Figure 3. This guideway design resulted from earlier PRT studies at Kansas State University (7) and meets all know code requirements for the area.

[FIGURE 3]

Financing of PRT at the Legends West

Finance options, like the structural guidelines, are affected by state and local legislation. The State of Kansas instituted a unique system to encourage the development of significant projects that can impact the economic vitality on a region in Kansas or the State as a whole. The system, called STAR bonds for Sales Tax and Revenue bonds, enables eligible government units to bond infrastructure improvements that will enhance the economies in their communities, and retire the bonds through state and local sales tax revenue. STAR bonds were instrumental in the financing of the road and parking infrastructure that serves the existing Village West development.

A similar scheme could be used for financing automated transit. The monorail proposal put forward by Schlitterbahn contained such an analysis. Typically the first 10 years of sales tax are used to retire STAR bonds. After 10 year the tax revenue reverts to the corresponding municipality. The financing analysis previously performed for the monorail proposal is shown in Table 1 (courtesy of Schlitterbahn development group). The analysis estimated a potential increase of sales tax revenue $8.8 million USD per year, providing $88 million in bonding potential over the course of a 10 year period. (This was based on a 10% increase in visitation as a result of the transit system.) Note that it reflects only increases sales tax revenue. Existing sales tax revenue could also be applied given proper approval. Although developed for a monorail concept, the financing analysis is independent of transit technology.

Table 1 Transit Funding Analysis from Monorail Proposal (courtesy of Schlitterbahn)

Potential Transit Riders and Impact on Sales Tax Revenue

	Visitors / Year	Autos/Year (1.8 riders/auto)	Gross Sales (in $millions)	Sales Tax (in $millions)	Gross Sales Per Cap per cap ($/person)
Cabelas	4,000,000	2,222,222	$72	$5.4	$18.0
Nebraska Furniture Mart	4,000,000	2,222,222	$400	$29.9	$100.0
Great Wolf Lodge	300,000	166,667	$25	$1.9	$83.3
Area Restaurants	3,650,000	2,027,778	$60	$4.5	$16.4
Legends	3,000,000	1,666,667	$100	$7.5	$33.3
Kansas Speedway	1,000,000	555,556	$15	$1.1	$15.0
Schlitterbahn	4,000,000	2,222,222	$500	$37.4	$125.0
T-Bones Stadium	350,000	194,444	$2	$0.1	$5.7
Totals	**20,300,000**	**11,277,778**	**$1,174**	**$87.8**	**$57.8**

Percentage of vistors to ride transits	25%	5,075,000 Total riders
Percentage of adult riders	70%	14,210,000 Total Adult Riders
Percentage of children riders	30%	6,090,000 Total Child Riders
percent increase in visitors due to transit	10%	2,030,000 vistors
Increase in sales due to increase visitation		$117,400,000
Yearly increase in sales tax		$8,781,520 Tax Rate 7.48%

Proposed PRT Route

Layout of a PRT route to serve the complex, both existing and planned development, was developed through collaboration of professions from multiple disciplines including planners, architects, engineers, advanced transit consultants. The issues considered include:

- Circulate patrons easily between all existing attractions and planned development
- Make efficient use of land area, minimize the need for additional parking
- Open up land currently dedicated to parking for denser development
- Link remote parking facilities
- Plan for future expansion to regional transit, and expansion of system to different land uses (for example residential and office space)
- Allow the development to grow from a one to two day event, to a 5-7 day resort destination by seamlessly linking attractions

Although several concepts were rendered, a final layout was adopted that took into account the issues listed above, plus feedback from stakeholders. This is shown in Figure 4. Notable aspects of the layout include:

- Links all existing development
- Provides for future development (areas in dark green)
- Constructed in a series of loops, conforming to PRT design concepts
- Layout encompasses 10 miles of guideway and 26 stations
- Links to planning regional transit (BRT station shown in yellow)

- Race track is not served directly (as per feedback from developers on nature of race day traffic.)

The proposed layout as shown in Figure 4 links all existing attractions. Stations are designated either directly at the entrance of existing facilities, or with the assumption that the station would be integrated into the current structure. The routes were chosen to not only link attractions with parking, but also to efficiently open up additional areas for development. The layout provides for efficient access of the entire complex with stations at all major attractions, allowing patrons to park once and enjoy multiple attractions with relative ease.

[FIGURE 4]

Environmental Issues

A review of applicable environmental and related code compliance issues was conducted by University of Kansas to explain the process that would need to be addressed in order to complete any needed Environmental Assessment, Environmental Impact Statement, as well as any local permits that might be needed in order to get approval to construct an automated small vehicle transit system at the Kansas Speedway area. A synopsis of the report follows.

The primary law governing the environmental protection process is the National Environmental Policy Act of 1969, or NEPA (1). The NEPA meets compliance with each federal law and regulation by requiring preparation of an Environmental Impact Statement (EIS) for all major federal projects that may significantly affect the environment. Core provisions of the NEPA include three primary mandates,
- "To the extent possible, policies, regulations, and laws of the federal government must be interpreted and administered in accordance with NEPA;
- Federal agencies must use an interdisciplinary approach in planning and decision making that impacts the human and natural environment; and
- The preparation of an EIS is required on all major federal actions that may significantly affect the human or natural environment."

Any federal assistance will likely result in the need to develop an Environmental Impact Statement (EIS). The EIS covers three primary areas: Air Quality, Noise and Vibration, Social and Economic Impacts. Note that when a proposed project does not include significant displacement of housing, is located on a single site, doesn't disrupt major business activity, and is compatible with area land use, economic impact will be minor and will not require extensive economic analysis. No environmental impacts were identified as potential contentious for an EIS, in fact the environmental benefits in most areas are obvious. However, the EIS process itself is a time consuming and costly process which could add significantly to the time line and budget of any proposed PRT system.

Ridership Estimates

Ridership estimates were established by Dean Landman, P.E., of LTR, P.A., Transportation Research & Planning. Ridership estimates on each portion of the system were based on the predicted number of persons boarding and departing at each station as well as the distribution of planned trips. Estimates were based on the projected number of annual visitors as originally estimated by Schlitterbahn in Table 1. Daily estimates were made for each station, resulting in a total estimate of daily visitors.

The ridership and revenue table from the Schlitterbahn report assumed that 25% of the visitors would use the then-proposed mono-rail. For this report, the 25% figure was used for those stations that served the Schlitterbahn complex but remaining stations were modeled at 10% and 20%. Ridership from the Park and Ride facility on State Avenue, was assumed to be 100%.

Other assumptions:
- Boarding time was assumed to be one minute
- Off-loading time was assumed to be 30 seconds
- The running speed on the system was assumed to be 23 mph as a weighted average between a design speed of 25 mph for most of the system and 20 mph for some of the sharper curves.
- The projected boarding as the Park and Ride Lot was 1,000 riders per day.
- Peak demand would be 150% of the average hourly demand

Using these parameters, the model estimated daily ridership of the system between 8200 and 16400 trips per day (for the 10% and 20% visitor ridership model respectively), and the average wait of 1.7 minutes.

Next Steps

The project is schedule to be completed in mid-2009. In addition to the results presented herein, the study also seeks to capture:
- Cost estimates for construction and operation of the proposed system based on the layout and taking into account local construction practices and codes.
- Energy consequences of the project
- Possible redevelopment of existing land dedicated to parking,
- Consequences if PRT had been integrated from the beginning.

Since the study commenced, Wyandotte County Unified Government has considered the proposed system for further study and possible implementation. At the time of this writing, the county government is investigating possible funding through the economic stimulus package.

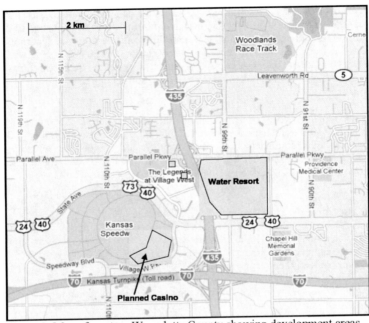

Figure 1 Map of western Wyandotte County showing development areas

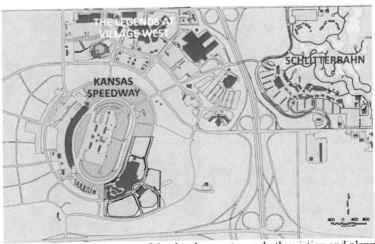

Figure 2 Detailed site map of the development area, both existing and planning

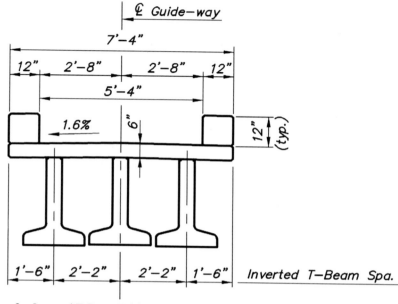

Figure 3 Inverted T-Beam Bridge Detail

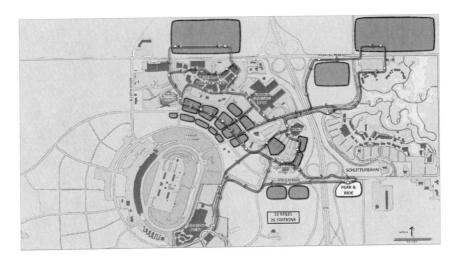

Figure 4   Proposed PRT Layout

REFERENCES

1. Kerr, A.D., P.A. James and A.P. Craig. *Infrastructure Cost and Comparisons for PRT and APM.* ASCE APM05 Special Sessions on PRT.

2. Ghosn, M. and F. Moses. *NCHRP Report 406: Redundancy in Highway Bridge Structures"*, Transportation Research Board, National Research Council, Washington, D.C., 1998.

3. *Fracture Critical Inspection Techniques for Steel Bridges,* Publication No. FHWA-NHI 02-307, U.S. Department of Transportation, Federal Highway Administration, March 2006.

4. *AASHTO LRFD Bridge Design Specifications, Third Ed.,* American Association of State Highway and Transportation Officials, Washington, D.C., 2006.

5. *Guide Specifications for Fracture Critical Non-Redundant Steel Bridge Members,* American Association of State Hightway and Transportation Officials, Washington, D.C., 1978.

6. Connor, R.J., R. Dexter and H. Mahmoud. *NCHRP Synthesis 354: Inspection and Management of Bridges with Fracture-Critical Details*, Transportation Research Board, National Research Council, Washington, D.C., 2005.

7. El-Aasar, M.G., R. Willard, and C Hibbs, *Kansas DOT Research Report No. KS-06-5 : Automated Small Vehicle Transit System Structural and Architecture Research Study for a University Campus*, 2006

PROJECT FUNDING OPPORTUNITIES

Sanjeev N. Shah [1], P.E., M.ASCE (Primary Contact)
Larry Coleman [2]

(1) Principal, Lea+Elliott, Inc., 5200 Blue Lagoon Drive, Suite 250, Miami, FL 33126; PH 305 500 9390; Fax 305 500 9391; snshah@leaelliott.com
(2) Manager of Planning Projects, Lea+Elliott, Inc., 5200 Blue Lagoon Drive, Suite 250, Miami, FL 33126; PH 305 500 9390; Fax 305 500 9391; lcoleman@leaelliott.com

Abstract

The traditional pay as you go project finance mechanism, which has been prevalent in the US, is becoming more and more untenable for many project owners due to several factors, including the hyper-inflationary pressures on project costs as well as the owners' financial capacity to issue long term debt due to the downturn in economic conditions. These pressures are common to transit agencies that do not have viable and dependable local revenue sources to leverage against limited state and/or federal funds, as well as airports where high fuel costs have negatively impacted the airline industry and the passenger traffic which is the key source to back long term debt.

Public-private-partnerships (P3) and other innovative finance/project structuring strategies are used extensively in the overseas market and offer an approach for funding projects in the US. This paper examines the key factors that can make a public-private-partnership approach viable and attractive including project structuring options, revenue stream opportunities and other tangible and intangible factors such as local economic impacts and provides examples of how these factors are being considered and applied on some projects in the US.

Background

Efficient and effective transportation systems have long played a vital role in advancing economic prosperity and growth in the United States. Unfortunately, in recent times, many communities and agencies across the US find themselves with constrained funding resources and financing capacity available through traditional sources to address pressing needs to modernize and enhance existing transportation infrastructure. Under the present circumstances, it may be prudent for project owners/sponsors to begin early in the planning process to consider project financing strategies as this can influence how a project might be phased and delivered later in the process.

In the sections that follow, traditional and alternative project financing mechanisms will be discussed with the focus on approaches for fixed guideway/transit improvements followed by a review of the factors to consider in establishing if a project is a candidate for a traditional or alternative financing approach.

FINANCING OPPURTUNITIES

Federal Funding Source

Some of the Federal funding sources typically available to cities, counties, states and/or authorities to finance a large capital project are summarized below. From a strategic standpoint, note that these funding sources can used either individually or in combination with other sources of capital as part of a project financial plan.

- **Federal Transit Administration's (FTA) Section 5309 New Starts** – This program provides funding for the development of new rail/fixed guideway transit systems and improvements or upgrades to exiting systems. Eligible systems include light rail, rapid rail (heavy rail), commuter rail, automated fixed guideway systems (such as a "people mover"), or a busway/high occupancy vehicle (HOV) facilities. Also, New Starts projects can involve the development of transit corridors and markets to support the eventual construction of fixed guideway systems, including the construction of park-and-ride lots and the purchase of land to protect right-of-ways. To become eligible, project sponsors must complete the major capital investment planning and project development process. Funding is provided on a discretionary basis and competition is considered very intense. FTA's evaluation criteria emphasizes travel time savings, costs and support for transit-oriented land use. Under the program, FTA will fund up to 60% of project cost with the balance covered by local sources.

- **Federal Highway Administration (FHWA) Flexible Funding for Transit/Highway Improvements** – Several FHWA Federal-aid highway programs have direct transit funding provisions including:

    - **Surface Transportation Program (STP)** – Provides funding eligibility for transit capital projects, vehicles, and facilities publicly or privately held, and for transit safety improvements.
    - **Congestion Mitigation and Air Quality (CMAQ)** – Provides funding eligibility for transit capital and operating expenses for new services in non-attainment areas only. Projects must demonstrate benefits to air quality and operating uses are limited to three years.
    - **National Highway System (NHS)** – Transit improvements within a National Highway System Corridor are eligible.

In addition, per the Intermodal Surface Transportation Efficiency Act (ISTEA), a State may transfer funds from Federal-aid highway programs that do not provide for transit related funding to ones that do provide for such eligibility. The fund transfers between programs are managed through the metropolitan and statewide transportation planning processes and eligible projects must be included in the regional Long Range Plan (LRP), the short-term transportation improvement program (TIP), and the approved Statewide Transportation Improvement Program (STIP).

Under these programs, FHWA will fund up to 80% to 90% of project cost with the balance covered by local sources.

- **Federal Aviation Administration Airport Improvement Program (AIP)** - The Federal Aviation Administration (FAA) administers the AIP program which provides grant assistance to public-use airports for capital improvements that enhance safety, capacity, security or the environment. The two primary categories of AIP funds airport operators receive are entitlement and discretionary funds. Entitlement funds are apportioned by the FAA based on airport passenger activity. Discretionary funds are distributed by the FAA based on their ranking of an airport sponsor's project relative to other competing projects under consideration. AIP eligible projects have included landside access improvements and fixed guideway conveyance systems such as APMs including those that connect to intermodal facilities off-airport and are used exclusively by airport patrons.

State/Local Sources

- **State Grants** – Many states through their respective Departments of Transportation provide grant programs for transportation infrastructure.

- **Tax Revenues** - Depending on the taxing authority of the governing owner/sponsor, there are a variety of tax methods that have been used to provide revenue to cover capital and operating costs or secure debt for a transportation project. Common forms include sales, income, property and gas taxes.

- **Special Tax District** – In this approach property owners within a particular district would be assessed a tax to reflect the access benefits associated with the provision of transit facilities within or to the district. In monetary terms, these benefits could be measured in several ways including the increased property values realized through the provision of transit improvements for the district or the cost savings developers may realize through the reduction in on-site parking requirements made possible by the improved access to the district.

- **Facility operating revenues** – System owner/operators can generate revenues from a variety sources including fare box revenue from direct operations and vehicle parking, concessions and leases from ancillary facilities. The net income from these sources can be applied to cover debt service, fund capital projects and/or cover operating and maintenance of costs.

- **User fees** – In some cases, owner/operators can apply user fees from associated facility operations to support financing for transportation improvements and operating costs. For example, airport operators have used fees paid by patrons of on-airport rental car facilities (typically called Customer Facility Charges or CFCs) to cover a portion of on-airport APM system operating costs. Passenger Facility Charge (PFCs) paid by airport patrons is another source funds that have been used to finance debt-service on eligible airport projects. PFC fees range from $3.00 to $4.50 per passenger. Under the program, the airlines collect the fee from each enplaning passenger as part of the ticket cost and the funds are transferred to the airport operator to invest in capital improvements at the airport that are approved by the FAA.

Financing Options

- **General Obligation Bonds (GOB)** – GOBs are a common form of finance for public projects in which tax revenues of a city, county or state are pledged as a source of repayment for a bond issue.

- **Revenue Bonds** - Proceeds from the sale of revenue bonds are the most common form of financing used by airport operators for large capital improvement projects. Debt payments can be supported and/or secured through general airport revenues, PFCs for eligible projects and revenues from the facility constructed or some combination thereof.

- **The Transportation Infrastructure Finance and Innovation Act (TIFIA)** authorizes the U.S. Department of Transportation (DOT) to provide Federal credit assistance to nationally or regionally significant surface transportation projects, including highway, transit and rail. Credit assistance is awarded through a merit based system to project sponsors, which can include public and private entities in one of three forms - secured (direct) loans, loan guarantees, and standby lines of credit. Loan cannot exceed 33% of the eligible project costs, are made at favorable U.S. Treasury rates, can be repaid up to 35 year term and require a favorable credit rating.

- **State Infrastructure Bank Program.** – This Federally authorized program enables States to capitalize Federal transportation grant assistance to provide loans, credit enhancement and other forms of assistance to eligible surface transportation projects.

Innovative Approaches

- **Public Private Partnership (P3)** - P3s are a growing method of implementing transportation infrastructure in which a private venture in partnership with a public agency will typically finance, design, build and operate a facility in exchange for a guaranteed revenue stream and/or land development rights from the public entity to cover debt service and operating costs. The partnership can be structured through a variety of mechanisms including concession or operating agreements and/or land leases. The revenues streams or financial incentive afforded to the private partner can come in many forms including fare box revenue, user fees and concession or parking fees from associated development. The P3 partner may also be granted development rights to adjacent parcels through a long-term land lease and develop the property to realize additional revenue from the development program. This latter approach to P3 partnering can be a challenge for transit agencies who have limited adjacent property to package in a P3 partnership. Also, not all States in the US have authorized local jurisdictions to enter into a P3 arrangement.

FINANCING STRATEGY AND PROJECT STRUCTURING

Under traditional models for large transportation capital projects, governmental owners/ sponsors such as states, cities or authorities typically finance projects through grants from the Federal Government. If they have the authority, the governing owner/sponsors raise additional capital to cover their local share through the sale of bonds which are secured by a stream of revenues or taxes. Under this approach for project financing, local owners/sponsors typically manage all phases of the planning, design and construction of the project through a design-bid-build approach.

Before embarking on a pursuit for Federal funds, local owners might consider the following:

- The timetable and resources necessary to fulfill the requirements to be eligible for the Federal funds – To be eligible, local sponsors have to fulfill a prescribed series of steps from project planning to obtaining environmental approvals. These steps require a commitment of local resources and often take several years to complete particularly if there is controversy associated with a project. The time needed to complete such a project can be an issue in situations where improvements are urgently needed.

- Competition for Federal funds and the likelihood of success – While the project may be a priority at the local level, there is much competition for limited Federal funds at the national level.

- The conditions which the Federal government may impose on the local entity to receiving project funding – These limitations can pertain to how a project is bid and implemented, how a system can be operated and how revenues generated from operations may be used by the owner thereafter.

- The availability of funds to provide for the local share – Local shares can vary but are generally in the range of 20% to 50% of project costs and coverage of the local share is typically dependent on the ability to leverage revenues from taxes and/or operations.

- The ability to cover operations and maintenance costs after implementation.

Bond sales also present a host of considerations for the owner/sponsor including,

- The credit rating for the selling entity

- Existing or potential sources of revenue to secure the debt service for the bond issue such as through existing or new tax revenues, the taxing authority of the sponsor to raise new revenues and the political viability of a new tax.
- Competing uses of funds and financing capacity – a plan of finance should be integrated within the sponsor's overall capital plan to clearly demonstrate that the sponsor is making the highest and best use of available funding and has the capacity to finance the overall capital program.
- Multi-tiered debt structure – Interest rates and debt coverage requirements can vary for different forms of finance. (Debt coverage is the ratio of revenue to annual debt service typically in the range of 125% to 135%) So it might be might be advantageous to pursue a tiered approach to debt financing to reduce overall interest and financing costs.

Finally there are project phasing and project structuring considerations:

- In cases where funds are limited but the need is great, the sponsor may consider implementing the minimum operable segment of a system that provides the greatest benefits from a level of service perspective and/or is the most feasible from a cost and financial point of view.
- If the initial capital is lacking but the project presents opportunities to generate a long-term of stream of revenue, then the pursuit of private investment capital might be a viable option.

After conducting such an evaluation, the project owner/sponsor may choose one of the following paths:

1. Pursue traditional approach - the project may have national significance and has a strong likelihood of receiving Federal funds and/or local revenue sources are available to secure debt service for bond financing
2. The project can generally be supported through public sector financing available to the owner/sponsor but multiple sources are needed to cover the local share of projects costs and/or secure bond debt service.
3. The project can not be fully supported through traditional public sector sources. In this case, the owner/sponsor may consider phasing implementation of the project if viable or may consider pursuing an alternative structure that draws in private sector financing through a public private partnership or P3 assuming that the that authority to pursue same has or is likely to be granted at the State level.

Project Delivery Considerations

If the sponsor concludes that P3 approach is desirable then the next consideration is the approach to project delivery. Traditional project delivery systems such as the design-bid-build approach noted above are generally not well suited for P3 applications, wherein the concessionaire has a financial interest in completing the project as early as possible to facilitate revenue generation. These traditional approaches require extensive interfaces and management of the different aspects that introduces schedule and budget risks which will typically lower the attractiveness of the project for investors who are likely to participate in a P3 concessionaire team.

The preferred mechanism would be a single Design/Build/Finance/Operate/Maintain (DBFOM) contract with the P3 partner as it provides the selected contractor more flexibility in managing and completing the Work. Benefits include a quicker project completion, less schedule and budget risks, and lower costs. The owner/sponsor also realizes cost savings as a smaller/leaner program management team will suffice for project/contractor oversight (compared to the traditional approaches).

A DBFOM contract for a transit/fixed guideway system could be arranged in two distinct phases:

- Phase 1 of the contract will incorporate the capital project, including the design and construction of the project infrastructure and installation of operating system equipment
- Phase 2 of the DBOM Contract will include the Operations and Maintenance (O&M) of the system and the fixed facility infrastructure by the same contractor for a period defined by the owner,

At the conclusion of the O&M period, the assets (developed under Phase 1 of the Contract, and maintained under Phase 2 of the contract), would revert to the owner with conditions that the assets be in good repair and require no major overhaul/maintenance for a specified period after the hand-over.

FUNDING STRATEGY CONSIDERATIONS: A CASE STUDY

As part of their long-term strategic plan issues in 2001, Broward County, Florida has been considering the development of an Intermodal Center and a People Mover. Under this plan, the Intermodal Center would be located on a site between the County's airport, Fort Lauderdale Hollywood International Airport (FLL) and the seaport, Port Evergaldes (PEV) with connections to planned regional transit and commuter rail and direct vehicle access to the regional highway system. The People Mover would link the four unit terminals at FLL with FLL's rental car center, potential remote airport parking at the Intermodal Center and the cruise ship terminals at the Port.

From a financing strategy point of view this proposed project offers a number of unique opportunities for consideration:

- The large volume of cruise passengers traveling between the airport and seaport would provide a captured market for the APM from a fare box or user fee perspective
- Revenues generated by associated airport properties including the rental car facility could be applied to cover operating costs for the on-airport portion of the APM system
- The potential for parking and concessions at the Intermodal Center offer another potential source of revenue for project financing
- The potential to leverage Passenger Facility Charges (PFCs) collected at FLL to cover capital financing costs for the on-airport portion of the project.
- The option to pursue Federal funding available through FTA, FHWA and or FAA.
- The County and airport's favorable credit rating
- The support demonstrated by the Florida DOT due to the traffic mitigation potential offered by the project and the resources they could bring to bear in the form of grants and the State Infrastructure Bank program.
- The project is not controversial from an environmental point of view which would keep project planning and review costs low. (This fact was later borne out by a Federal determination that the project would only require an Environmental Assessment and not a full Environmental Impact Statement.)

As the planning for the project took shape, the County conducted a preliminary analysis as outlined below to examine if the project could be self sustaining financially or would external fund be required.

Preliminary Financial Analysis

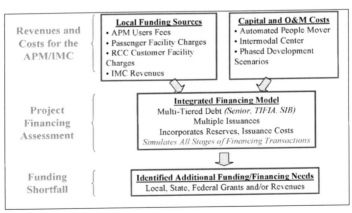

The findings issued in late 2004 indicated that external funding would be needed. At that juncture the County made the following go forward strategic decisions:

- Pursue Federal funding through the FHWA rather than FTA as there would less competition for transit funding with FHWA and the FHWA offered a higher Federal match.
- Proceed with a State sponsored Project Development and Engineering Study (PD&E) to fulfill the environmental review requirements of the National Environmental Policy Act (NEPA) which is required to obtain eligibility for Federal and State funding.
- Examine how the project might be more effectively phased from a financial perspective.

The basic strategy then was to position the project for funding consideration during the next cycle of Federal transportation legislation.

A second set of financial analyses was then conducted based on the following set of factors:

- The two most viable transportation system alternatives identified were an Elevated Busway or an Automated People Mover System.
- The project implementation was divided into four phases
- Capital costs were estimated by phase and escalated to the anticipated midpoint year of the implementation as summarized below.

Phase of System	Period of Development	Elevated Bus		APM Alternative	
		Cost in 2007$	Escalated to YOE	Cost in 2007$	Escalated to YOE
On-Airport	2016-2020	$82M	$126M	$173M	$267M
Extend to Midport	2018-2022	$227M	$378M	$410M	$683M
Extend to N. Port	2020-2022	$110M	$184M	$177M	$295M
IMC	2020-2022	$79M	$132M	$79M	$132M
Totals		$498M	$820M	$840M	$1,377M

- O&M costs were similarly escalated up to the start-up year of operations for each phase.
- Due to changing priorities in FLL's capital program, PFCs were no longer available for consideration as part of the financial plan.

- Local revenue sources were limited to user fees paid by cruise passengers to cover debt service and a portion of Customer Facility Charges (CFCs) paid by airport rental car customers.

- By agreement between the County and the rental car companies, rental cars users are assessed a Customer Facility Charge as part of the rental fee at FLL to cover costs for the on-airport rental car center. A portion of this fee presently covers the cost of the consolidated shuttle bus and would revert to cover a portion of on-airport costs of the APM when it came on line.

- About a half of the multi-day cruise passengers arriving through FLL were projected ride the APM system on the inbound leg to the Port and about two-thirds would use the system on the return from PEV to FLL at the end of their cruise. Per available information passengers currently pay about $10 per direction to be transported between the Airport and Seaport and this was fare level assumed in the financial analysis.

- To reduce costs, the project would be financed though a multi-tiered debt structure with general revenue bonds providing the senior debt and a TIFIA loan would be the subordinate debt.

- The project debt financing was assumed to have the following set of conservative characteristics:

  a. Senior Debt
     - Bonds issued by the County for this project would be "BBB" Rated
     - Interest rate: 6.5%
     - 30 Year Maturity
     - 1.85 Minimum Debt Service Coverage
     - 1.5% Financing costs

  b. TIFIA Loan
     - Loan issued through U.S. Dept. of Transportation
     - 30 Year Maturity
     - Limited to 33% of total project costs
     - Interest rate: 6.5%
     - 1.15 Minimum Debt Service Coverage
     - 1.5% Financing cost

Comparing cumulative costs to revenues over the bond repayment period, the findings indicated that the cruise passenger user fee would cover about 40% of the Alternative APM Alternative project costs and about 50% of Elevated Busway Alternative project costs leaving the project with a shortfall which would have to be covered by other external Federal, State and/or P3 sources.

In the final analysis, the County's approach to pursue external financing was still valid and the options going forward are as follows:

- Federal approval of an EA will establish the Project's eligibility for potential Federal and/or State funding opportunities which the County may pursue, and it will enhance the Project's attractiveness for Public-Private-Partnership (P3) funding opportunities.

- The cruise passenger ridership still offers a secure revenue source to attract private investments and a possible DFBOM approach with a P3 partner as does the development potential of the Intermodal site

- In view the cost and financial considerations, another option is that the County may consider initially constructing portions of the system as a lower cost elevated busway which could later be converted to APM system technology. The conversion from bus to APM could be accomplished by constructing the supporting elevated guideway for the busway with the dimensions and structural capacity required to accommodate the operation of APM system technology in the future.

EVOLVING CLARK COUNTY APM CODE REQUIREMENTS

David Mori, P.E.[1], Eric Troy[2]

[1]President, Jakes Associates, Inc., Jakes Plaza, 1940 The Alameda, Suite 200, San Jose, California, 95126, USA, Tel: (408) 249-7200; Fax: (408) 249-7296; E-mail: jakes@jakesassociates.com

[2]Associate, Jakes Associates, Inc., Jakes Plaza, 1940 The Alameda, Suite 200, San Jose, California, 95126, USA, Tel: (408) 249-7200; Fax: (408) 249-7296; E-mail: jakes@jakesassociates.com

Abstract

APM code requirements within the Clark County (Las Vegas, NV) environment are continuing to evolve. New sections of the ASCE code have been adopted as have liability and responsibility requirements. While designed to implement new APM systems having greater public safety, their imposition has increased the cost threshold of new APM development which, in turn, has had a potential impact on their marketability beyond the marketplace of Las Vegas. This paper further explores likely future code requirements within the Clark County jurisdiction.

Introduction

Over the past four years, the American Society of Civil Engineers (ASCE) along with the American National Standards Institute (ANSI) and the Transportation and Development Institute (T&DI), has updated and refined a set of Automated People Mover standards that serves as a benchmark for the entire industry. The latest standards include:

Part 1 (ANSI/ASCE/T&DI 21-05), revised in 2005 with the following scope:

- Operating Environment;
- Safety Requirements;
- System Dependability;
- Automatic Train Control (ATC);
- Audio and Visual Communications;
- System and Safety Program Requirements.

Part 2 (ANSI/ASCE/T&DI 21.2-08), revised in 2008 covering the following:

- Vehicles;
- Propulsion and Braking.

Part 3 (ANSI/ASCE/T&DI 21.3-08), revised in 2008 covering the following:

- Electrical;
- Stations;
- Guideways.

Part 4 (ANSI/ASCE/T&DI 21.4-08), new for 2008 and including the following:

- Security;
- Emergency Preparedness;
- System Verification and Demonstration;
- Operations, Maintenance, and Training;
- Operational Monitoring.

For most North American APM system applications, both public and private, these standards are often used as a representative guideline for APM design and development, but are not necessarily required to be enforced to the letter. Indeed, in the Foreword section of each ASCE standard it is stated that "the overall goal…is to assist the industry and the public by establishing standards for APM systems", but "[t]his standard has no legal authority in its own right". The Foreword goes on to clarify that it "may acquire legal standing" by any of the following or a combination thereof:

1. Adoption by an authority having jurisdiction;
2. Reference to compliance with the standard as a contract requirement;
3. Claim by a manufacturer or manufacturer's agent of compliance with the standard.

Such is typically the case that all or part of the standards are adopted by a customer seeking proposals for a new APM project, while adding further requirements as necessary to cover specific needs of the project in question.

For Clark County in Nevada, which because of the casino/entertainment industry has seen some of the most active project development of privately-owned APM systems in the world, system regulation falls not on Federal oversight committees such as the Regional Transportation Commission or the State of Nevada, but by the Clark County Building Department. Oversight applications include resort installations such as the Bellagio-Monte Carlo-CityCenter people mover, Circus Circus rubber-tired people mover, Mandalay Bay Express, MGM Grand-Bally's monorail, Primm Valley monorail, and the Las Vegas Monorail, as well as the series of APMs already operational and in process at McCarran International Airport in Las Vegas. Clark County has opted to adopt the ASCE APM standards directly as part of its Amusement/Transportation System Code for commissioning and oversight testing.

Case Study Examples

The two cases below represent examples of systems where different levels of oversight may be required, based on system characteristics.

Mandalay Bay Express

The Mandalay Bay Express People Mover tram system in Las Vegas, operating between a Las Vegas Boulevard/ Tropicana Avenue intersection station and a Mandalay Bay Resort station with intermediate stops at the Luxor and Excalibur Hotels and Casinos, is a fully automated cable-propelled transit system designed to provide transportation along a dual-lane, elevated steel guideway structure utilizing two (2), 5-car trains. The system was originally manufactured by Doppelmayer Cable Car (DCC) and represents an innovative, state-of-the-art People Mover and guideway design.

Figure: 'Mandalay Bay Guideway' illustrates the system dual-lane elevated guideway design. The system was designed without an emergency egress walkway between stations, as it was a.) not considered functionally necessary, and b.) not part of Clark County adopted ASCE code requirements at the time of system installation. In the event of a vehicle emergency and/or failure on the elevated guideway, where a vehicle is unable to return to a station for passenger unloading, a local fire department and system personnel would be dispatched to the train site. They would access the vehicles from hook-and-ladder trucks positioned below the guideway. A manual exterior door opening mechanism installed as part of the emergency exit doors for each vehicle would be utilized to reach the passengers and escort them to safety via the truck ladders.

Per ANSI/ASCE/T&DI 21.3-08, Part 3, Section 11.3, "[t]he APM guideway emergency evacuation and access shall be designed in accordance with the requirements of *Fixed Guideway Transit and Passenger Rail Systems*, NFPA 130, 2007 edition". These requirements imply a need for an emergency walkway along the entire guideway length outside of station areas. However, a walkway adds significant cost to an APM system, not to mention the additional cost to allow passengers manual access to the walkway from the vehicle interior. Further, a control system must be implemented to prevent door egress except in emergency situations. For the Mandalay Bay Express system (1/2 mile length), the additional capital costs of an emergency walkway and door control system could be upwards of $5 million. Therefore, for installations such as Mandalay Bay that can be accessed from the streets or parking areas below, it does not necessarily justify the requirement.

This does not mean that all APM installations (for Clark County or elsewhere) should not require an emergency walkway. For systems with all or part of an elevated guideway constructed taller than fire department ladders can easily reach, or a system that has sections not easily accessible from below, a walkway for passengers (along with appropriate means of access from the train interior) may be the only viable emergency evacuation option, and also a necessary and vital safety component. But an oversight process needs to be in place that allows functional interpretation of the ASCE APM standards for just such gray areas within the Code requirements.

Figure: Mandalay Bay Guideway

McCarran Airport T3

On the flip side, some APM installations are designed such that a greater degree of oversight may be necessary. As an example, the currently ongoing Las Vegas McCarran Airport Terminal 3 integration project includes provisions for an APM system to link Terminal 3 with Satellite Concourse D as part of an expansion effort. The APM system design consists of two 245 meter (803 ft) tunnels connected by ventilation shafts at each end. Adjacent to each of the stations is an emergency ventilation shaft, where the emergency fans will be installed. These shafts vent to atmosphere and will be grated at the interface between the ventilation shafts and atmosphere. Figure: 'Terminal 3 Station Ventilation Flow Concept' provides a rendering of the ventilation flow path for a fire event in a tunnel.

Figure: Terminal 3 Station Ventilation Flow Concept

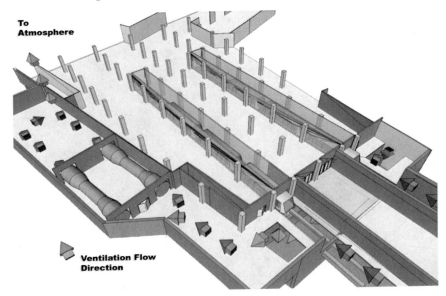

The control of smoke and fire suppression is necessary for an underground/ subterranean guideway to facilitate the evacuation of APM passengers to a point of safety. This is achieved in the APM System tunnels by providing tunnel emergency ventilation systems. Being a fixed guideway tunnel APM System, local Clark County NFPA 130 codes require that an emergency ventilation system be designed for the Terminal 3 APM System. Half of the APM System tunnels, stations and emergency ventilation shafts were already constructed. Some conceptual analysis of the emergency ventilation system design was performed by others prior to the construction of the existing facilities, however a detailed design analysis of the required airflows, ventilation equipment and ventilation control strategies had not been developed for the existing APM fixed facilities design.

Therefore, the ventilation system was designed in accordance with industry standard subway design principals within the constraints of the existing tunnel and ventilation shaft designs at Concourse D. In addition, an engineering analysis and simulation of the proposed APM ventilation system design was performed to identify design parameters and ventilation control strategies necessary to maintain a tenable environment for APM System passengers, maintenance staff and other people that may access the APM System tunnels in emergency and non-emergency situations.

Research of other similar transit system applications suggests that smoke control and fire suppression during emergency tunnel conditions, as well as the elimination of heat gain due to normal APM System operations, are best accomplished within the constraints of the existing facilities with reversible, axial-flow fans mounted in equipment rooms at each end of the tunnel system. Axial-flow fans are capable of providing the volume of airflow required to direct smoke flows away from evacuating passengers during tunnel fire events. Axial-flow fans can also be designed with a low-speed function for ventilating the tunnels during normal train operation. The T3 APM tunnel system will be equipped with fans in each tunnel emergency fan room located in each tunnel ventilation shaft. Such positioning will allow the ventilation system to utilize the ventilation shafts as a means of drawing fresh air into the tunnel and exhausting smoke out of the tunnel.

This fan arrangement allows a "push/pull" concept to be used during a tunnel fire event. Depending upon the location of a tunnel fire, it may be desirable to force smoke out of the tunnel through Concourse D ventilation shafts by operating the Concourse D emergency fans in exhaust mode ('pull'), simultaneously operating the emergency fans in Terminal 3 in supply mode ('push'). In this fan operating scenario, passengers would evacuate the tunnel system in the direction of ventilation air flow toward Terminal 3 station. In other tunnel fire scenarios it may be desirable to supply fresh air in through the Concourse D ventilation shaft and exhaust air from Terminal 3 ventilation shafts with passenger evacuation toward Concourse D station.

Although for this case study a comprehensive emergency evacuation and ventilation system has been developed, it may be prudent to expand ASCE/NFPA requirements and/or Clark County code requirements for similar tunnel systems given the extreme sensitivity of such a system to an emergency scenario and high potential for liability. Additional specific requirements to be addressed could include:

- Required airflows based on system size;
- Ventilation control strategies based on number of passengers evacuating the train(s) and emergency personnel entering the area;
- Fan size and/or quantity and blade speed based on emergency smoke removal rates, ventilation requirements and/or elimination of heat gain during normal operations;
- Fan operational requirements for "push/pull" arrangements based on location of incident and direction of nearest evacuation point.

ASCE APM Standard, Part 4

As referenced at the beginning of this paper, ASCE recently released Part 4 of the APM Standards (ANSI/ASCE/T&DI 21.4-08) with requirements for security, emergency preparedness, system verification and demonstration, operations, maintenance, and training, and operational monitoring. From a safety and security perspective, these additional standards certainly encompass a much more detailed and thorough action plan for verification of system security, emergency preparedness, coordination, training, recordkeeping, and operational monitoring programs than what was discussed previously in the other three parts of the APM Standards.

However, most of these requirements typically surface for larger public entity projects such as airport APMs. When an oversight agency such as the Clark County Building Department incorporates these standards as part of its code requirements, they may be taking on a level of regulation that is not necessary for the smaller-scale privately funded projects that are typical of Las Vegas and Clark County. Instead, consideration could be given to simply establishing an alternative framework for safety and security, operations and maintenance, training and auditing standards without requiring documentation such as:

- System Security Program Plan;
- Emergency Preparedness Program Plan;
- System Verification Plan;
- System Operations Plan;
- Service Restoration Analysis;
- Maintenance Plan;
- Training Plan;
- System Operational Monitoring Plan;
- Independent Audit Assessment;
- Other.

Conclusion

New sections of the ASCE Automated People Mover standards have recently been adopted which have shined a spotlight on APM security and safety. These recommendations are being or have been adopted by some jurisdictions (including Clark County) as Code. This could create situations where a broad enforcement of the Code affects the marketability of APMs for which viable and cost-effective alternatives exist to what is called out in the standards. In other instances, the standards may not be enough to adequately ensure public safety. An oversight process needs to be implemented that allows functional interpretation of the ASCE APM standards. Further, when adopting the standards for use in Clark County, special consideration should be given to the needs and requirements of smaller-scale, privately funded projects that are typical of Las Vegas and Clark County.

A CAMPUS TRANSPORTATION SYSTEM FOR MICHIGAN TECH

William H. Leder*, Frank W. Baxandall**, and William J. Sproule***

*Adjunct Professor, Michigan Technological University, Department of Civil and Environmental Engineering, Houghton, MI 49931; Ph 906-487-1647; bleder@mtu.edu

**Graduate Student, Michigan Technological University, Department of Civil and Environmental Engineering, Houghton, MI 49931; Ph 906-487-3583; fwbaxand@mtu.edu

***Professor, Michigan Technological University, Department of Civil and Environmental Engineering, Houghton, MI 49931; Ph 906-487-2568; wsproule@mtu.edu

Abstract

This paper explores the conceptual design of a high level of service transportation system for the Michigan Technological University campus in Houghton, Michigan. Like most colleges, Michigan Tech's campus has expanded during recent decades, student auto usage has grown dramatically, and parking is now a major problem. The proposed transportation system links the central campus with an athletic complex area about 850 meters (2,800 feet) south of and 53 meters (175 feet) above the main campus. The athletic complex area has available parking and land for additional campus housing and other development. Using urban planning "smart growth" principles, this transit link would help achieve several campus master plan objectives, including: (a) relocating parking from the central campus to an upper campus activity center, thereby enabling more efficient land use, creating new opportunities for development, and providing a more esthetically pleasing appearance; (b) providing a high level of mobility for a proposed campus housing development located in a Transit Village; and (c) promoting sustainability by helping to control commuter student vehicle miles traveled. Steep terrain, combined with very harsh winter weather, poses significant engineering challenges that rule out self-propelled APM technologies. Rope propelled APM systems, and a rope-propelled and supported aerial tramway technology, similar to a system operating at the Oregon Health Science University in Portland, are alternative solutions.

Project Context

Michigan Technological University is located in the City of Houghton, Michigan. Houghton, in the state's Upper Peninsula, is approximately 675 kilometers (420 miles) north of Chicago, Illinois and 570 kilometers (350 miles) northeast of Minneapolis, Minnesota. Figure 1 (on page 2) shows the location. The region is known for harsh and long winters that include lake effect snow created as cold air masses cross Lake Superior. Annual snowfall can reach eight meters (315 inches), and temperatures well below zero degrees F are common. The western half of the Upper Peninsula is rich in minerals, and Houghton is located in the heart of what was a very productive copper mining region beginning in the mid-1800s and continuing for nearly a century.

Figure 1 -- Location

Michigan Tech began in 1885 as the Michigan Mining School. The school, established by the mining companies to overcome a shortage of trained engineers, soon received a charter from the State of Michigan. The first classes were held on the second floor of the Continental Fire Hall in Houghton. By 1900 the renamed Michigan College of Mines had moved to the present campus location, about 0.8 kilometer (1/2 mile) east of the Houghton central business district. Hubbell Hall, shown in Figure 2, was the first building.

Figure 2 -- Hubbell Hall, first campus building

By 1931 enrollment had increased to 591 students, and the college offered degrees in several engineering disciplines, metallurgy, and chemistry. Figure 3 shows the campus during the 1930s. In 1964, with a broader range of degree programs in engineering and science and 3,400 students, the Michigan College of Mining and Technology became Michigan Technological University.

Figure 3 -- Campus circa 1930

Today Michigan Tech, one of four research universities in the state, has an enrollment of 7,000 students and offers more than fifty degree programs with an emphasis on Science, Technology, Engineering and Mathematics, referred to as STEM education [1].

As enrollment and degree programs grew, so did campus facilities. Today the core campus is located on a narrow glacial terrace overlooking Portage Lake to the north. The south side of the core area is bounded by steep terrain composed of rock outcroppings and glacial till rising steeply about 53 meters (175 feet). An upper campus area comprised of athletic facilities and fields and large automobile parking lots first appeared in 1980 with the opening of the Student Development Complex (SDC). Since then several other buildings have been added to the upper campus. MacInnes Drive is the arterial street linking the lower and upper campus activity centers. Many students use sidewalks flanking MacInnes Drive for trips between the two campus areas. Figure 4 shows the current campus with Portage Lake to the north.

Figure 4 -- Current campus

Campus Master Plan

During the past four decades Michigan Tech has undertaken several master planning efforts aimed at preparing and refining a comprehensive development plan for campus facilities. The most recent effort has been documented in the "Fresh Look Scenarios Plan Report" *[2]*.

The body of campus planning has been based on many parameters such as forecasts of
student enrollment (including mix of undergraduate and graduate students), status of existing facilities, research programs, changes in approaches to education, student housing needs, the best thinking about a wide range of related Michigan Tech activities, and the most effective approach to land use considering physical constraints posed by topographic features, the surrounding community, and the environment.

Historical campus development occurred along an axis that parallels U.S. 41, a state highway. In the early 1970s the segment of U.S. 41 bisecting the heart of the campus was realigned to the south, thus allowing the creation of a central pedestrian mall. The idea of developing a stronger, more focused "woods-to-water" axis, roughly perpendicular to the U.S. 41 corridor, has emerged in recent thinking and studies such as the "Fresh Look Report" previously cited.

The significant walking distance and elevation change between the lower and upper campus areas, combined with harsh winter weather, pose transportation issues that must be solved in order to unify the campus in a functionally effective and sustainable manner that will be acceptable to the University and community. If this problem is not solved, more buildings on the hill above U.S. 41 could be problematic, and future campus development may be constrained. Master plan work to date has not addressed the details of transportation alternatives that could link the upper and lower

campuses.

Figure 5 -- Campus with underutilized Lot 24 in foreground

Another issue at Michigan Tech that is typical of many college campuses concerns automobile parking. Michigan Tech has 37 parking lots with a combined capacity of 3,700 spaces. [3] Much of the parking is conveniently located around the perimeter of the core campus. Thus, considerable land that could be available for new facilities, pedestrian circulation, or landscaping is not being used for its highest purpose. The perimeter parking has created an undesirable suburban "shopping center" look as one views the campus from U.S. 41. A large inventory of parking in the Upper Campus, including a 660 space surface lot shown in Figure 5, has been constructed for use during major athletic events. The 660 car lot is almost completely vacant most of the time.

Senior Design

The Civil and Environmental Engineering curriculum at Michigan Tech requires all students to take a three credit, 14 week long course titled Senior Design. The general objective of Senior Design is to provide an opportunity for student teams to successfully collaborate on a major, semester-long assignment integrating a range of civil and environmental engineering disciplines and skills. This course provides a transition between traditional classroom instruction and professional practice. For most students it is the first time they have "owned" a complex, open-ended project for an entire semester.

Inspired by the "Fresh Look Report," the authors developed a scope of work for a Senior Design Project pertaining to future transportation between the Lower and Upper Campus activity centers that would address:

- Automated People Mover (APM) and Aerial Ropeway alternatives to walking or road based modes;
- Automobile parking reform;
- Student housing;
- Sustainable development; and
- A strong woods-to-water corridor.

While in most applications a wide range of APM alternatives would be considered initially and narrowed through technology analysis and evaluation, given the terrain and climate, it was decided to quickly focus on a rope (spun steel cable) propelled APM. Moreover, the Aerial Ropeway was a logical alternative to the APM believed worthy of consideration, partly because of many years of experience in operating ski lifts at Mont Ripley, the downhill skiing facility owned by Michigan Tech.

The rich industrial heritage of the local copper mining district includes many examples of rope hauled transportation systems. The mines were some of the deepest in North America, and rope hoists were used extensively to transport ore, construction materials and miners, and to remove water. Figure 6 (on page 6) shows the Nordberg steam hoist installed in 1918 at the Quincy No. 2 shaft rockhouse. Weighing 880 tons, the hoist could transport cars loaded with 10 tons of ore at a speed of 58 kilometers per hour (36 miles per hour) from the 9,260 foot deep mine. *[4]* As well, several mines had aerial trams as part of their surface works to transport ore.

The proposed project was reviewed with and endorsed by senior Michigan Tech officials in the summer of 2007. Michigan Tech became the "client" for the project. Cooperative
alliances were established with Doppelmayr Cable Car, GmbH & Co. for the APM and Doppelmayr CTEC for the Aerial Ropeway. Doppelmayr staff visited Michigan Tech in
the late summer of 2007 and agreed to provide technical support to the students and significant financial support to Michigan Tech.

Doppelmayr's generous gift of $4,000 enabled the class of 11 students and two faculty members to take a field trip to the Toronto International Airport, where Doppelmayr has an operating APM.

Figure 6 -- Nordberg double expansion steam hoist

The Senior Design Project was undertaken during the spring 2008 semester. An excerpt from the scope of work provided to the students on the first day of class follows:

"The scope of work is to prepare a complete plan and conceptual design for:

- A high level of service Campus Transportation System (CTS) that will link the Lower and Upper Campus areas; and
- A Transit Village in the Upper Campus based on smart growth, Transit Oriented Development (TOD) concepts.

Michigan Tech desires that the CTS be a showcase for sustainable development. To that end, top University officials are generally aware of Automated People Movers (Minneapolis/St. Paul and Detroit Airports) and rope suspended systems (Mont Ripley). Moreover, it is envisioned that the Transit Village will include a multi-story parking garage, medium rise dormitory/apartment building(s), supporting commercial space, and related urban infrastructure such as roads and utilities. No budget constraint has been specified, but the client expects that good judgment will be followed using customary standards for university facilities.

In order to accomplish the scope of work, the class will divide itself into three teams that will undertake planning and conceptual design:

- Team A will address the CTS based on Automated People Mover (APM) technology.
- Team B will address the CTS based on aerial tram technology.
- Team C will address (a) the CTS stations and maintenance facility, and (b) the Transit Village."

The remainder of this paper summarizes the work accomplished by the students and presents some general conclusions.

Transit Village

Team C prepared a conceptual design for a mixed use development on the Lot 24 site that includes 240 two-bedroom apartments for student housing, a 1,692 stall parking garage, and diverse retail space, all contained within a single five-story structure. *[5]* The upper CTS station is north of the larger mixed use structure and linked to it by a pedestrian bridge.

The Michigan Tech Strategic Plan envisions more emphasis on research and an increase of 400 graduate students. *[6]* Accordingly, based on the Strategic Plan and meetings with the Assistant Vice President of Housing and Student Life, Team C decided on apartments rather than undergraduate dormitory housing. The apartments are arranged around the perimeter of the parking garage, on floors 2 through 5. This design provides a variety of good views and enables residents to access their parked cars without the need for vertical circulation.

The parking garage capacity is based on parking for apartment residents, the elimination of five surface parking lots in the campus core, sufficient additional capacity so that visitors and sports events can be accommodated, and future growth in the number of students and staff commuting by automobile to the campus each day from the surrounding community.

One of the goals for the transit village is to function as a Transit Oriented Development (TOD) that will help control vehicle miles traveled. *[7]* Accordingly, 3,065 square meters (33,000 square feet) of retail space is located on the outside face of the parking structure on the ground level. Team C prepared a list of desirable retailers such as dry cleaning, a barber shop/salon, video rental, a convenience store, small restaurants, bicycle rental and storage, and other similar services.

Given the very significant snow fall, the roof was not designed for parked cars. Snow removal would be very problematic, requiring a design that could accommodate heavy equipment. Team C planned a green roof that will reduce the heat island effect and contribute to sustainability.

The design details of the CTS station adjacent to the mixed use structure are to some extent dependent on the CTS technology, either the APM or the Aerial Tram.

- For the APM, stationary traction equipment and some maintenance functions will be located on the ground level, there will be an intermediate level for under floor vehicle maintenance, and the passenger platform will be on the top level. The top level will be connected to the parking garage and apartments by a 38 meter (125 foot) pedestrian bridge.

- For the Aerial Ropeway, the stationary traction equipment and maintenance will be located on the ground level. The second level will contain a passenger platform and space to store detachable grip gondolas that will be explained later. Because of differences in the APM and Aerial Tram alignments, also explained later, a 9 meter (30) foot pedestrian bridge will connect the station and the parking garage and apartments.

APM Alternative

Team A investigated alternative APM guideway alignments within the corridor linking the upper and lower campus areas. The preferred aerial alignment, 850 meters (2,800 feet) in length, skirts the edge of the hockey arena, passes in front of the day care center, parallels MacInnes Drive, and crosses U.S. 41 into the lower campus, ending at the lower station adjacent to Rekhi Hall in a parking lot that will be eliminated as part of the overall plan.

Figure 7 — APM Alternative alignments (preferred in red)

Figure 7 shows alignment alternatives. Considerations included not exceeding a 10% grade, minimal horizontal curves, avoiding existing structures, column placement, and guideway accessibility from existing streets. Although the ASCE APM Standards permit a maximum 12% grade, Doppelmayr indicated a preference for

10%, which was adopted as a criterion. *[8]* Even with the rough terrain, the recommended guideway is never more than 20 feet above ground level, which permits maintenance and, in very rare events, access by mobile ladder equipment for emergency evacuation in accordance with NFPA-130 Section 6.2.3.2.1. A total of 34 columns support the guideway.

To provide a high level of service and availability, the APM is configured as a single lane with center bypass to permit two trains to operate in shuttle mode. Each train consists of two 30-passenger cars with the capability of adding a third car to each train in the future if warranted by increased peak demand. A Doppelmayr train is shown in Figure 8.

Figure 8 - **Doppelmayr train on guideway**

Maximum speed is 32 kilometers per hour (20 miles per hour). The round trip time is 5.4 minutes including 30 second station dwells. Headway is 2.7 minutes. System capacity with both two-vehicle trains operating is 1,300 passengers per hour per direction. In order to establish a recommended system capacity, Team A undertook field studies to measure the peaking characteristics of the current lower campus parking lots. *[9]*

The proposed two train operating schedule during fall and spring semesters is shown below:

Day	Hours
Monday-Friday	06:30 -- 22:00
Saturday	07:45 -- 20:15
Sunday	11:45 -- 20:15

During other times one train will be available in standby mode. During significant snow events, both trains will be operated to keep the guideway and rope clear.

Summer service, when the student population is low, can be tailored to fit demand patterns, and the schedule can be adjusted to accommodate special events.

Aerial Ropeway Alternative

In addition to the many aerial ropeways associated with downhill skiing, there are two public, commuter transportation systems in the United States that use rope technology. In New York City since 1976 the Roosevelt Island Tram has operated between Manhattan and Roosevelt Island in the East River. *[10]* The Portland Aerial Tram, opened in 2006 at a cost of $57 million, connects the Oregon Health & Science University Marquam Hill Campus with the Portland Waterfront. The Portland Aerial Tram, shown in Figure 9, is 1,000 meters (3,300 feet) long and rises 152 meters (500 feet). *[11]*

Figure 9 – Portland Aerial Tram

Taking advantage of the steep grade capability of aerial ropeway technology, the alignment Team C recommended is on a tangent directly between the Transit Village station and a station in the lower campus adjacent to the Memorial Union Building. The maximum height above ground is 15 meters (50 feet), and only 13 towers will be required. *[12]* The alignment passes through a residential area. It is unclear at this point if Michigan Tech would need to acquire all the land under the alignment or if an air rights arrangement could be established, but for purposes of their project, Team C assumed that these issues could be resolved.

Team C conceived their system with detachable grip gondolas, each with a capacity of 8 passengers. Although the haul rope speed maintains a constant 21 kilometers per hour (13 miles per hour) in a loop operating configuration, at each station the gondolas detach from the haul rope and slow to 0.8 meters per second (2.5 feet per second) for boarding and deboarding passengers. An off-line design feature permits disabled passengers to board and deboard stationary gondolas. Unlike the APM alternative, the Aerial Ropeway requires continuous station attendants.

Trip time between the stations is 2.75 minutes. With 28 gondolas in operation, the headway is 11 seconds, and system capacity is 2,400 passengers per hour per direction. The operating schedule planned by Team C is similar to the APM alternative.

The aerial ropeway will be equipped with a diesel auxiliary drive for temporary operation should there be an extended electrical power failure. This auxiliary drive will be used only for evacuation of the system. This would permit movement of all occupied gondola cars into either the upper or lower station. Should there be a rope derailment or some other event that would preclude operation, an emergency evacuation would be undertaken. For gondolas that are less than six meters (20 feet) above the ground, evacuation would be accomplished using ropes that are strung across the haul rope and belayed from below. An evacuation seat attached to the rope would allow safe lowering of the gondola's occupants to the ground. Evacuation of persons with disabilities would be accomplished using the rope/evacuation seat procedure no matter the height of the car. The ski industry and the National Ski Patrol have developed effective emergency evacuation procedures for aerial ropeways. An emergency evacuation procedure for the Michigan Tech system will be developed following these procedures and in accordance with ANSI B77.1-2006. [13]

Costs and Funding

Capital cost estimates were prepared by the teams in 2008 dollars using cost data from RSMeans and Doppelmayr. [14] Each CTS is approximately 850 meters (2,800 feet) long. The estimates appear below:

	Cost (millions of U.S. dollars)
Parking garage, including roughed in commercial space	35
Apartments	39
2 CTS Stations (approximately the same for each technology)	3
APM Alternative (fixed facilities and operating system)	20
Aerial Ropeway Alternative (fixed facilities and operating system)	7

Although there is no user charge for current parking except for large athletic events, it might be feasible to charge apartment residents and commuters to use the parking garage. Desirability factors include the advantage of sheltered parking in the winter and a very high level of service trip to the lower campus compared to current parking and walking conditions.

Student housing at Michigan Tech is financed through the sale of revenue bonds with debt service recovered through rent. There are various federal and state grant programs that potentially could pay for most of the capital costs of either the APM or Aerial Ropeway.

Conclusions

The authors believe that Michigan Tech can build on campus transportation successes like the APM at the University of West Virginia in Morgantown and the Oregon Health & Science University in Portland to achieve master plan objectives.

The project described in this paper eliminates 682 lower campus parking spaces, enabling the former lots to be redeveloped for higher uses and removing a visual impairment that has existed for decades.

The Transit Village has been conceived as a TOD smart growth project that will contribute to sustainability by reducing student automobile trips to and from the campus and to retail and service establishments located off the campus.

The key to making this project feasible is a high level of service transportation link capable of ascending and descending steep grades and operating in adverse winter weather conditions.

The Senior Design student reports were forwarded to University Facilities Management and Administration officials and are being reviewed and considered.

Further studies involve preparation of a detailed site plan and traffic impact study for the Transit Village and more detailed engineering investigations of the two transportation technology alternatives, including advanced facilities design work. Perhaps these tasks will form the basis of future Senior Design Projects at Michigan Tech, and participating Senior Design students will be able to return to their alma mater and ride the CTS they helped conceive and engineer.

References and Notes

[1] *Michigan Tech Website*, http://www.mtu.edu

[2] *HGA*, "Fresh Look Scenarios Plan Report," December 2006.

[3] *Carl Walker, Inc.*, "Parking Study for Michigan Technological University," January 2000.

[4] *Wikipedia contributors*, Wikipedia the Free Encyclopedia, "Quincy Mine," accessed November 1, 2008.

[5] *TSH Tech*, "CTS Stations and Transit Village Design, Michigan Tech Civil and Environmental Engineering Senior Design Project," May 2008.

[6] *Michigan Technological University,* "Michigan Tech Plan Progress Committee Final Report," January 13, 2006.

[7] Transit Oriented Development (TOD) involves mixed use residential or commercial areas that are intentionally designed to maximize use for public transportation, walking, and bicycles as alternatives to automobiles.

[8] *American Society of Civil Engineers,* "Automated People Mover Standards, Part 3," ASCE 21-00, Section 11.8, 2002.

[9] *Michigan Tech Transit,* "Automated People Mover Transit Link," Michigan Tech Civil and Environmental Engineering Senior Design Project, May 2008.

[10] *Wikipedia contributors,* Wikipedia the Free Encyclopedia, "Roosevelt Island Tram," accessed October 31, 2008.

[11] *Wikipedia contributors,* Wikipedia the Free Encyclopedia, "Portland Aerial Tram," accessed October 31, 2008.

[12] *Granite Engineering*, "Michigan Tech Aerial Ropeway," Michigan Tech Civil and Environmental Engineering Senor Design Project, May 2008.

[13] *ANSI B77.1-2006 - American National Standard for Passenger Ropeways,* American National Standards Institute, Inc., 2006.

[14] RSMeans, a provider of construction cost information in North America, produces 27 annually updated cost data books.

Photo Sources

Figure 1	Michigan Technological University
Figures 2 and 3	Michigan Technological University Archives
Figures 4 and 5	Michigan Technological University
Figure 6	http://www.galen_frysinger.com
Figure 7	Michigan Technological University Senior Design, May 2008
Figure 8	Doppelmayr Cable Car, GmbH & Co.
Figure 9	Doppelmayr CTEC

California University of Pennsylvania (CALU)—Maglev Sky Shuttle

Thomas E. Riester, P.E.* and Dr. Husam ("Sam") Gurol **

*Vice President, Transportation Services, Mackin Engineering Company, 117 Industry Drive, Pittsburgh PA 15275; PH 412-788-0472; ter@mackinengineering.com
**Director of Maglev Systems, General Atomics, 3550 General Atomics Court, San Diego, California 92121-1194; PH (858) 455-4113; sam.gurol@gat.com.

Abstract

This paper discusses the California University of Pennsylvania (CALU) – Maglev Sky Shuttle Project, the first Demonstration Project for Urban Maglev. The paper includes the history and background of the project, discussion of California University Projects 1, 2, and 3, project funding, discussion of Project 1 design, advancements in Guideway design and construction currently under consideration for the California University Project, and a summary of advantages of the Urban Maglev System.

Introduction

The CALU Sky Shuttle Project was initiated in 2001 by the Urban Maglev Team and California University of Pennsylvania and its President Angelo Armenti, Jr.

California University of Pennsylvania is situated on the banks of the Monongahela River in California Borough, Washington County, approximately 35 miles south of Pittsburgh. The University is divided into 2 Campuses – the lower, or Main Campus, which includes all class room and administration buildings and the upper campus (sometimes called "The Farm") which includes James Adamson Stadium (football field), various other sports fields and recreation areas, and student housing constructed between 2000 and 2006.

The University plans to implement a new transportation system, in this instance – Urban Maglev, to transfer students between the campuses.

The University was, in 2001, and currently is, using a shuttle bus system to transfer students between the two campuses – the University desired to replace the shuttle bus system with a more environmentally friendly system with greater capacity. (Shuttle busses run every 20 minutes between 2 campuses)

The University currently has extensive parking on the lower campus – some located between train tracks and the Monongahela River. Parking in this area constitutes a hazard for students – at least 1 fatality has occurred due to use of at-grade crossings.

The University has developed a new master plan. The plan eliminates a large portion of parking from the lower campus – currently, the majority of parking on the campus proper, excluding areas adjacent to the river, has been eliminated. The University envisions construction of a large capacity parking garage on the upper campus, with the urban maglev system utilized to transfer students from the parking garage on the upper campus to classrooms and the student center on the lower campus.

Project Description

Following initial meetings, including enlistment of support from local political leaders, the Maglev Team developed a program plan, with estimated construction costs and design/construction schedules, and development of conceptual alignments. The Project was divided into 3 individual projects to spread required funding over a period of years, and to facilitate environmental clearance. The project will be constructed in 3 phases as follows: (Refer to Figure 1 which follows).

Project 1

Upper Campus – James Adamson Stadium to the Mid-Mon Valley Transit Authority (MMVTA) Intermodal Center, including stations at Adamson Stadium, Student Housing, and the Intermodal Facility. Project 1 will be constructed as a Demonstration Project.

Project 2

Intermodal Facility to the future Convocation Center Station on the lower campus. This project includes a one-mile, 7% grade. Project 2 will also be constructed as a Demonstration Project.

Project 3

Main campus system including the extension to California Borough adjacent to the lower campus and the extension from James Adamson Stadium on the Upper Campus to the Center in the Woods, a senior citizen facility located in California Borough at the southern end of the upper campus. The future parking garage will be located on the west side of State Route 88, across from the Center in the Woods.

Status

Due to lack of funding, the majority of the project is still in the preliminary stages of design; however, some progress has been made.

Project 1

An Environmental Assessment (the required environmental document required by the Federal Transit Administration) was prepared in early 2006, and underwent several reviews by the FTA in 2007. At this point, all comments have been addressed, and Final Environmental Approval has been issued by the FTA. The conclusion of the Assessment was that there are no permanent environmental impacts – the only impacts are temporary and largely consist of noise impacts due to construction equipment. It is important to note that noise impacts due to maglev operation, which passes in close distance to student housing, is less than 70 dB, or the level of soft music.

The study also concluded that there are no ill effects due to magnetic fields.

Under $1,000,000 in funding provided by the Pennsylvania Department of Transportation in 2004, Mackin Engineering Company conducted final alignment design, preliminary guideway and station design, and final pier and foundation design. Additional funding is required to complete the guideway design, as well as to complete vehicle, magnetics, and communication and signaling design.

Projects 2 & 3

Preliminary Planning, Conceptual Guideway Alignment Design, and right-of-way planning and pre-acquisition have been completed. It is important to note that right-of-way acquisition will be confined to 7 or 8 properties within Project 2. All other property in Projects 1, 2, and 3 is owned by either California University or the California University Student Association. Environmental studies and environmental clearance documents have not been initiated. No funds are currently available to advance either Project 2 or 3.

Design/Construction Costs and Funding

The estimated program cost for Projects 1, 2, and 3 are as follows:

Project 1:	$50,000,000
Project 2:	$75,000,000
Project 3:	$125,000,000
Total:	$250,000,000

Funding to Date

The project has received no federal funds to date. The following funding has been provided by the Pennsylvania Department of Transportation (PennDOT) or the State of Pennsylvania.

Project 1: $1,000,000 provided for the Alignment (PennDOT) and Guideway Design discussed above.

Federal Match: $40,000,000 was authorized in the Commonwealth of Pennsylvania Capital Budget to Match Federal Funds. Approximately $3,000,000 of the total has been appropriated as State Match for FTA funds utilized by the Maglev Team for research and development. The remainder of the $40M authorization will have to be appropriated to match Federal Funds, if they are allocated for Projects 1, 2, and 3.

Project 1 – Preliminary Design Summary

Under funding provided by the Pennsylvania Department of Transportation, Mackin Engineering Company completed final alignment design, preliminary/pre-final guideway design, and final pier and foundation design for Project 1.

Project 1 begins at the Adamson Stadium Station, proceeds through Student Housing to Station 2 at the Housing Clubhouse, and terminates at Station 3 at the Intermodal Transit Facility, for a total length of 578 M (1,900 feet). The alignment utilizes a minimum horizontal radius of 50 M (164 feet) and spiral curvature to weave its way from Adamson Stadium through the Vulcan Village Housing Complex, culminating at the Intermodal

Transfer Facility (Refer to Figure 2). The vertical alignment consists of a combination of vertical grades and curvature to accomplish changes of elevation between Adamson Stadium and the Intermodal Parking Facility (Refer to Figure 3). Maximum grades are 4.5% and 5.8%, in combination with vertical curvature (vertical curves are constructed with parabolic geometry, resulting in large equivalent radii which facilitates transition between vertical grades). The horizontal and vertical alignment clearly demonstrates the feasibility of Urban Maglev to operate at the minimum level of curvature (50 M, or 164 feet). The maximum 5.8% grade of Project 1 is very near the maximum 7% grade anticipated for the entire project, and demonstrates the vehicle's ability to handle steep grades. The only design requirement that the alignment does not demonstrate is speed – due to the short alignment, number of stations, and curvature, vehicle speed will be limited to a maximum of 20 mph.

Vehicles

The standard maglev vehicle will be utilized at CALU – the vehicle will be capable of carrying a maximum passenger load of 100. One vehicle will be used for Project 1 Demonstration (4 vehicles, ultimately) for Projects 1, 2, and 3. Vehicles are designed by Hall Industries of Pittsburgh (Refer to Figure 4). It is important to note that the second guideway may be constructed under Project 1 to be utilized as a test track for a Cargo Maglev vehicle. The guideway section (Section 5.3) can carry either vehicular or Cargo Maglev.

Figure 4. CALU Maglev Vehicle

Pre-Final Guideway Design

Guideway structural design was essentially completed under the PennDOT, Project 1 Funding. In order to design piers and foundations, it was necessary to substantially complete the guideway design, and detailed geometry.

The guideway consists of 19 spans of varying lengths and curvature: minimum length 18.4 M (60.3 ft.): Maximum length 36.3 M (119.2 ft.). Geometry varies between straight to a minimum radii of 50 M (164 ft.). Total length of the guideway for Project 1 is 584.3 M (1,916.9 ft.),

Guideway Section

The Typical Guideway Section (Refer to Figure 5) is composed of precast pre- or post-tensioned box beams, a concrete leveling slab, and the Steel Guideway Module. The Post-Tensioned Sections are utilized on curved sections of the alignment and consist of three prestressed box girder sections which are delivered to the site, then post tensioned and erected in place. The concrete leveling slab, which is constructed after the beams are erected, is utilized to provide a flat surface on which to mount the Guideway Module and accommodate upward camber in the beam which cannot be completely predicted.

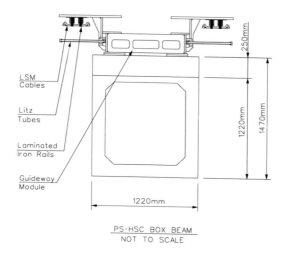

Figure 5. Baseline Box Beam/Guideway Module Cross Section

The Guideway Module is composed of stainless and carbon steel (top plate only) and carries the vehicle, LSM Cables, and Litz Tubes. The permanent magnets on the vehicle react with the Litz Tubes to achieve levitation and with the LSM Motor to propel the vehicle. The Guideway Module is attached to the concrete leveling slab with anchor bolts cast in the concrete. Final adjustments to achieve the exact alignment are made with the leveling nuts.

The guideway has several advantages. The beams are fabricated offsite, delivered, then quickly erected in place. This is particularly advantageous in congested urban areas – impacts to pedestrians, traffic are minimized, and construction can proceed quickly. The concrete leveling pad corrects for all variations in beam camber and geometry, and allows the guideway and guideway module to be constructed to less stringent tolerances, reducing project costs. The Guideway Module does not require special fabrication techniques, and can be fabricated and erected quickly. However, it is important to note that Guideway Modules were utilized for General Atomics Test Track. Costs were significantly more than estimated, and distortion due to welding was a problem on some modules – we anticipate that higher volume production would reduce cost, and that automated fabrication and fabrication software/ processes currently under development would reduce or eliminate the distortion.

Piers and Substructure

Geotechnical Engineering, and Pier and Foundation Design were completed for Project 1 of California University, along with the Pre-Final Guideway Design. The construction of the piers and foundations can begin almost immediately, following appropriation of construction funds.

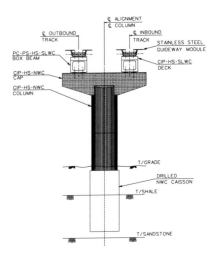

Figure 6. T-Shaped Pier with Hammerhead Cap

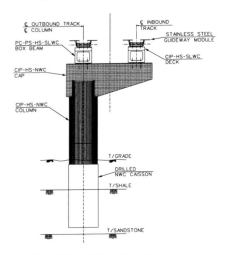

Figure 7. L-Shaped Pier with Cantilever Cap

The piers are composed of reinforced oncrete, and are either T-shaped or L-shaped (Refer to Figures 6 & 7). T-shaped piers are generally used; L-shaped piers are utilized when space is not available for T-shaped piers – minimizing impacts to buildings, right-of-way, city streets, etc. At CALU, the piers (and foundations) are designed to carry Project 1 loads (1 vehicle/guideway only) and Project 3 loads (dual guideway system). Note that the second guideway may be utilized to demonstrate Cargo Maglev, in which instance the second guideway would be constructed under Project 1. The Project 1 loads control stability and bending stress in the pier column and foundation. Piers at CALU are designed to be cast-in place, as preferred by Trumbull/P.J. Dick, the construction partner of the Urban Maglev Team. However, precast piers can also be utilized, and may be preferable in true urban areas. The piers vary in height from 3.0 M (9.9 ft.) to 14.7 M (48.2 ft.).

Foundations

Urban Maglev systems generally use concrete caissons to support the pier and guideway – concrete caissons can be constructed without the noise and vibration of steel pile foundations, and with a lesser footprint. At CALU, a single concrete caisson is utilized to support the pier. This is possible due to the presence of a rock layer (shale) located between 0.9 M (3.0 ft.) and 5.3 M (188 ft.) below the ground surface. Use of the single caisson results in several advantages over multiple caissons, which may be necessary depending on the location of the soil/sand/rock layers below the piers.

Faster and more economical construction.

Minimal footprint and effect on existing streets, utilities, etc.

Plans for all substructure units were completed in 2004 – construction, as indicated earlier, is dependent on funding.

Advances in Guideway Technology

Mackin Engineering Company and General Atomics, over the past 2 years have been evaluating an improved Guideway Concept – the Hybrid Girder. While the Pre-Cast Girder with Guideway Module has all of the advantages identified, there is a desire to 1) Improve Aesthetics, and 2) Combine the prestressed beam and Guideway Module into 1 unit, the Hybrid Girder (Refer to Figure 8). The hybrid girder consists of a pre-or post-tensioned high-strength concrete box beam, with stainless steel fiber reinforcement instead of standard steel reinforcement. The hybrid girder will be fabricated to include all components present on the current prestressed girder, Guideway Module combination, but in a much more aesthetic manner. The hybrid girder includes carbon steel top plates (to accommodate vehicle landing wheels), Litz Tubes, and LSM cables all attached to embedments cast into the concrete girder.

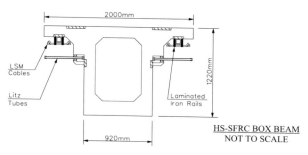

Figure 8. Alternative Hybrid Girder Cross Section

Original Concept

The original concept considered use of high strength steel fiber reinforcement only, without pre- or post-tensioning. Initial and follow-up testing was conducted by Mackin Engineering Company and General Atomics. The results of the testing indicated a brittle failure of the girders (a sudden failure due to a loss of bonding between the fiber reinforcement and the concrete (Refer to Figure 9)) at strengths well below those required. As a result, it was determined that pre- or post-tensioning would be necessary.

Figure 9. Failed Hybrid Test Beam Due to Brittle Fracture

There are several challenges associated with the hybrid girder which need to be addressed – solution of those challenges will affect both the cost and feasibility of use, and include:

The Litz Tubes and LSM Cables must be constructed at a constant distance from the alignment profile grade – adjustability must be provided to account for beam camber, beam superelevation, and vertical curvature.

Girder camber, due to pre- or post-tensioning cannot be completely predicted. Different methods of design/fabrication of the girders must be evaluated.

Fabrication of the girders to meet the vertical geometry must also be evaluated. The current beam/guideway module combination utilizes the concrete leveling slab to account for the variations in the camber. The guideway module accounts for the superelevation and vertical curvature.

The hybrid girder is a very promising concept. Additional design and fabrication development, and associated funds, are needed to finalize the concept, including evaluation of cost effectiveness.

Advantages of the CALU Urban Maglev System

The California University of Pennsylvania Project clearly illustrates the numerous environmental and operational advantages of an Urban Maglev System.

Environmental Advantages

The Urban Maglev System has significant environmental advantages, and is a truly "Green" system.

- Urban maglev is an all electrical system (in the case of California University, it will replace gas/diesel engine shuttle busses), with no emissions.
- The elevated system and tight turning radii minimize or avoid environmental impacts, as well as construction impacts.
- The system is very quiet – measurements taken at General Atomics' test track in San Diego show noise levels equal to 70 dB (equivalent to soft music).
- There are no ill effects due to the magnets – magnetic field levels are less than the natural magnetic field levels of the earth.
- The vehicle and guideway are aesthetically pleasing and will blend with the surrounding environment.
- Safe and secure – on-board close circuit cameras will be placed in all vehicles and stations. The elevated guideway avoids any interference with vehicular traffic.
- The system will eliminate bus traffic and most vehicular traffic between the campuses at California University.

Urban Maglev is truly a "Green" Technology.

Operational Advantages

Urban Maglev Systems, due to the large air gap between the guideway and vehicles, elevated guideway, and permanent maglev technology provide many advantages which minimize cost, impacts, and maintenance.

- There are no moving parts, with the exception of the doors and air conditioning system.
- The one-inch air gap permits the vehicle to make horizontal turns on radii as little as 18.3 M (60.0 ft.), permitting guideways to be placed in urban areas (i.e., city streets) with minimal building and right-of-way impacts and displacements, and avoiding impacts to park and recreation areas.
- The system can be placed on alignments with vertical grades of 10% or greater, providing additional operational flexibility.
- The system can be operated under all weather conditions.
- The elevated guideway is significantly less expensive than underground systems, and absolutely minimizes impacts to streets, utilities, etc.
- The elevated guideway is significantly safer than an at-grade system, and can be placed in areas where at –grade systems are not feasible.
- There is no friction between the vehicle and guideway, and no moving parts – as a result, maintenance costs will be minimized.
- The vehicle is automated – no driver – and is controlled at the system control center.

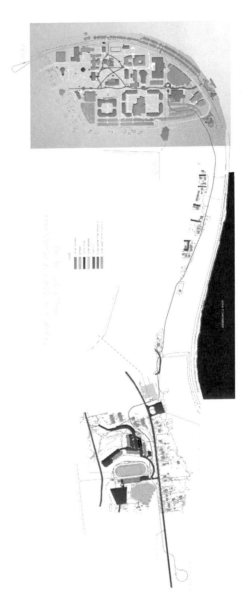

Figure 1. Plan of Projects 1, 2 and 3

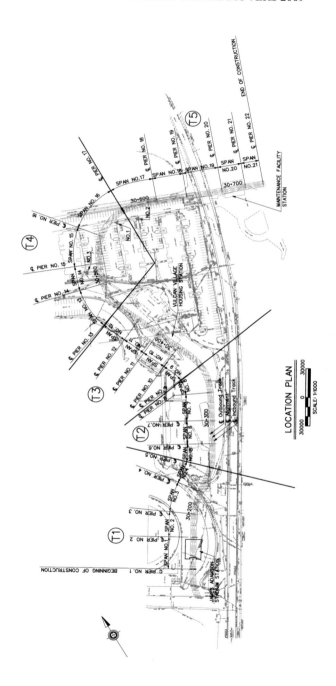

Figure 2. Plan of Project 1

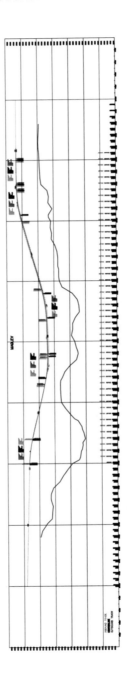

Figure 3. Profile of Project 1

Market trends and comparative study of economic and technological parameters of APM systems

E. Todt[1], A. Gehlen de Leão[2], L. A. Lindau[3], E. Bortolini[2] and B. M. Pereira[3]

[1]Universidade Federal do Paraná, Departamento de Informática. Curitiba, Brasil.
email: todt@ieee.org
[2]Pontifícia Universidade Católica do Rio Grande do Sul, Faculdade de Engenharia. Porto Alegre, Brasil. email: gehleao@pucrs.br; bortolini@pucrs.br
[3]Universidade Federal do Rio Grande do Sul, Laboratório de Sistemas de Transportes. Porto Alegre, Brasil. email: lindau@producao.ufrgs.br; brenda@producao.ufrgs.br

ABSTRACT

This paper analyses the technological evolution and the trends of Automated People Movers (APM) in recent years. It establishes a baseline of the technological and economic aspects in order to support the development process of an enhanced version of the Aeromovel system (APM that uses air flow for propulsion on an elevated track). For the definition of the new Aeromovel guidelines, a significant sample of systems already developed worldwide was analyzed using an operational research tool based on a multivariate quantitative model, known as Data Envelopment Analysis (DEA). Several quantitative analyses taking into account implementation investment, total line length, number of boarding stations, passenger capacity per hour, and maximum speed were performed. As a preliminary result, it is observed that technologies that use external propulsion, for instance by means of cables, currently offer the best efficiency indicators when compared to other technological solutions for propulsion and power transmission, based on the performed DEA analysis.

1 INTRODUCTION

Since middle of the decade of the 70 until the end of 80 years the Aeromovel Automated People Mover system had a significant evolution. Initiating with a stretch of 30 meters of length, with a rudimentary vehicle designed for just one passenger, passing for the construction of a pilot line in Porto Alegre, it evolved until the implantation, in 1989, of a ring of 3.2 km in a park in Jacarta (Figure 1), with six passenger stations (Coester, Soehartono, Pinto, and N., 1989).

Figure 1 Aeromovel in Jacarta (Aeromovel, 2008)

After a gap in the research activity, in 2006, with founding from FINEP (Brazilian governmental agency that supports projects and research), it was structured a new R&D project in collaboration with two universities (Pontificia Universidade Catolica do Rio Grande do Sul – PUCRS and Universidade Federal do Rio Grande do Sul - UFRGS), a private company, Aeromovel Brasil S.A., and a technological center (Centro Tecnologico de Mobilidade Urbana).

The project was structured in two stages. The first stage encompasses a set of actions in several areas, from motor control to architectonical design. The second stage aims at to the construction of an experimental circular line at PUCRS campus, with 2.7 km and six boarding stations (Figure 2). The final objective of the project is to get the international certification of the Aeromovel, with its consequent accreditation to compete in the international market of Automated People Movers.

Figure 2 Experimental line planned to be built at PUCRS campus (Aeromovel, 2008)

One of the actions of the first stage of the project consists of an evaluation of technical and economical parameters of APM technologies. In this context, this work establishes a baseline of technological and economic aspects in order to support the development process of the Aeromovel system. In the following section the Aeromovel system is presented.

2 AEROMOVEL SYSTEM

Automated People Movers, APM, is the denomination used for non-conventional systems characterized for electric, guided and fully automated transport modalities used in short lines operating as shuttles or loops in sites as airports, campi, centers of conventions and leisure parks (Vuchic, 2007). However, a consensus does not exist on this classification, and some works as Trans 21 (2008) includes in this category even automated subways operating regular lines of collective transport, a system more classically classified by other authors (Jakes, 2003; Vuchic, 2007) as Automated Guided Transit, AGT. Also, autonomous vehicles (AVGs) in protected environments can function as low-capacity, low-speed APMs.

The Aeromovel differs from other AGT/APM technologies by the use of pneumatic principle for the propulsion. Day and Wilson (1957) describe pioneering systems of pneumatic propulsion introduced in Europe in 19th century that, differently of the Aeromovel, used high air pressures.

The Aeromovel has an elevated guideway with a duct inside it. Aeromovel stationary blowers propel air (under low pressure) through the duct. The pressurized air pushes a propulsion plate attached to the bottom of the vehicle. This propulsion plate acts like an upside down sail, propelling the vehicle forward and helping to stop it when the airflow is reversed. The plates are fixed to the vehicles through connecting rods and, therefore, it is required a ridge longitudinal throughout all structure extension. Throughout this ridge sealing elements allow the passage of the plate rods and prevent the excessive air leakage. The vehicles have steel wheels supported and guided by conventional tracks settled to the superior face of the structure. The control of the system requires pressure sensors along the guideway and acts on the static air blowers and control valves.

The Aeromovel presents some advantages over other APM systems (Lindau and Furtado, 1987):

(i) The possibility of safe use of non-energized elevated track in the case of the necessity to evacuate the passengers of a vehicle;

(ii) The propulsion concept has the intrinsic safety feature of an air buffer between propulsion plates which helps to prevent collision between vehicles;

(iii) Air propulsion eliminates the problems of heavy rail traction; wear on wheels and tracks is reduced;

(iv) The light weight of vehicles ensures that energy is not wasted moving heavy deadweight; the extreme simplicity results in reduced maintenance requirements.

Taking into account that the construction of the guideway answers for about 60% to 80% of the investments required by elevated systems (Jakes, 2003), one of the main technological contributions of the Aeromovel is the use of vehicles with low deadweight per carried passenger .

3   MARKET TRENDS

Throughout last the two decades the APM/AGT market quadruplicated, although passed for a strong decline in the middle of the decade of 90. Jakes (2003; Jakes, 2005) and Warren (2000) report the existence of a present potential market of the order of 6 the 7 billion dollar. The main market for APM is at airport shuttle applications, but also leizure centers, institutions, like universityes and hospitals, and local connections between conventional public transport and specific areas, like shopping centers, represent important market segments. Vuchic (2007) reports APM systems at 26 airports, with line lenghts ranging from 700m (Pittsburs, USA) to 10km (San Francisco, USA).

While the market of big systems (typically AGT, with wide area operational range) presents certain stability, the amount of projects regarding small and medium systems oscillates significantly throughout the time. Currently, the market of medium size systems, where APMs are inserted, is around 2 billion dollars.

Jakes (2003) reports that, although more than 200 types of APM technologies have been developed until now, only some remained in the market. Also, the market of airports has been dominated by just one technology, provided by Bombardier.

It is remarkable the recent participation of the cable-traction technology in the APM market, introduced by the Doppelmayr-Garaventa group (Lindau, Gehlen de Leão, Todt, and Pereira, 2007). The interest for a technology also marked by the use of an off-vehicle propeller system strengthens the potential of simple but smart technologies, hopefully cheaper than that dominated the market until nowadays.

Warren (2000) remarks the necessity of the APM costs to be low enough in order to induce its adoption by incorporators interested in developing urban areas, revitalizing obsolete industrial districts, and promoting links between office or commercial centers and parking lots or public transport stations. Jakes (2003) reports research that disclosed the willingness of incorporators to compromise until up to 5% of the total enterprise costs with internal transport systems. This represents, in the general case, a budget of the order of some few millions, instead of tens of millions of dollars required by current APM technologies.

4 METHODOLOGY

In order to get parameters that make possible the development of a technical and economical study comparing distinct implanted APM systems around the world, it has been carried through a hard work of data collection. From the 129 systems listed in the Airfront guide (Trans21, 2008), and with the definition of parameters traditionally adopted for the characterization of urban passenger systems, a survey that contemplates technical publications, supplier manuals, and operators and manufacturers web sites have been performed. Until now, with varied levels of detailing and precision, information have been gathered from 122 systems used in airports, urban transport, centers of leisure and institutions, such as university, shopping centers and hospitals.

It is very difficult to obtain all desirable information, as already pointed by Jackes (2003), specially those regarding implementation, operational costs and energy source. The most relevant available parameters for our study are listed in the following, together with the amount of systems for which we could get the respective information:

- Line extension [km], available in 86 systems
- Number of stations, available in 80 systems
- Total cost of implantation [US$ millions], available in 33 systems
- Capacity [passengers/hour/direction], available in 36 systems
- Area of the vehicle [m2], available in 35 systems
- Vehicles for composition, available in 35 systems
- Deadweight per passenger [kg/passenger], available in 13 systems
- Acceleration [m/s2], available in 15 systems
- Maximum speed [km/h], available in 46 systems
- Operational speed [km/h], available in 26 systems
- Passengers/day, available in 19 systems

It can be observed that there aren't comprehensive information for most of the systems. However, the gathered information is sufficient to proceed a general analysis comparing the systems.

First of all, it was observed which parameters were available in relevant amount of cases to be significantly analyzed. Table 1 presents 19 systems that possess enough information for the accomplishment of a comparative analysis. It contemplates systems implanted in airports (code AIR), in lines of urban transport (LT), in the local transport (LC), for leisure (LZ), and other private enterprises (IT). The investments in these systems vary from millions to billions of dollars, depending on their application, technology, and extension, among others.

Table 1 APM/AGT systems taken in the comparative analysis

Case	Manufacturer	Propulsion	Ext. [km]	Stations	Cost [MUS$]	Line type	Capacity [pass/h/dir]	Max speed [km/h]	Nickname
Dallas-Ft Worth	ADtranz (Bombardier)	Rubber-tired self-propelled	8	10	864	2-way	5,270	56	AR2
Toronto	Doppelmayr	Cable	1,422	3	40	2-way	1,090	43.2	AR7
Birmingham	Dopplelmayr	Cable	0.588	2	16	2-way	804	36	AR11
San Francisco	ADtranz (Bombardier)	Rubber-tired self-propelled	4.27	9	104	2-way	6,000	48	AR10
Morgantown PRT	Alden/ Boeing	3rd Rail Rubber-tired self-propelled	13.9	5	319	1-way	1,500	48	IT4
Porto Alegre*	Aeromovel	Pneumatic	3.2	6	12	1-way	10,000	80	LZ20
Aichi HSST	Chubu HSST	Linear Induction Motor	8.9	9	955	NA	4,000	100	LZ1
Mandalay bay	Dopplemayr	Cable	0.82	4	16	2-way	1,480	36	LZ2
Ina	NA	NA	12.6	13	266.5	1/2-way	3,480	60	LT20
Taipei-Brown	IHI-Niigata	Rubber-tired self-propelled	10.5	12	920	2-way	30,000	80	LT1
Tokyo Yukarimome	IHI-Niigata	Rubber-tired self-propelled	11.9	12	1,456.5	2-way	7,200	60	LT21
Yokohama	NA	NA	10.8	14	570	2-way	4,320	60	LT22
Hiroshima Skyrail	MHI	Maglev/ Rubber-tired self-propelled	18.4	21	1,542.7	2-way	5,720	60	LC20
Kobe Portliner	KHI	Rubber-tired self-propelled	6.4	9	383	1/2-way	3,840	60	LC22
Kobe Rokkoliner	KHI	Rubber-tired self-propelled	4.5	6	367.3	2-way	972	63	LC23
Tokadai (Nagoya)	NA	NA	7.4	7	283	2-way	965	55	LC2
Yukarigaoka	NA	NA	4.1	6	18.4	1-way	1,630	50	LC24
Miami Metromover	Bombardier	Rubber-tired self-propelled	7.1	21	424	2-way	3,200	33.3	LC25
Denver	ADtranz (Bombardier)	Rubber-tired self-propelled	1.48	4	102	NA	6,000	48	AR20

* System not implemented, design data, NA=Not Available
(Aeromovel, 2008; Bombardier, 2007; Dopplemayr, 2007; Jakes, 2003; Trans21, 2008; Vuchic, 2007; Yamamoto and Nakazumi, 2001)

In cases of comparative analysis of systems where there is a great amount of inputs and outputs to be considered, it is difficult to get a single index of performance. In this sense, the Data Envelopment Analysis (DEA) is a technique based on linear programming that contemplates multiple inputs and outputs which comes to be an effective tool for evaluation of systems relative efficiency (Coelli, Rao, O'Donnell, and Battese, 2005; Cooper, Lawrence, and Tone, 2007; Zhu, 2003).

The *efficiency* for a system with multiple entries and exits can be defined as:

$$efficiency = \frac{weighted\ sum\ of\ outputs}{weighted\ sum\ of\ inputs}$$

Thus, the efficiency of a system i is given by:

$$efficiency_i = \frac{\sum_{k=1}^{s} v_k y_{ki}}{\sum_{j=1}^{m} u_j x_{ji}}$$

where i is one of the considered systems, s is the number of outputs, m is the number of inputs, v_k is the weight given to output k, y_{ki} is the value of k output of system i, u_j is the weight given to input j, and x_{ji} is the value of input j of system i.

Assuming that there are n systems, each with m inputs and s outputs, the *relative optimal efficiency score* θ^* of a system p is the result of the solution of the following model (Cooper, Lawrence e Tone, 2007):

$$\theta^* = \max \frac{\sum_{k=1}^{s} v_k y_{kp}}{\sum_{j=1}^{m} u_j x_{jp}}$$

$$\text{such that } \frac{\sum_{k=1}^{s} v_k y_{ki}}{\sum_{j=1}^{m} u_j x_{ji}} \leq 1 \; \forall i, \, v_k \geq 0, u_j \geq 0, \, \forall k, j$$

This model is computed n times in order to identify the relative efficiency score of each system. Each system has the inputs and outputs weights adjusted in such a way that it has the maximum possible score in relation to the other systems. The solution of the problem is denoted by the tuple (θ^*, u^*, v^*), where the asterisks indicate the weights that produce the maximum relative efficiency. The systems with relative efficiency score near 1.0 are considered the most efficient ones, while the scores near zero indicate the least efficient systems.

Since it is desired to analyze the benefit obtained by each implemented system as a function of the investment done, just the cost of implantation is taken as input variable. By the other hand, the line extension, the number of stations, the hourly passenger capacity, and the maximum speed are taken as output variables.

In the analyzed systems it is evidenced the existence of single (1-way) and double lines (2-way), where the compositions can pass through in just one direction or both directions, respectively. This is a complex point that has implications in the results obtained. The approach adopted in this work doesn't take into account the benefit to the users that a double line brings because the necessary data to do it is not available in most cases. Although the area covered by a double line is the same as with a single line system, the global performance of a double line is better, if composition occupation, energy and travelling and wait times are considered. If the passenger wants to travel to a station that precedes the current station, in single lines, either he/she has to wait for the reversion of the direction of the compositions, in linear lines, or to travel a complete loop, in circular lines. This extra travelling in circular lines potentially reduces availability of installed capacity, and wastes energy. Here circular means that the line path is closed, not that the line shape is circular.

In the following the application of DEA to the analysis of the systems enumerated in Table 1 is described. To simplify and clarify the text nicknames are used to identify the different systems.

5 APM SYSTEMS DATA ANALYSIS

The DEA analysis was performed using two softwares, distributed freely for academic pourposes: DEA Solver, delivered with the book of Cooper at al. (2007) and the EMS - Efficiency Measurement System (Dortmund, 2008).

The DEA computing conditions were assumed as convex, to assure convergence, constant and variable returns to scale, to consider linear and non-linear output-to-input relations, radial distance, for simplicity, and input oriented model, that aims at reducing the input amounts by as much as possible while keeping at least the present output levels when searching for optimal weights for each case.

All Decision Making Units from **Table 1** were taken and processed with DEA solver. It is important to observe that Aeromovel data (LZ20) are the only theorethical data, all others are obtained from implemented systems.

Considering as input the implementation cost, and as outputs the total line length, number of stations, maximum speed and passenger capacity, the systems were ranked according to DEA efficiency as shown in Figure 3.

According to the DEA analysis shown in Table 2, which assumes constant returns to scale, the most efficient systems are the LZ20 (Aeromovel) and AR7 (Toronto), which serve as references with efficiency 1.0, LC24 (Yukarigaoka), LZ2 (Mandalay bay) and AR11 (Birmingham). Three of them (AR7, LZ2, and AR11) are cable-propelled from Doppelmayr, one not specified (LC24) and one is the pneumatic Aeromovel (LZ20). The systems with lower efficiency score are the LT21 (Tokyo Yukarimome), LZ1 (Aichi HSST) and AR2 (Dallas-Ft Worth). All of these are based on self-propelled vehicles, LT21 and AR2 rubber-tired self-propelled and LZ1 with linear induction motor.

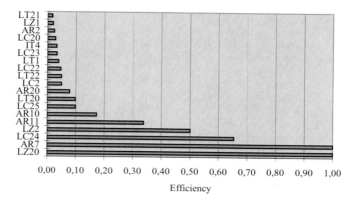

Figure 3 DEA relative efficiency of the systems, considering constant returns to scale

Table 2 DEA relative efficiency of the systems, constant returns to scale, input and output weights for each case (Decision Making Unit)

Nickname	Case	DEA efficiency score	Computed output weight	Computed input weights			
			Cost	Line length	Stations	Max. speed	Capacity
LZ20	Aeromóvel	1,000	0,0833	0,0020	0,1656	0,0000	0,0000
AR7	Toronto	1,000	0,0250	0,0007	0,0000	0,0000	0,0000
LC24	Yukarigaoka	0,653	0,0543	0,0013	0,1080	0,0000	0,0000
LZ2	Mandalay bay	0,500	0,0625	0,0000	0,1250	0,0000	0,0000
AR11	Birmingham	0,338	0,0625	0,0000	0,0000	0,0094	0,0000
AR10	San Francisco	0,173	0,0096	0,0000	0,0192	0,0000	0,0000
LC25	Miami Metromover	0,099	0,0024	0,0000	0,0047	0,0000	0,0000
LT20	Ina	0,098	0,0038	0,0001	0,0075	0,0000	0,0000
AR20	Denver	0,078	0,0098	0,0000	0,0196	0,0000	0,0000
LC2	Tokadai(Nagoya)	0,050	0,0035	0,0001	0,0070	0,0000	0,0000
LT22	Yokohama	0,049	0,0018	0,0000	0,0035	0,0000	0,0000
LC22	Kobe Portliner	0,047	0,0026	0,0001	0,0052	0,0000	0,0000
LT1	Taipei-Brown	0,039	0,0011	0,0000	0,0000	0,0000	0,0000
LC23	Kobe Rokkoliner	0,033	0,0027	0,0001	0,0054	0,0000	0,0000
IT4	Morgantown PRT	0,032	0,0031	0,0001	0,0062	0,0000	0,0000
LC20	Hiroshima Skyrail	0,027	0,0006	0,0000	0,0013	0,0000	0,0000
AR2	Dallas-Ft Worth	0,023	0,0012	0,0000	0,0023	0,0000	0,0000
LZ1	Aichi HSST	0,019	0,0010	0,0000	0,0021	0,0000	0,0000
LT21	Tokyo Yukarimome	0,017	0,0007	0,0000	0,0014	0,0000	0,0000

It is interesting to observe that in general the outputs *maximum speed* and *capacity* received weights zero, meaning that just the total length and number os stations were relevant in the ranking. Recall that DEA maximizes each DMU computing the appropriate weights to the respective input and output values.

If we relax the DEA analysis allowing for variable returns to scale, meaning that the relationship between inputs and outputs can be non-linear, than the rank of systems change, as can be seen in Table 3. Variables which weren't relevant in the constant returns to scale analysis now receive weights that promote some DMUs, like the maximum speed for LZ1 (Aichi HSST). Thus, many systems appear with relative efficiency near the optimal value, for instance, the first eleven DMUs having relative efficiency greater than 0.7. From the seven least efficient systems, under this analysis, six of them are based on rubber-tired self-propelled vehicles and one is not specified respective the propulsion technology.

Table 3 DEA relative efficiency of the systems, variable returns to scale, input and output weights for each case (Decision Making Unit)

Nickname	Case	DEA efficiency score	Computed output weight Cost	Computed input weights Line length	Stations	Max. speed	Capacity
AR7	Toronto	1,000	0,0250	0,0000	0,0007	0,0000	0,0000
LZ20	Aeromóvel	1,000	0,0833	0,0000	0,0020	0,1656	0,0000
LZ1	Aichi HSST	1,000	0,0010	-3,1750	0,0011	0,1235	0,0305
LT1	Taipei-Brown	1,000	0,0011	-0,4803	0,0003	0,0000	0,0000
LC20	Hiroshima Skyrail	1,000	0,0006	-3,0851	0,0011	0,1172	0,0262
LC25	Miami Metromover	1,000	0,0024	-0,3604	0,0000	0,0648	0,0000
LT20	Ina	0,991	0,0038	-11,378	0,0038	0,4426	0,1094
AR10	San Francisco	0,908	0,0096	-1,4705	0,0007	0,2639	0,0000
AR11	Birmingham	0,750	0,0625	0,7500	0,0000	0,0000	0,0000
LZ2	Mandalay bay	0,750	0,0625	0,7500	0,0000	0,0000	0,0000
LT22	Yokohama	0,667	0,0018	-5,3196	0,0018	0,2069	0,0512
LC24	Yukarigaoka	0,656	0,0543	-8,3113	0,0042	1,4917	0,0000
LC22	Kobe Portliner	0,247	0,0026	-0,3993	0,0002	0,0717	0,0000
AR2	Dallas-Ft Worth	0,141	0,0012	-0,1770	0,0001	0,0318	0,0000
LC2	Tokadai(Nagoya)	0,141	0,0035	-0,5404	0,0003	0,0970	0,0000
LT21	Tokyo Yukarimome	0,122	0,0007	-0,1050	0,0001	0,0188	0,0000
AR20	Denver	0,118	0,0098	0,1176	0,0000	0,0000	0,0000
IT4	Morgantown PRT	0,038	0,0031	0,0374	0,0001	0,0000	0,0000
LC23	Kobe Rokkoliner	0,033	0,0027	-0,4164	0,0002	0,0747	0,0000

6 DISCUSSION AND CONCLUSIONS

The review presented in this paper shows that APM systems constitutes a promising worldwide market for the next years. The implementation costs and the efficient utilization of resources with high return to society are fundamental to the success of new technologies. This point is very critical due to the high absolute implementation costs typical of public infraestructure systems.

The analyses performed in this work, although preliminary and based on 15% of the systems that we could identify (19 of 129), show that technologies based on alternative propulsion methods, for instance by means of cables or pneumatics, represent a strong potential to become benchmarks.

The recently introduced propulsion external to the vehicles seems to be promising in terms of cost/benefit relationship. It is still early to know about operational costs, energy consumption and ambiental impacts.

The Data Envelopment Analysis method showed to be a simple and interesting method to evaluate relative efficiency of systems. A critical to the model adopted in this work is that it allows for arbitrary large or small individual weighs for the variables, allowing the result to give excessive relevance to those items were each system performs better. By the other hand, the analysis of the weights gives information about the composition of the global efficiency of each technology. More sophisticated DEA methods should be considered for future works in order to refine results and to restrict the influence of an excellent parameter over the estimated total efficiency of a system.

Finally, it is important to increase the amount of available information in order to get a more comprehensive analysis, and to replace theoretical data from Aeromovel with real data. However, the early evaluation of Aeromovel helps designers to adjust project guidelines toward more efficient results.

REFERENCES

Aeromovel. (2008). *Aeromovel*. Retrieved, 2008, from the World Wide Web: http://www.pucrs.br/aeromovel/index.php

Bombardier. (2007). APM catalog.

Coelli, T. J., Rao, D. S. P., O'Donnell, C. J., and Battese, G. E. (2005). *An Introduction to Efficiency and Productivity Analysis*. New York, U.S.A.: Springer.

Coester, O. H., Soehartono, Pinto, C. F. S., and N., A. (1989). *Technical aspects of Porto Alegre's Aeromovel*. 2nd International Conference on Automated People Movers, Miami, FL, U.S.A., pp.686-695.

Cooper, W. W., Lawrence, M. S., and Tone, K. (2007). *Data Envelopment Analysis: a comprehensive text with models, applications, references and DEA-Solver software.* New York, U.S.A.: Springer.

Day, J. R., and Wilson, B. G. (1957). *Unusual Railways.* London: Frederick Muller Limited.

Dopplemayr. (2007). DCC Dopplemayr Cable Car GmbH & Co. Catalog.

Dortmund. (2008). EMS Data Envelopment Analysis Software (Version 1.3). Dortmund, Germany, http://www.wiso.uni-dortmund.de/lsfg/or/scheel/ems/.

Jakes, A. S. (2003). *Reasons why people movers are underutilized in solving traffic problems.* 9th International Conference on Automated People Movers, Singapura

Jakes, A. S. (2005). *Trends in Airport Automated People Movers.* Automated People Movers 2005, Florida, U.S.A.

Lindau, L. A., and Furtado, S. M. L. (1987). Aeromóvel: uma contribuição tecnológica nacional no campo dos sistemas não convencionais. *Revista dos Transporte Públicos*(37), 33-45.

Lindau, L. A., Gehlen de Leão, A., Todt, E., and Pereira, B. M. (2007). *Tendências de mercado e estudo comparativo de parâmetros técnicos e econômicos de tecnologias APM no contexto do Sistema Aeromóvel.* XIV Congresso Latinoamericano de Transporte Publico y Urbano (CLATPU), Rio de Janeiro, Brazil

Trans21. (2008). *APM Guide on-line.* Airfront, Boston, U.S.A. Retrieved December 6, 2008, from the World Wide Web: http://airfront.us/

Vuchic, V. R. (2007). *Urban Transit Systems and Technology.* Hoboken, U.S.A.: John Wiley.

Warren, R. (2000). *Automated People Movers. A1E11 Committee on Major Activity Center Circulation Systems.* Washington, U.S.A.: Transportation Research Board.

Yamamoto, K., and Nakazumi, S. (2001). *An Update of the ATS and AFC Systems for Kobe Portliner.* Automated People Movers, California, USA

Zhu, J. (2003). *Quantitative Models for Performance Evaluation and Benchmarking: data envelopment analysis with spreadsheets and DEA Excel Solver.* New York, U.S.A.: Springer.

Financing Transit Usage with Podcars in 59 Swedish Cities

Göran Tegnér * and Elisabet Idar Angelov **
* Göran Tegnér, M Sc Econ., WSP Group, Arenavägen 7, SE-121 88 Stockholm-Globen, Sweden. Phone:+46-8-688 76 77; Fax: +46-8-688 77 32; E-mail: goran.tegner@WSPGroup.se
** Elisabet Idar Angelov, M Sc. Econ. WSP Group, Arenavägen 7, SE-121 88 Stockholm-Globen, Sweden. Phone:+46-8-688 76 64; Fax: +46-8-688 77 32; E-mail: elisabet.idar.angelov@WSgroup.se

Abstract

The main question that will be addressed and answered by this paper is how a doubled transit ridership by podcars could be financed?

This paper summarizes a Swedish research project financed by the Swedish Institute for Transport and Communications Analysis (SIKA). It deals with several analytical comparisons between bus, LRT and podcars, based on a city data base with 59 Swedish cities, and four more in-depth case studies, the cities of Kiruna, Södertälje, Linköping and the Commercial Area of Kungens Kurva in Stockholm and Huddinge cities. The comparisons comprise the following aspects:
1. Generalized travel times for the bus and the podcar modes for 59 cities
2. Market shares and total transit ridership with bus and podcars for 59 cities
3. Financial costs (investment, operational & maintenance costs) and ticket revenues with bus and podcars for 59 cities
4. Various financial solutions will be discussed. Calculations for the case study cities for podcars will be presented with the (Public Private Partnership) solution.

The analyses show that it would be possible to double the transit ridership in cities with bus or LRT traffic when shifting to podcars. The cost per trip is showed to be lower by podcar than with LRT and - in some cases - than with bus.

1 Transit Mode Shares – Goals and Reality

1.1 Swedish goals - double transit share

The public transport mode share is one of the key indicators used in transit planning but also at the political level, when setting goals for transit investments and operations. In many cities and regions in Sweden the political goal is formulated to double the transit mode share, with the main reason to reduce greenhouse gases from the car traffic.

1.2 Swedish transit mode share

Swedish transit mode shares is usually and officially measured in terms of the number of trips by the car, the transit and by the walk and bicycle modes. However, as trip distances vary across these modes, a better measure would be in terms of passenger-kilometers.

On average, the transit mode share is 10 % in all Sweden. In the metropolitan Stockholm is has dropped from 22 % in 1999 to 20 % in 2006. In the rest of the other 59 largest cities it is around 16 %.

Figure 1. Transit mode shares in Sweden 1999-2007

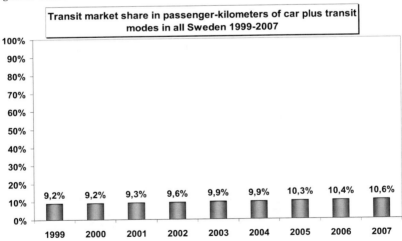

In spite of an impressive extension of transit supply (in terms of vehicle-kilometers) during the latest 20 years, transit ridership per inhabitant has dropped by 1 %, while transit supply has increased by 13 %. At the same time car ownership has increased by 26 %, see Figure 2 below:

Figure 2. Development of car and transit trips 1985-2007 in Sweden

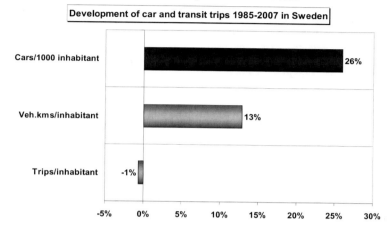

1.3 International transit mode shares

From the UITP Millenium 52 cities data base the figure below is derived. It plots the transit market share against the city density (in terms of population plus employment. Hong Kong and Singapore are outstanding in both high density and high transit market shares, well above 40 %. Also four eastern European cities, Moscow, Warsaw, Prague, and Budapest show higher market shares than 40 %. Of all other 46 cities, only Vienna shows a higher market share than 30 %.

Figure 3. Transit market shares in 52 cities as a function of city density (jobs and inhabitants)

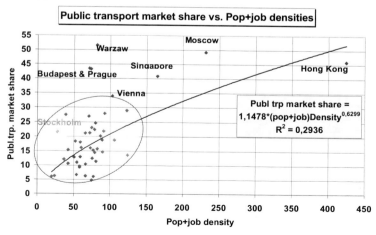

Most western European cities are not that densely populated, and partly due to this fact, also show rather low transit market shares. Stockholm, in an international comparison, has a low transit market share, also being rather sparsely populated.

1.4 A better transit system is needed

In Sweden, transit supply has increased both absolutely and per inhabitant, while, at the same time, transit demand has stagnated absolutely and dropped per inhabitant. My conclusion is that the transit industry faces some fundamental problems as regards Transit supply has increased both absolutely and per inhabitant, while, at the same time, transit demand has stagnated absolutely and dropped per inhabitant. My conclusion is that the transit industry faces some fundamental problems with and its service attractiveness. A better transit system will be needed if the goal to double the transit mode share should be achieved.

1.5 Meta-analysis of modal split with podcars: 15 %-units higher than with bus

A common experience from many urban podcar studies, in which travel demand models have been applied, is that bringing podcars to the customers in the cities would affect the modal split in favor of more public transport trips in a substantial way.

Figure 4 summarizes 10 cases with the modal split without podcar networks as compared to a forecasted situation with podcar networks:

Figure 4. Public Transport Modal split without and with PRT (Podcars). Results from British and Swedish Case studies in which travel demand models have been adopted

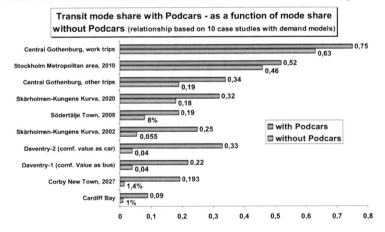

The result is also presented in Table 1 below:

Table 1. Public transport Modal split without and with PRT (podcars). Results from British and Swedish Case studies in which travel demand models have been adopted:

Marked asssessment based on demand models of PRT ridership	Public Transport Modal split without Podcars	Public Transport Modal split with Podcars	Higher modal split By PRT number of times
Cardiff Bay	1%	9%	9,0
Corby New Town, 2027	1,4%	19,3%	14
Daventry-1 (comf. Value as bus)	4%	22%	5,5
Daventry-2 (comf. value as car)	4%	33%	8,3
Skärholmen-Kungens Kurva, 2002	5,5%	25%	4,5
Södertälje Town, 2008	8,0%	19%	2,4
Skärholmen-Kungens Kurva, 2020	18%	32%	1,8
Central Gothenburg, other trips	19%	34%	1,8
Stockholm Metropolitan area, 2010	46%	52%	1,1
Central Gothenburg, work trips	63%	75%	1,2

As can be seen (also from Figure 5 below) the augmentation in modal split is substantial.

On average it might increase by 15 percentage units, when podcars will be introduced. The improvement in the modals split is higher when the modal split is lower without podcar networks. This relationship between the transit mode share without and with podcars can also be illustrated by Figure 5 below:

Figure 5. Relationship between transit mode share without and with podcar networks in 10 case studies, in which travel demand models have been used.

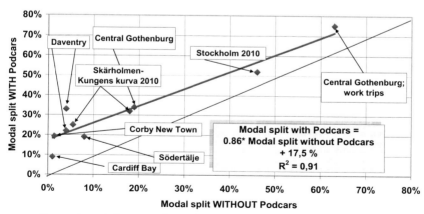

To our knowledge, very few other urban public transport projects yield the same magnitude in increasing the public transport modal split as podcars tend to do.

2 The Swedish 59 Cities Data Base

A Swedish Company, Stadsbuss & Qompany, has collected bench-marking data from some 60 Swedish city transit companies. For the year 2006 the following variables were reported:
1. Line length
2. Vehicle-kilometers
3. Annual trips (boardings)
4. Annual costs
5. Annual ticket revenues
6. Subsidies
7. Population
8. Population density

The data base was completed by us at WSP with the following variables:
9. City area
10. Street length
11. Annual operating costs for the street network

This data base allows us also to calculate:
12. Transit network density (i.e. line length per street length)
13. Population density
14. Frequency of transit service (vehicle-kilometers per line-kilometers, assuming 18 hours transit service)
15. Average walking distance to stops (by combining line length and city area data)

The average bus speed has been assumed to be 24 km/hour (data from the transit industry), which yields the bus travel time, once the average trip length is known. Trip length and market share data was obtained from the national travel survey data for the various cities.

3 Podcar networks in 59 cities

To be ale to compare today's bus network (and partly, in Göteborg and Norrköping, tramway network) with a podcar network, a "synthetic" podcar network has been suggested, designed as a grid network with 250 meters walking distance in the origin and in the destination area, respectively. The size of the podcar network depends on the city size and population density, see Figure 6 below:

Figure 6. A structure of a podcar track system

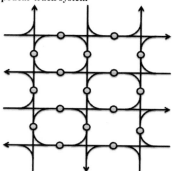

The present bus network is very dense, with an average walk time of 4 minutes, but with very low service frequency – over the day with an average waiting time of 18 minutes. The podcar network provides very short waiting times, between 0 and 1 minute for the vast majority of riders, but the average walking time will be 6 minutes (3 minutes at each of the origin and destination station).

Figure 7. Network length of the bus and the podcar networks in 59 Swedish cities

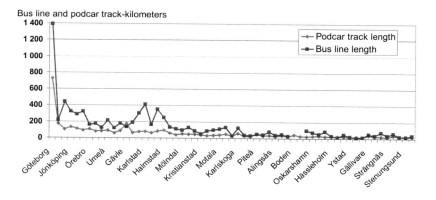

The average network size will be 80 km for the podcar network, compared to 144 km for the bus network in the 59 cities. The total bus network length is 8,162 km and the total podcar network is 2,835 km long.

4 Generalized Time and the Demand for Podcars

To calculate the future demand for podcars in 59 Swedish cities, we need first to calculate the generalized time and cost. This was done for each of the 59 cities with the database described above for the bus/tram mode. For podcars the following assumptions were made:

- Walk speed: 5 km/hour
- Wait time: 0,5 minutes
- In-vehicle speed: 40 km/hour
- Number of transfers: 0 within the city-wide system
- Podcar fare: the same as for bus/tram, i.e. 0,78 € per boarding on average

In calendar time the total door-to-door travel time is 32-33 minutes by bus or tram, but would be 11-12 minutes by podcar, which is one third of the bus/tram travel time. The generalized time will be 44 minutes by bus/tram and 18 minutes by podcar, i.e. less than half the travel time. These figures are averages for the 59 Swedish cities, see Table 2 below:

Table 2. Travel time component comparison between bus/tram and podcar. Averages for 59 Swedish cities

Minutes of travel time components	Bus/tram	Podcar
Walk time	4	6
Wait time	18	0,5
Transfer time	2,5	-
In-vehicle time	8	5
Total calendar time	32,5	11,5
Generalized time	44	18

The generalized time for the bus/tram and the podcar modes is shown in Figure 9 below:

Figure 8. Generalized travel time with bus/tram and podcar in 59 Swedish cities

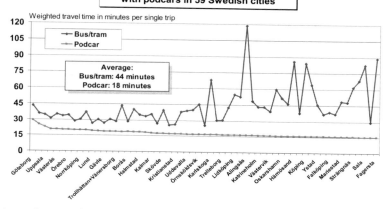

As can be seen, the variation in (weighted) travel time between cities would drop from a range of 25-120 minutes to a range of 15-29 minutes.

A simplified demand model, called ELMA, has been used to derive the new demand for transit trips by podcars. The model is an elasticity model, with variable elasticities (such as the logit model). The model is based on the generalized cost (thus including not only the above mentioned travel time components, but also the fare), and also the original market share for transit.

In most demand model applications, the podcar mode is treated as a public transport mode, with "a mode specific constant term" that resembles the negative mode specific perception of the bus mode. As most stated preference studies and the pilot tests in Cardiff Bay with the ULTra system clearly has shown, the podcar travelers regard the comfort and convenience in riding the podcar is much more like going by taxi. Therefore we have tested to treat the podcar journey as something in between going by bus and going by the private car. This has been achieved by inserting "a half car mode specific constant" into the demand model. In Daventry, a similar approach has been carried out, shown in Table 3 below:

Table 3. Car, transit trips and modes shares in Daventry, UK, at varios PRT penalties

Option	Highway trips	PT Trips	Highway mode share (%)	PT Modes share (%)
Base 2004	3 617	157	96%	4%
DDC Preferred Bus option 2021	8 023	911	90%	10%
PRT – modal penalty as car	6 354	3 110	67%	33%
PRT – modal penalty as bus	7 214	1 978	78%	22%
PRT – modal penalty as car £1.60 fare	7 186	2 014	78%	22%

Source: Daventry Development Transport Study. Draft Stage 1B and 2 Report. Daventry District Council; November 2006, By Malcolm Buchanan, Colin Buchanan.

The Daventry study shows that using the "car constant" instead of the "bus constant" increases the share of PRT trips from 22 % to 33 % or by 50 percent.

Our similar tests refer to the City of Linköping, where bus traffic has a market share of 12 %. The walk and bike modes make up 40 % and car traffic 48 %. With a podcar network for the city of Linköping, the transit market share would double to 23 %, assuming a bus constant. With a "half car constant" it would augment up to 28 % and up to 41 % if we apply "a full car constant". According to the Kungens Kurva site assessment study of the EDICT project, a podcar system would generate some 17 % new transit trips, besides from the diversion from previous trips made by car and walk/bike. Both options as regards the choice of constant terms are presented below in terms of transit market shares for the bus/tram mode, and the podcar mode:

Figure 9. Transit market shares in 59 Swedish cities with bus/tram and with podcars

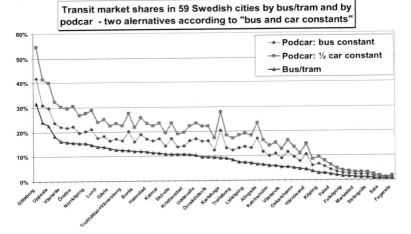

On average over the 59 cities the market share might double from 17 %, with the present day bus/tram networks up to 32 % transit trips by podcar (high estimate), or to 23 % (low estimate), see Figure 10 below:

Figure 10. Average transit market shares in 59 Swedish cities with bus/tram and with podcars at two alternative assumptions about the constant

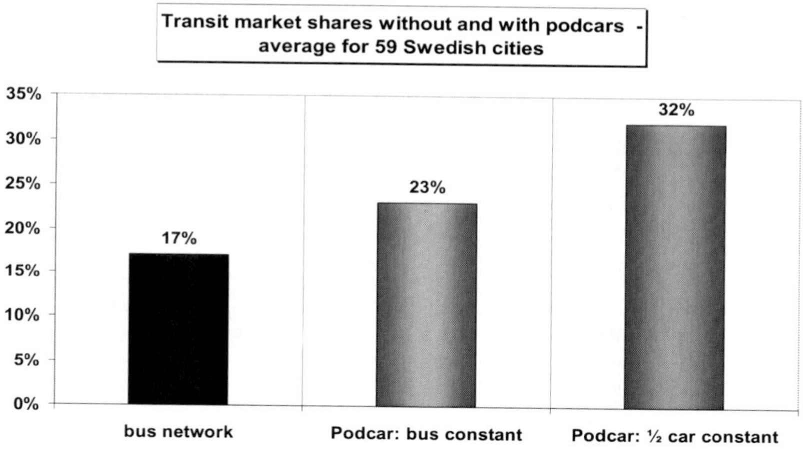

Transit market shares without and with podcars - average for 59 Swedish cities

- bus network: 17%
- Podcar: bus constant: 23%
- Podcar: ½ car constant: 32%

5 Cost and Revenue Comparisons between Bus, LRT and Podcars

In a previous paper, the costs per trip for bus, LRT, PRT, metro and commuter rail were presented (see: **"PRT Costs Compared to Bus, LRT and Heavy Rail – Some Recent Findings"**[1]).

In this project a different approach was taken. The starting point has been the actual annual costs for the existing bus network in the 59 cities (in Göteborg and in Norrköping, part of the transit network is provided by tramway). Then these costs have been compared with the costs for a 100 % percent tramway network in the 10 biggest cities and for a 100 % podcar network in all 59 cities. For the 10 biggest cities, we can, therefore, compare bus, LRT and Podcar networks, and - for all 59 cities, we can compare the bus and Podcar costs.

5.1 Bus costs

Traditionally, bus costs include capital costs for the vehicle, and operating and maintenance costs (vehicle-hour and vehicle-kilometer costs). But very seldom the road/street infrastructure cost will be included, unless we speak about BRT, Bus Rapid Transit or dedicated bus only lanes. Assume a new car free city (such as Masdar City in Abu Dhabi). Then, if one wants to introduce a bus network, all the street infrastructure costs should be accrued to the bus cost.

We have calculated the replacement costs for all city streets in the 59 Swedish cities, and then calculated the annual installment costs for this. For the 59 cities

---

[1] Paper presented at the AATS European Conference in Bologna 7-8 Nov, 2005: "Advanced automated transit system designed to out-perform the car", by Göran Tegnér, TRANSEK Consultants

with 163,420 km of street length this replacement cost amounts 634 billion €. The corresponding annual installment cost would be 32 billion €. But what would the fair share for the bus network be? A starting point could be the bus share in terms of the number of line-kilometers per street-kilometer in the city. This amount is - on average for the 59 cities - about 5 %, with a variation of 1.3 % and 12.7 % between smaller and bigger cities. But the share of vehicle-kilometers differs, with only 0.7 % bus-kilometers of the total bus plus car-kilometers. Finally we have considered buses as corresponding to 3 to 3.5 vehicle-equivalents as big as the private car. Therefore, our suggestion is to accrue a share of 2.5 % of the annual total street costs (both investment and operating street costs) as a fair share for the bus network.

The official annual Swedish cost for the bus network amounts to 428 million €. (with 3,835 line-kilometers) in the 59 cities An estimate from bus operators' show that some 89 % of these costs are operating costs and only 11 % is capital costs for the vehicles. Adding the infrastructure cost for the road network (with the transit share of 2.5 %) adds another 229 million € to the total bus cost, thus amounting 726 million €.

We have also considered the infrastructure costs for the bus network in terms of bus stops, bus terminals and bus depots. These bus infra costs adds another 69 million € per annum, which corresponds to 14 % of the total annual official bus costs of 428 million €. Even when we include the more un-traditional street infrastructure costs for the bus system, these bus stop, bus terminal and bus depot costs add another 9 % to the total annual costs of 726 million €. Per trip these costs can be compared to the ticket revenue and the corresponding subsidy rates:

Table 4. Bus cost, ticket revenue and subsidy in € per trip

Bus cost alternative	Cost in €/trip	Ticket price in €/trip	Subsidy in €/trip	Subsidy rate
Official	€ 1,46	€ 0,78	€ 0,68	47%
Incl stops, term's, depots	€ 1,70	€ 0,78	€ 0,92	54%
Incl. also street costs (bus share of cap.& Oper costs)	€ 2,48	€ 0,78	€ 1,70	69%

Therefore, the official subsidy rate of 47 % can be regarded as low, when we also include the full infrastructure costs for the bus system. With all such infrastructure costs the subsidy rate is estimated to be 69 %, and the corresponding cost to be 2.48 € per trip for bus.

5.2 Light Rail Transit costs varies substantially

In the HiTrans-report[2] an international comparison of 37 Light Rail Transit (LRT) projects has been made. The LRT cost per double-track (all infrastructure costs included) vary between 6 and 101 million €, and with an average of 23 m€ per track-km. This is indeed a substantial variation in the unit costs.

Figure 11. Light Rail Transit (LRT) cost per track-kilometer in m€

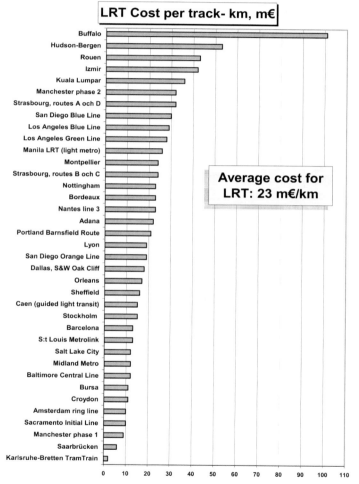

Source: HiTrans Best Practice Guide: Public transpoprt - Mode options and Technical Solutions, Intereg North Sea

The tangential LRT line built in Stockholm around year 2000 had a cost of 21 million € per double track-kilometer, while extensions of existing tramway lines in Göteborg and in Norrköping are much cheaper, with costs around 3 to 5 million € per double track-kilometer.

---

[2] HiTrans Best Practice Guide: Public Transport – Mode options and technical Solutions.

Newly planned LRT lines in Northern Stockholm now show cost estimates in the range from 13 -76 million € per double track-kilometer. Since 1994, building cost index has increased by 46 % in Sweden. Thus, it is a substantial variation in unit costs for LRT.

5.3 Podcar costs lower than LRT costs

Also the unit cost for podcar network per track-kilometer varies between different studies by various suppliers.

A natural starting point is to refer to ATRA's Status Report from 2002[3]. In a summary table they summarized the unit costs as follows:

Table 5. Podcar (PRT) cost components, according to ATRA in US 2002 k$

Component	Unit Cost	Number	Total (k$)
Guideway – straight	2,300 k$/km	8	18,400
Guideway – curved	3,400 k$/km	2	6,800
Vehicle	38 k$ each	100	3,800
Stations @ 2/km	250 k$ each	20	5,000
TOTAL			34,000

In US 2002 $, the costs were estimated to amount 34 m$ for a 10 kilometer long podcar system (single track).

I have collected information from 18 various sources regarding investment cost estimates for Podcar systems. They do by no means reflect all possible PRT systems costs that might be available after a much deeper research, but only what has been known to the author. The cost estimates have been adjusted to the 2007 year price level, with the following results:

---

[3] "Personal automated Transport – Status and Potential of Personal Transit – Technology Evaluation, By Advanced transit Association (ATRA), September 2002

Figure 12. Investment costs for podcars (PRT) per track-km from 18 studies 1998 -2 008

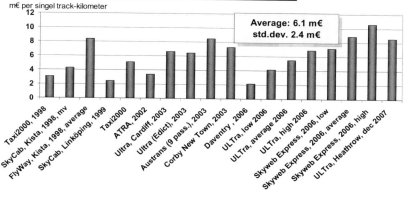

As can be seen, there is a substantial variation in Podcar cost. The costs vary between 2.1 and 10.6 m€ per track-kilometer. The average is 6.1 m$, and the standard deviation is 2.4 €/km. as the observations area arranged along a time scale, one can calculate if there is a time trend in costs. There is such a tendency, with an annual increase of 0.24 m€/km and year.

In a study by Booz Allen Hamilton for New Jersey in the US, Paul Hoffman argues for much higher PRT costs, in the range 6.5 m€ to 21.8 m€ per track-km. His lowest estimate corresponds well with my own findings. However, his higher estimates refer to large scale systems in dense and complex large cities, such as New York.

The ULTra Heathrow podcar system, with 4 km, 5 stations and 18 vehicles, is estimated to cost 25 m£, or 8.5 m€ per track-kilometer[4].

---

[4] In a recent study: "The Viability of Personal Rapid Transit in Virginia: Update" Virginia, 18th Dec. 2008 , the ULTRa cost is lowered to 20 m£, or to 6 m€ per track-kilometer

ULTra Picture

Vectus, with its Podcar test track in Uppsala, Sweden, has recently confirmed that it is a tricky task to give accurate and general costs estimates, But, for 10-20 kilometer long Podcar network an estimate between 7.5 m€ to 10 m€ might be realistic. This cost level corresponds well with the ULTra cost for the Heathrow installation.

Vectus picture

5.4 Three cost estimates for Bus, Podcar and LRT

As costs vary substantially both in Sweden, and on the international scene, we have decided to present three cost estimates, one low, one high and one average cost estimate for both LRT and for Podcars.

The cost estimates have been chosen according to reflect the substantial variation in costs per track-km for the LRT and the podcar modes. The high cost estimate for podcar reflects the ULTra Heathrow cost level, which also is the average Vectus' cost estimate. The low podcar cost estimate corresponds to ULTRa's cost estimate at a lower utilization rate (i.e. 100,000 annual trips per track-km) from 2002, adjusted to the 2007 price level.

Figure 13. Investment cost for Buss, Podcar and LRT in m€ per track-km at three cost alternatives

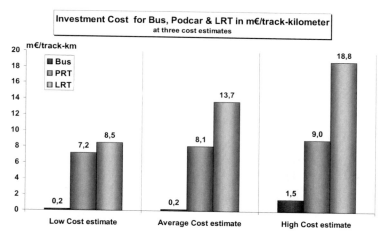

For the high cost estimate, the full annual transit share of the street replacement cost (including also the corresponding transit share of the street operating cost) has been accrued as the bus infrastructure cost. For the average and for the low cost estimate, these street infrastructure costs are not included, but only the infrastructure costs for bus stops, terminals and depots. In Figure 14 below the corresponding unit costs per trip is shown:

Figure 14. Unit cost (capital and operating) in € per trip for bus, podcar and LRT: average for 10 cities in Sweden at three cost levels

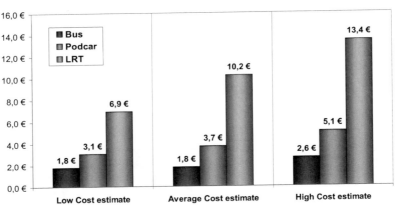

At the average cost estimate level, an average Swedish city bus trip (with a weighted n average trips distance of 7.8 kilometer) costs 1.8 € by bus, 3.7 € by podcar and 10.2 € by LRT. The relationship between bus, podcar and LRT does not change when the cost levels shift from low to high. Thus, podcars can be regarded as less costly than LRT-systems. However, the bus network is cheaper than the podcar network, when all infrastructure costs are included (even for the bus network).

5.5 Double ticket revenues with podcars yields lowest operating deficit

As the podcar mode will yield up to twice as many transit trips as the traditional bus an tramway modes, even the ticket revenues will augment by a factor proportional to the number of trips, provided we adopt the same pricing policy for podcar trips.

5.6 Comparison of operating results

The operating deficit (or surplus) is defined as the difference between the ticket revenue minus the operating cost. This deficit can be calculated in totals and per trip.

Figure 15. Operating deficit/surplus in € per trip for bus, podcar and LRT: Average for 10 Swedish cities at three cost levels

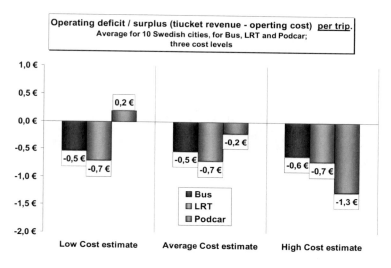

Figure 15 shows the operating deficit per trip for the bus, the LRT and the Podcar modes. At the average cost estimate, the bus and the LRT modes show an operating deficit of 50 and 70 cents, respectively, while the deficit by podcar will be only 20 cents. At the low cost estimate the podcar mode even show a positive operating surplus by 20 cents per trip. At the high cost estimate level, the deficit will be highest for the podcar mode, -1.3 5 per trip.

The uncertainty as regards the "true costs" for Podcars, explains the big difference in these operating results.

However, an average for the ten cities hides the details. As a matter of fact, the podcar mode shows a negative operating deficit only for a few large cities. Only eight bigger cities out of the 59 Swedish cities show a negative deficit, while the rest that is 51 cities, yield a positive operating surplus by Podcar. For the bus mode there is a negative deficit in all 59 cities, se Figure 16 below:

Figure 16. Operating deficit/surplus in € per trip for bus, podcar and the difference between podcar and bus for 59 Swedish cities

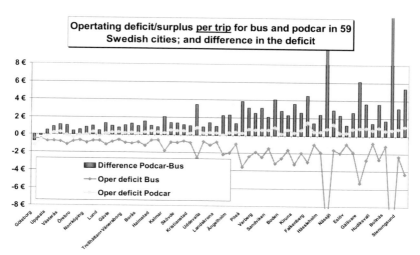

The bars in Figure 16 above show the difference in the operating deficit between the podcar and the bus mode.

6 Cost-Benefit Analysis and Environmental Impacts of Podcars
6.1 Benefits are 27 % higher than cost with podcars in 59 Swedish cities and towns

A cost –benefit analysis as seen carried out for the 59 cities and towns, with podcar networks replacing the existing bus/tram networks. On the benefit side the following aspects are considered:
- Travel time gains
- Ticket revenues
- Traffic safety gains
- Environmental gains (reduced CO_2 exhausts from private cars and from buses

The extra comfort and convenience by podcars is not considered

And on the cost side the following aspects are considered:
- Investment costs
- Operating and maintenance costs
- Reduced gasoline tax revenue from less car traffic

The main result is that the overall benefits amount 2.85 billion €, while the total costs amount 2.24 billon € in present value. The net benefits amount 0.61 billion € and the benefit-cost ratio is 1.27. This means that one € spent on podcars yield 1.27 € in return in terms of benefits to the society.

Figure 17. Benefit and cost components for Podcar networks compared to bus networks. Annualized present values over a 40 year period (at 4 % discount rate) in billion €

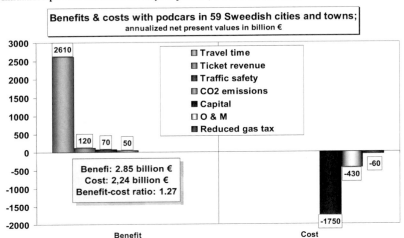

Travel time gains make up more than 90 percent of the total benefits. Increased ticket revenues, traffic safety and environmental gain add up to the rest. Podcar networks are clearly worth any cent. But in how many cities and towns will a podcar network be economically justified from the social surplus point-of-view? To answer this question, I have calculated costs and benefits for each of the 59 cities and towns, with the following result:

Figure 18. Relationship between benefit-cost ratio and city population size. Statistics for 59 Swedish cities, at podcar capital cost of 8 m€ per track-km

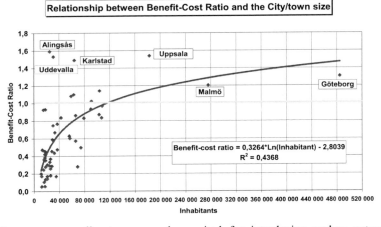

On average smaller towns are less suited for introducing podcar networks compared to the larger towns and cities. Even if there is certain variation from the regression line, on average, one might conclude, that podcar networks seem to be suitable to introduce in cities and towns down to a size of approximately 100,000 inhabitants. From the Figure 18 above it can be seen that at least seven smaller cities than 100,000 inhabitants still show a positive benefit-cost ratio for podcars.

The social profitability (i.e. when the social benefit-cost ratio is greater than 1.0) is highly sensitive to the capital cost per track-km for the podcar system. A sensitivity analysis has been carried out, for the capital cost per track-kilometer in the range between 4 – 8 m€ per km, with the following results fro the 59 cities Swedish data base:

Table 6. Benefit cost ratio, city size, share of profitable cities, total benefits and cost in m€: Results from 59 Swedish at capital costs of 4 – 8 m€ per track-km

Podcar capital cost/track-km, in m€/km	B/C ratio, all 59 cities/towns	City size limit in no. of inhabitants for profitability	No of profitable cities	Share of profitable cities	Total annual benefits in m€	Total annual costs in m€	Share of total population in profitable cities
4 m€/km	2,15	20 000	31	53%	2 036	946	82%
5 m€/km	1,67	30 000	27	46%	2 036	1 218	76%
6 m€/km	1,32	45 000	22	37%	2 036	1 547	68%
7 m€/km	1,03	70 000	15	25%	2 036	1 972	59%
8 m€/km	0,9	100 000	10	17%	2 036	2 220	43%

At 8 m€ per track-km the city size limit for profitability is around 100,000 inhabitants. At this cost level only 10 out of the 59 cities are profitable, but they carry 43 % of all citizens in the 59 cities group of Swedish cities.

Figure 19. Relationship between City size and Social profitability (B/C ratio > 1.0)

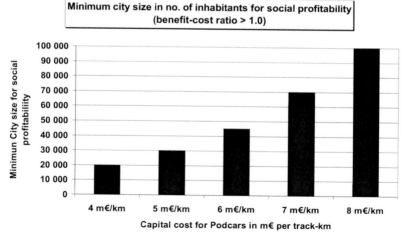

At 6 m€ per track-km, the overall benefit/cost ration becomes 1.32, i.e. benefits are 32 percent higher than the costs. The minimum city size for a podcar network, that covers the whole city drops to 45,000 inhabitants. Approximately one third of all 59 cities fulfill this criterion, and these 22 profitable cities carry two-thirds of all citizens (3.2 million) in all the 59 cities.

If the podcar capital could be reduced to 4m€ per kilometer, then the benefits would be more than twice as high as the costs, and podcars would be profitable down to a town size of only 20,000 inhabitants. Of all 59 Swedish cities and town, 31 cities and towns fulfill this criterion, and they make up more than 80 of all inhabitants.

Thus, the conclusion is that costs matters and that the profitability of podcars in cities is highly dependent on the unit costs.

In reality, I recommend to carry out detailed cost-benefit analysis for each town and podcar case in order to draw the correct conclusion if the podcar project will be economically justified in terms of benefits and costs.

6.2 Reduction in carbon-dioxide emissions by 25 % with podcars

If a system of podcars would replace the urban diesel bus, then the local exhausts from diesel buses would be eliminated. Also, the modal shift from trips made by the private car to podcar trips would contribute to reduce the local air pollutions exhausts substantially.

Figure 20 below the carbon dioxide emissions per vehicle-.kilometers is presented according to Swedish calculations for nine modes. The figure refers to urban traffic conditions; and for the electric modes, LRT, Metro and rail as well as PRT (podcars) we have based lour figures on the average Swedish electricity production system (with high proportions of hydro and nuclear electric power).

For the podcar mode the energy consumption (as the basis for CO_2 emissions) are derived from the Ultra and Vectus podcar systems.

Figure 20. Estimated carbon dioxide emissions per vehicle-kilometer for 8 transit modes and for car

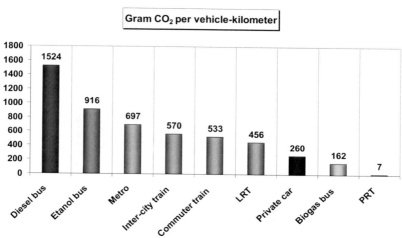

The diesel and ethanol bus modes show high CO_2 emissions per vehicle-kilometer. Also the rail modes show higher exhausts then the private car, partly due to the bigger size of the vehicles. The podcar mode is estimated to produce 7 grams of carbon dioxide per vehicle-km. Adopting average passenger loads per vehicle type, gives us the following carbon dioxide emissions per passenger-kilometer.

Figure 21. Estimated carbon dioxide emissions per passenger-kilometer for 8 transit modes and for car, at average vehicle occupancy

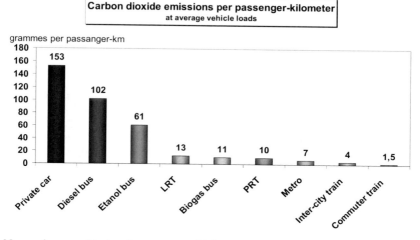

Now, when considering the average vehicle occupancy, the private car becomes the highest emitter of CO_2 gases, followed by the diesel bus. The podcar mode has 10 grams of CO_2 gas per passenger-km, which is of the same magnitude as by

biogas bus, and a little more than by metro (however, the energy consumption in the building process of each transport mode has not been estimated here).The resulting impact on carbon dioxide is presented in the figure below:

Figure 22. Impacts on carbon dioxide emissions in kiloton per annum from replacing bus networks into podcar networks in 59 Swedish cities

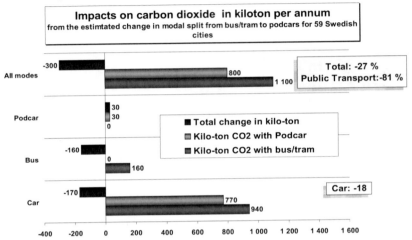

Exhausts from the car traffic would be reduced by 18 % when podcars compete as the local public transport system for these 59 Swedish cities. Exhausts from diesel buses will then be eliminated, and replaced by a much smaller exhaust from the new podcar mode (only 19 % of the diesel bus exhausts). The overall reduction in carbon dioxide is estimated to become 27 % or 300,000 tons per annum. The CO_2 exhausts from the public transport system will be reduced by 81 %.The total carbon dioxide emission from road traffic in Sweden amounts 13.2 million tons annually. Therefore, a replacement of bus networks for podcar networks in 59 Swedish cities would reduce the CO_2 road emissions by 2.3 percent of all surface transportation exhausts. Even if this is positive, there are other cheaper ways of reducing CO_2 gases.

7 PPP Solutions for Financing Podcar Systems - Case Studies

In Sweden several pre-feasibility studies have been undertaken lately. In cities and towns like Värmdö, Kiruna, Linköping, Östersund, Eskilstuna, Södertälje and Uppsala local podcar networks have been assessed. Also the Kungens Kurva Area in the Municipality of Huddinge was analyzed within the EDICT-project[5].

7.1 City of Linköping Case Study

The PRT double track network for Linköping (104,000 inhabitants) was designed by Beamways AB.

---

[5] EDICT is an acronym for European Demonstration of Innovative City Transport

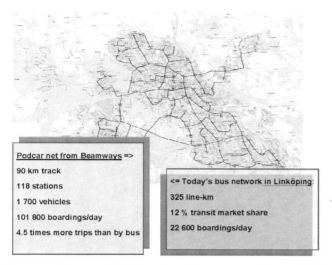

Beamways has simulated the demand for podcar trips on double track network with 500 meters of station spacing and assuming 1.5 podcar trips per person and day. The average vehicle occupancy is assumed to be 1.5 persons.

The estimated impacts of an area-wide podcar network in Linköping would be:
- The transit mode share would increase from 12 % to 28 % (assuming a bus constant in the mode choice model) or 40 % (assuming a "half car constant" in the mode choice model)
- The annual costs are estimated to increase from 14.4 to 45.3 m€.
- The ticket revenues are estimated to increase from 6 to 33.6 m€, i.e. more than five times.
- The Annual ticket revenues will exceed the operating costs with the podcar system by 9.4 m€, which corresponds to 44 % of the annualized capital costs.
- The net cost (total cost – ticket revenue) per boarding will be reduced from 1.14 € by bus to 0.35 € by podcar.

7.2 City of Södertälje Case Study

An area-wide podcar network has been designed for Södertälje (82,000 inhabitants) and analyzed by LogistikCentrum and WSP in 2008. A regional nested logit travel demand model for the entire Metropolitan Stockholm area was used to calculate the regional podcar demand, for the year 2030; that was later simulated into more detail by the PRTsim model, in order to calculate the waiting and travel times within the podcar network, but also to estimate the size of the vehicle fleet. At present a first phase of a podcar network is assessed by an engineering design. No decision is yet taken to start building a podcar system

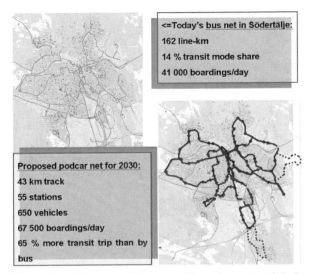

The estimated impacts of an area-wide podcar network in Södertälje would be:
- Transit demand would increase from today's 25,000 trips by bus to 69,500 trips in 2030 by podcar
- The transit mode share is estimated from 14 % to 18 %.
- The annual ticket revenues would balance the annual operating costs, if the same ticket price would be adopted.
- The net cost ((total cost − ticket revenue) per boarding would increase from 0.6 € to 1.0 € per boarding by podcar.
- The social benefit-cost ratio for the first phase is calculated to be betrween 1,45 − 2.00.

7.3 Area of Skärholmen - Kungens Kurva Case Study (EDICT)

This Study was carried out in 2002-2004 under the European Commission project called EDICT - European Demonstration of Innovative City Transport. In 2002 Kungens Kurva Shopping Mall had 42,000 daily visitors. Around 2015 this number is estimated to increase to 63,500 visitors. The proposed podcar network was designed by Transek and LogistikCentrum in collaboration with the municipality of Huddinge, to be connected to the metro station at Skärholmen, in the city of Stockholm. Also two remote parking houses in each of the two entrances to the area was proposed to carry a podcar station inside the parking area. A regional nested logit travel demand model for the entire Metropolitan Stockholm area was used to calculate the regional podcar demand, for the year 2030; that was later simulated into more detail by the PRTsim model, in order to calculate the waiting and travel times within the podcar network, but also to estimate the size of the vehicle fleet.

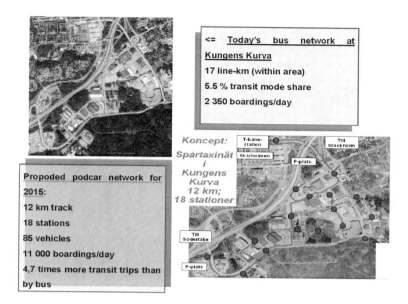

The estimated impacts of an area-wide podcar network at Kungens Kurva, Huddinge, would be:
- The number of transit trips would augment almost fivefold
- The operating costs would increase three times
- The ticket revenues would increase by a factor of ten, as the ticket price was proposed to double. The tax subsidy could then be reduced from 61 % to 24 %.
- The operating cost per boarding could be reduced from 1.06 € to 0.7 € by the podcar system.

7.4 The City of Kiruna Case Study

The city of Kiruna (23,000 inhabitants) is a fairly small town with a very low transit mode share. A pre-feasibility study was made by WSP in 2006 for a podcar network. The reason for this study is that the entire city has to be moved from its present position within the next two decades, due to the fact the iron ore, on which the town is based, causes cracks in the soil. The demand for podcar trips is based on a meta analysis of the transit modes share for bus and podcar networks (based on previous Swedish and British PRT demand model studies[6].

---

[6] See section 1.5 above about the meta analysis

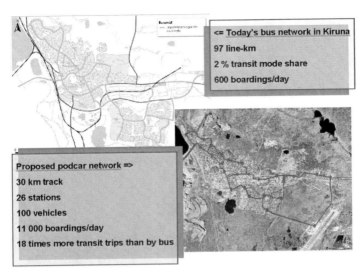

The estimated impacts of an area-wide podcar network in the town of Kiruna would be:
- The number of transit trips would increase from 600 to between 7,000 and 15,000 daily boardings by the podcar network
- The transit mode share would increase from 2 % to between 12 and 24 %.
- The annual costs would increase from 0.8 m€ to 19.5 m€.
- The ticket revenues would increase from 0.14 m€ to 2.5 m€, or by a factor of 18
- The annual ticket revenues will exceed the annual operating costs by some 1 m€
- The net cost ((total cost − ticket revenue) per boarding would be reduced from 5.2 € by bus to 3.2 € by podcar in Kiruna.

7.5 Conclusions from four case studies

The following conclusions can be drawn from the podcar case studies in Linköping, Södertälje, Kungens Kurva (Huddinge) and Kiruna:

- **Four to five times more trips with podcars compared to bus**
  Compared to bus, with podcars transit ridership will increase by four to five times in Linköping and at Kungens Kurva. At Kiruna the effect is even greater, but in that town the analysis is more coarse

- **3,5 times higher transit mode share with podcars**
  The transit mode share augments by 3.5 times in Linköping and at Kungens Kurva. In Södertälje the impact is estimated to be smaller, or from 14 % to 18 % with a podcar network for the built up area.

- **Increased annual costs, but a positive operating surplus with podcars**
  The total annual costs will increase with podcars, and they vary with the size of the city and the podcar network. The ticket revenues exceed the

operating and maintenance costs in Linköping, Kiruna and at Kungens Kurva. This is explained by the substantial increase in ridership. In Linköping and at Kungens Kurva we have also assumed a higher ticket price with podcars.

8 Public-Private Partnership – a recommended financial solution with examples from six Swedish case studies

Public transport has since long been a case for various public private partnerahip solutions. And there is still room for further improvements towards an even better collaboration for the local, urban and regional public transport sector. Investments in podcar systems would mean any exception from such collaborations. A closer Public Private Partnership (PPP) solution might not change the financial burden dramatically, but it can bring other advantages, e.g. in terms of a more efficient transport system. A negotiated annual fee from the service provider could also ease the planning of the annual budget for the system.

BOT stands for Build-Operate-Transfer and is a form of public-private partnership. In its most common form, a BOT project implies that a private actor receives a concession from the public sector to finance, design, construct, and operate a facility for a specified period of time, normally between 20 and 30 years. After the concession period ends, ownership is transferred back to the public sector. Within the transportation sector, large road investments have so far been the most common and talked-about BOT projects. However, it can also be used for implementing investments in podcar systems.

There are several advantages to the public actor:
- Costs are spread out over the concession period.
- The risks are divided between the private and public actors and thus lower for the public sector.
- The project has good chances to be cost effective, since the private actor is forced to optimize maintenance.
- An advantage to the private actor is that so far, only a few actors are strong enough to offer such long-term commitment.

The financial means are usually gathered through borrowing on the international financial market. Revenues normally come from ticket sales, and/or a yearly payment from the public actor.

BOT is suitable for podcar investments, since they are characterised by relatively large initial investments but rather low operating and maintenance costs. Podcar systems are compared to other public transport, such as bus with no or low investments costs and high operating and maintenance costs. The comparison is made easier through BOT, since both alternatives seem to have only "operating and maintenance costs" in the eyes of the public actor.

A podcar system may be a private transportation alternative comparable to taxis and may be financed through tickets sales. In that case, the private actor will want to be free to set the ticket price that he chooses. In such an alternative, a public actor will most likely keep the existing public transportation, seeing the podcar system as a separate alternative.

In Sweden it is more suitable to see podcars as a part of the existing public transportation system. In that case, the municipality is in charge of planning, procurement, and ticket sales. A podcar system may transfer travellers to and from other public transport and can in some cases replace this locally. The same ticket is supposed to be valid on different types of public transport. The private actor's revenues will in this case be fees from the public actor.

A BOT consortium may include the following parties:

- Podcar system supplier
- Property developer
- Public transport operator
- Host municipality

Below we show illustrative examples of what a yearly fee and fee per trip may be for six different podcar investments organised as BOT. We assume a rate of interest of 5 %, that the concession period is 30 years, that there are no subsidies, and the unit costs shown below. The examples are simplified compared to reality. In real projects, the design of the project is an important cost.

Table 7. Assumed unit costs for different parts in a podcar system

	Unit cost in m€*
Track per km	4.0
Station	0.3
Vehicle	0,075
Control system	1,0
Depot*	3 – 6

* Depending on system size and localization. Source: LogistikCentrum

The total investment cost per system km is dependent on the size that control system and depot are distributed over, as well as on the number of stations and vehicles per track km. In the case studies, the total investment cost varies between 5.8 and 8.7 m€ per km, with an average value of 7.5 m€.

Operating and maintenance costs include salaries for employees, and costs for spare parts, maintenance, and power supply. The number of employees is varied depending on system size. The energy consumption per vehicle is assumed at 7 kW (incl. AC), 9 hours per day, 300 days per year. The energy cost is assumed to be 0.1 € per kWh.

The operating costs might probably by underestimated somewhat, since no real wage increases have been included. Also, costs for land and encroachment compensations have not been included. However, a 20 % cost increase have been made to cover project management, unforeseen costs, and profit.

The result is seen in
Table 8 below:

Table 8. Example of six different podcar investments organised as BOT and where the private actor is responsible for initial investment. The rate of interest used is 5 %, and the systems are to be repaid after 30 years.

Alt. without any subsidy	Linköping	Kiruna	Kungens Kurva	Södertälje 1st phase	Södertälje areawide	Östersund
Track lenth (km)	90	30	8,3	11	43	25
No of stations	118	26	15	18	55	11
No of vehicles	1 700	100	85	140	731	150
Trips per annum (million)	33.1	3.3	4.0	4.1	22.2	4.2
Capital cost (m€)*	780	175	65	89	361	161
Capital cost per track-km	9	6	8	8	8	6
O & M Cost (m€)	16	2	2	2	8	2
Annual BOT Fee (m€)**	69	14	6	8	33	13
BOT Fee per podcar trip (€)	2,1	4,2	1,5	1,9	1,5	3,0
Corresponding Cost by bus (€)	*1,4*	*3,8*	*1,8*	*2,2*	*2,2*	*5,8*
Cost difference podcar - bus in %	53%	10%	-14%	-14%	-34%	-48%

* Including reinvestment for vehicles year 11 and 21.
** Including 20 % for project management, and profit.

The podcar systems in Södertälje and Kungens Kurva are the cheapest per trip with: 1.5 € per trip. The podcar cost per trip is lower than for bus in four of the six case study areas, namely in Kungens Kurva, Södertälje (both alternatives), and in Östersund. In Kiruna, a podcar trip would be 10 % more expensive than a bus trip, while it would be 53 % more expensive than a bus trip in Linköping. The primary reason for this in Linköping is that the study included double tracks for the podcar system, which would not be necessary. In Figure 23 below, costs per trip are compared between podcar and bus.

Figure 23. Costs per trip with podcar in a BOT solution with (dark green) and without (light green) 50 % state subsidies as well as with bus (red).

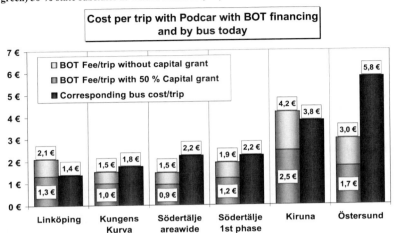

Figure 23 shows that BOT solutions for podcar investments in all five case study cities would imply lower costs per trip, provided that 50 % of investment costs are paid by state grants. Also without grants, the cost per podcar trip is lower than per bus trip in four of the six cities.

9 Other ways of financing Podcar networks

When public funds are scarce, there are a number of possibilities for financing public transport. Successful alternatives are distinguished by the following criteria:

- high socio-economic efficiency
- large flows of revenue
- small demands on the judicial system
- low administration and system costs
- political acceptance
- public acceptance

Unfortunately, few sources of finance combine all of this, but some are nevertheless well suited for podcar systems.

Agreements on co-financing with the private sector
An "Agreement on co-financing" is a politically attractive solution that may give positive external effects through strengthening the feeling of cohesion between public and private in a municipality. However, agreements presuppose that the concerned private actors appreciate the benefits they would get, enabled by the public transit investment in question, and that they accept that the municipality or

– the Public Transit authority cannot finance the investment by itself. When many private actors are involved, the transaction costs may become large.

Local fees or taxes
If the interest in co-financing is not large enough among local and regional private actors, local or regional fees or taxes may offer alternatives. The fees can be regarded as compensation for increased accessibility. In already developed areas, they can be implemented in the form of temporary supplements to the ordinary corporate taxation or as supplements to the existing employment tax. In expanding areas, it can be implemented as an earmarked exploitation fee imposed on developers.

A local fee or tax may be unusually well suited for investments in podcar systems since

- the investment offers something entirely new, rather than an upgrading of a (possibly neglected) public transportation system.
- the investment offers a higher standard of service in the local public transportation.
- the prospect of decreasing local levels of combustion, increasing traffic safety, improving accessibility for some groups and in the long run increasing land supply, all with one investment, is likely to be interesting to many.
- making stations elevated and putting them on side tracks makes it possible to integrate them with buildings, such as hospitals and malls, which increases benefits and probably also public acceptance.
- while the initial investment need of a podcar system is large, operation costs are low, making single payments sufficient.

Congestion and environmental charges
Congestion and environmental charges may be considered for several reasons, not the least because they may at the same time guide car traffic and finance public transportation. But while they may be implemented as a long-term strategy for financing public transit, they are not well suited for financing single investments.

Services connected to public transportation
Other alternative sources of finance that are already used today are revenues from various services that may be connected with the supply of public transportation, such as revenues from commercial advertising, and services connected to the public transportation payment system. The potential for this type of financing is estimated to be far from exhausted, although the single revenues are usually small. However, incomes from these types of services should be seen as a possible addition to public transportation funds in general, rather than to single investments. As an alternative, the municipality – as planned by the consortium behind the railroad Norrbotniabanan in northern Sweden – build tracks and sell station rights and vehicle kilometers to private companies that want to supply the traffic.

Government subsidies and EU-funds
The Swedish government subsidies' programme for local and regional public transportation is applicable to local podcar systems, but not easily so. The

investment should either be very small, or exist in the long-term planning schemes of the national road or rail agency. These are normally updated every four years, meaning that it may take four years before an application is considered. In other countries, other rules apply, for better or for worse.

Within the European Union, EU's regional funds, as well as the Seventh Framework Programme, seem to fit very well with investments in podcar systems. However, when applying for an EU subsidy one must be aware of the large load of administrative work that an application implies. The Seventh Framework Programme is aimed at funding research programmes, rather than ordinary investment projects. Considering the relative novelty of podcar systems, however, there is most likely research to be done, related to system implementation that may fit within the Programme.

10 Conclusions and recommendations

10.1 Conclusions about demand and mode share

The basis for this paper has been a data base over public transport in 59 Swedish cities. Two of the cities (Göteborg and Norrköping) have LRT as a backbone as a complement to the traditional line-haul scheduled bus system. A synthetic podcar network has been assumed for the 59 cities with the typical podcar (PRT) features (less than 1 minute waiting time, 45 km/hour in cruising time, no intermediate stops).

The main conclusions as regards the podcar performance, mode shares and travel demand compared to line-haul bus service for the 59 cities are:
- Generalized time: reduction from 44 to 18 minutes (- 59%) on average
- Mode share: increase from 17 to 32 %
- Number of trips: Up by 94 %
- Benefit and costs: Net present value of benefits are 27 % higher than the costs
- Environmental impacts: Carbon dioxide emission down by 27 % by replacing diesel bus traffic to podcar traffic

10.2 Conclusions about the economy with podcars compared to bus and LRT

- As regards **capital costs**, podcar is cheaper than LRT, but more expensive than an urban bus system. This is true even when we include the fair share for the bus mode for the infrastructure costs for the street network that the busses use. This holds true for all the three costs levels adopted in this study.

- As regards **operating and maintenance costs** the picture becomes less evident, at the cost levels we have found. There is a substantial variation not only for capital costs, but also among various studies about the O & M costs. At our Average cost level, podcars have the lowest O & M costs compared to both bus and LRT, while LRT show the highest O & M costs per passenger-kilometer. This is also true at the Low cost level alternative, except for the City of Göteborg (because of a large podcar network there, with a very high utilization). However, the cost comparison results are highly sensitive to variations in the unit cost assumptions. For podcar networks one might argue that the O & M costs should vary either according to the number of passenger-kilometers (demand) or to the number of track-kilometers (supply), or to both. In this study we have assumed dependence on the demand only, and not to the supply side.

- **The operating deficit or surplus** is defined as the difference between ticket revenues and the O & M costs. For bus and LRT there is an operating deficit throughout all 59 Swedish cities, which means that the O &M costs always are larger than the ticket revenue at the present low price levels. At the <u>Low cost level</u>, the podcar networks yield a positive operating surplus (except for the two largest cities in the data base, Göteborg ands Malmö). At the both other cost levels, the <u>Average and the High cost levels</u>, podcars yield a negative operating deficit than by bus for the 11 largest cities. At the <u>Average cost level</u> the podcar operating deficit will be much lower than by bus. Podcar networks yield an operating deficit only in some 8 larger cities, while it yields a positive operating surplus in the other 51 cities.

- If one wishes to further reduce the uncertainty as regards the capital and O & M costs that are accrued to podcar networks, we recommend to carry out more detailed civil engineering studies and also to let build pilot podcar networks in various areas in order to gain experiences both about the demand for podcar service and about capital and operating costs.

10.3 Conclusions about financing Podcar networks

This study summarizes a variety of financial solutions, among the most inportant I would like to mention:
- Agreements on co-financing with the private sector
- Local fees or taxes
- Congestion and environmental charges
- Services connected to public transportation
- Government subsidies (grants) and EU Funds

The method recommended in this report is a specific form of agreements on co-financing with the private sector, called PPP: Public Private Partnership. And even more specific: the BOT method, Build, Operate and Transfer. At reasonable assumptions, based on six Swedish Case Studies, we have show that a BOT Fee per trip for an area-wide podcar network might be even lower than the actual bus cost per trips in four of the six cases. With a traditional Swedish Governmental capital grant at 50 % of the capital cost, a BOT podcar fee would be cheaper than the actual bus cost per trip in all of the six case studies.

10.4 Recommendations - an Action Plan

For a local town or city interested in implementing a podcar network, we recommend the following action plan:

1. Delimit a dense settlement area with multi-storey housing, workplaces, commercial areas and major transit hubs
2. Undertake out a feasibility study for a potential podcar network, containing level-of-service, trip forecasts, benefits and costs as well as comparisons with today's public transportation and with a step-by step development.
3. If there is still an interest for a podcar network, the next step would be to undertake an engineering design phase that answers the questions about the feasibility and the economy
4. Seek public acceptance with inhabitants, land owners and with local transport authorities.
5. Negotiate about co-financing with developers, landlords and with relevant authorities and (in Europe) with EU Regional Funds,
6. Invite potential suppliers together with the responsible transportation authority to offer the remaining financial capital, installment and operations of a first phase for a podcar system at a fixed annual BOT Fee.
7. Demand a safety certification from the relevant public authority and financial guarantees from the consortium of suppliers.
8. An agreement with the consortium of suppliers ought to contain payment conditions for operations and service levels with a specified accessibility during the entire period of operations.
9. Later developments of the podcar network ought to be procured under competition, based ion specifications from the first supplier.
10. Include the podcar project in the relevant local and regional planning process and undertake the necessary political decisions.

Acknowledgements

This work has been made possible to carry out thanks to a substantial research grant commissioned to WSP Sweden AB by The Swedish Institute for Transport and Communications Analysis (SIKA). Without SIKA's trustful support this work would not have been possible to materialize. However, the authors is solely responsible for all facts, views and conclusions. For all potential mistakes, we are solely responsible.

References

1. Andréasson, I., et al., Research and development in advanced transit systems - Survey of academic and industry efforts, Report, Chalmers Industriteknik. 1996
2. Andréasson, I. " Innovative Transit Systems – Survey of Current Developments". Vinnova Report VR 2001:3.
3. Badger, Charles, M.: "The Viability of Personal Rapid Transit in Virginia: Update": Commonwealth of Virginia; 18 Dec. 2008
4. Buchanan, M: "Daventry Development Transport Study – Draft Stage 1B & 2 Report". Daventry District Council, Nov. 2006. By Malcolm Buchanan, Colin Buchanan consultants.

5 Dahlström, K, et al, 2005, A General Transport System – A Case Study within the project: The Value of Alternative Transport systems. SIKA Report 2006:1 (including two reports in Swedish)
6 Daventry District Council: "Daventry PRT Scoping Study". Febr. 2008
7 EDICT, European Commission, 2004, DG Research, 5th Framework Programme, Key Action: "City of Tomorrow and Cultural Heritage", EDICT
Final Report, Deliverable 10, December 2004.
8 Hoffman, P.: "Personal Rapid Transit - Innovative Transportation Technology; Overview and State-of-the Industry: Boos Allen & Hamilton, Febr, 2006.
9 Hoffman, P.: "Personal Rapid Transit - Strategies for advancing the state of the industry. Transportation Research Board; and Boos Allen & Hamilton, January 23rd, 2007.
10 Hoffman, P. and Carnegie, Jon A: "Viability of Personal Rapid Transit Study in New Jersey - Final Report". By Jon A Carnegie, Alan Vorhees Transportation Center, Rutgers, The State University of New Jersey and Paul.S.Hoffman, Booz Allen Hamilton, Inc., Febr 2007
11 Johansson, O. "Are PRT systems socially profitable?" Gothenburg University; 1997
12 Lowson, Martin, Advanced Transport Systems Ltd and University of Bristol:
"Service Effectiveness of PRT vs Collective – Corridor Transport", in Journal
of Advanced Transportation - Vol. 37. No 3 Sep 2003.
13 Poskey, J: "Report on the Feasibility of Personal Rapid Transit in Santa Cruz,
California", Draft Report prepared for the City of Santa Cruz, by Jeral Poskey,
Da Vinci Global Services.
14 Raney, S. & Young, S.: "Morgantown People Mover - an updated Description". http://www.cities21.org/morgantown_TRB_111504.pdf
15 SIKA: "Evaluation of a PRT System". Report 2008:5 (in Swedish)
16 SIKA: "Infrastructure planering towards an enhanced goal fulfilment in major cities" Report 2008:6 (in Swedish)
17 Tegnér, G., "Market Demand and Social Benefits of a PRT System: A Model Evaluation for the City of Umeå, Sweden". Infrastructure, Vol. 2, No. 3, pp. 27-32, 1997, John Wiley & Sons, Inc.
18 Tegnér, G.: "Benefits and Costs of a PRT system for Stockholm". 7th APM Conference. Copenhagen, May 1999.
19 Tegnér, G and Fabian, L.: "Comparison of Transit Modes for Kungens Kurva, Huddinge, Sweden"; 8th APM Conference in San Francisco, July 2001.
20 Tegnér, G. and Andréasson, I: "Personal Automated Transit for Kungens Kurva, Sweden - a PRT system evaluation within the EDICT project". 9th APM 2003 Conference, Singapore, Sept. 2003
21 Tegnér, G.: "EDICT – Comparison of costs between bus, PRT, LRT and metro/rail". Transek, February 2003

22. Tegnér, G.: "PRT Ridership Analysis; in: Personal Automated Transportation- Status and Potential of Personal Rapid Transit". The Advanced Transit Association, January 2003. (ATRA). http://www.advancedtransit.org/doc.aspx?i=1015
23. European Commission, 2004, DG Research, 5th Framework Programme, Key Action: City of Tomorrow and Cultural Heritage: "EDICT Final Report, Deliverable 10", December 2004
24. European Commission, 2004, DG Research, 5th Framework Programme, Key Action: City of Tomorrow and Cultural Heritage, "EDICT Huddinge Site Assessment Report", June 2004 (Ed: G Tegnér, I. Andréasson, N.E. Selin)
25. Tegnér, G., 2005, "PRT Costs compared to Bus, LRT and Heavy Rail, Some Recent Findings". Paper presented at: AATS European Conference in Bologna 7-8 Nov, 2005"Advanced automated transit systems designed to out-perform the car"
26. Tegnér, G, et.al. "PRT in Sweden - From Feasibility Studies to Public Awareness"; Paper presented at 11th International Conference on Automated People Movers, Vienna, 22-25 April 2007.
27. UITP: "Ridership and Market Share". : www.uitp.org

SUSTAINABILITY, PRT and PARKING

Shannon Sanders McDonald, AIA

Architect, Shannon Sanders McDonald, AIA, 1375 Harvard Rd. NE, Atlanta, GA 30306, 404-394-2501, s1027arch@aol.com

Abstract

Energy, time-efficiency, air-quality, efficient land-use, cost-effectiveness, utilizing existing infrastructure and appropriate transportation choices for user needs are the key points in defining transportation sustainability. PRT is a leader in addressing all of these sustainable transportation choices as this system has the benefits of combining and maximizing all of these issues. This changing transit paradigm allows a networked system to occur that provides for all of these issues including what is most important to the user: less distance traveled and user time shortened.

Linking to existing parking facilities can provide a realistic approach to first integrating a PRT system into our current automobile dependant urban fabric. Today's parking facilities can allow for a transfer point of personal rapid transit pods to other transit systems and form of movement. By utilizing the parking facility in this type of linkage with PRT an end user will be able to park a car, interface with the transit system as well as easily access the rest of a building or urban fabric. The PRT's ability due to its small size, similar or smaller than a car, to be able to move vertically, change levels and direction is another key to maximizing the time and distance potential of the networked system that PRT provides. With other-mixed use spaces the parking facility can now be a sustainable multi-modal mixed-use place accommodating many forms of movement (walking, bicycle, new movement forms such as segway) engaging with the urban fabric and the people that it serves.

Introduction

The missing link in creating a sustainable system of movement is how to integrate the personal rapid transit device into existing urban fabric and the current automobile society providing greater flexibility of use for people. A first step to creating seamless sustainable movement is efficiently connecting all of the available ways to move in space, new technologies and currently proven parking strategies. These technologies and existing strategies in combination can reduce the amount of land required, shorten the time for the user encouraging appropriate movement choices, create easier access for the user, and assist in providing points of transfer that share required systems and spaces. These positive benefits along with alternative sources of power for the PRT system, due to its size, in turn can positively impact air-quality and energy use further maximizing the sustainability of the system.

Existing Facilities

Parking lots and existing parking facilities can be retrofitted to provide different types of connections to PRT. Depending on the locations and position of a parking facility on its site several approaches to integrating PRT can be accomplished. Due to the basic layouts of parking lots and garages, minimal existing parking spaces could be now utilized to connect to and for the movement of a PRT system while still allowing for parking. This would provide a cost effective way to initially link an area with parking and PRT. Using existing parking within existing urban environments allows the use of all the other infrastructure pieces such as water, sewer, electricity and schools for development to occur more cost-effectively, inherently creating more sustainable living environments.

One of the original visions of PRT in the United States was as a dual mode vehicle called the Starr Car, designed by William Alden. A dual mode vehicle is one that can travel on the typical highway allowing access into low density environments and as well become a part of a larger transit system when appropriate density exists. This far sighted vision allows PRT to become the personal system of choice for individual ownership opening up many opportunities to create more efficient travel scenarios. PRT systems could also be rented for point to point use such as currently starting to occur with car sharing and point to point car rental.

The dual mode vehicle concept is still alive and well and currently continues under study. Mr. Alden as well continues the work with his vision looking at new ways to connect with parking facilities. His new vehicle is called the DAVe (dual-mode autonomous vehicle system). Moving PRT and automobiles on top of livable space is another approach being studied in the Netherlands that could allow access where none was imagined before. **(Image 1)**

Image 1: Parkhouse Carstadt, NL Architects

A recent conference held in Ithaca, NY with a Swedish/Korean PRT collaboration has proposed a podcar which continues this fully integrated transportation vision of the future linking cities and the country side in Sweden. **(Image 2)**

The same could occur within parking garages. Mr Alden's DAVe system arrives directly to your parking spot to take you via a prescribed route to your final destination. What a better way, especially in an airport situation, to manage transportation/parking service to the user. This can now be possible combined with cell phone linkages and real-time messaging. **(Image 3)** These approaches in combination allow for shorter transfers in time and distance allowing PRT to become a preferred transit mode.

Image 2: Podcar: Swedish Institute for Sustainable Transportation, Vectus, http://www.vectus.se/eng_press.html

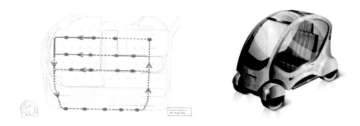

Image 3: DAVe, Alden DAVee systems. Wla07@comcast.net

Ramp Facilities

The ramp garage can provide sustainable solutions when linked with PRT. PRT has the ability to link architecturally as part of the city urban fabric in many types of location such as: underground, as part of other buildings, or as a stand-alone structure. The location of PRT and PRT stations within a parking facility is very achievable in the design process. The PRT system due to its similarity to the automobile can navigate ramps. In the design process of new parking facilities integrating transit stations has already been accomplished in many ways and forms. Detroit provides excellent examples of these types of connections with the downtown people mover system as well as within the Detroit Metro Airport. Minneapolis also provides another approach – the skywalk system - currently being used for human connections that is very applicable to PRT, parking facilities and sustainability.
(Image 4 & 5)

Image 4: Personal Photo Skywalk Minneapolis

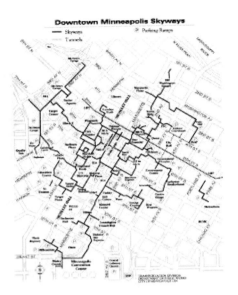

Image 5: Skywalk Map Minneapolis

In Detroit the downtown people mover connects buildings, stations and the places of the city; its only limitation is its scope. The system does not connect all parts of this spread out city the way the original trolley system did, however a system of this type allows for greater density to be developed within the core downtown area where other infrastructure exists and does and could allow for a preferred transit mode allowing for shorter time and distance connections. At the Detroit airport we see transit fully integrated within a building so the concept of moving transit through buildings is quite safe and viable. The skywalk system in downtown Minneapolis that connects parking facilities and the downtown on a second level provides another approach for building, transit and automobile linkage while providing more transportation and development options for a growing and aging society with limited additional infrastructure investment. A combination of all of these approaches would create a fully functional networked system for a new activity center or even a limited urban area.

In the 1950's and 1960's Victor Gruen, Paul Rudolph and Ulrich Franzen began to imagine parking integrated throughout the community along with small vehicles that could take people all around an area quickly and easily. **(Image 6)** It was from Victor Gruen's background in shopping center design that he understood the complexity of developing accessible interconnected. The parking strategy that he was most known for was the parking as a podium, which is parking under the entire

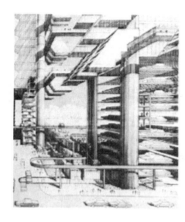

Image 6: Paul Rudolph and Ulrich Franzen, *The Evolving City* (New York: American Federation of the Arts, 1974), 74-75.

built area, linking the parking with small vehicles to better access the entire center. Parking linked with other movement devices still continues today as a future solution for success. Now more than ever as we want to limit the amount of driving we do and consolidate our shopping and work trips; envisioning a way that we can provide this level of access in a timely way is something that we could actually now build due to modern advanced technological computer advances such as fuzzy logic.

Today we can further expand upon Paul Rudolph, Ulrich Franzen and Victor Gruen's idea and use elevators, people movers, PRT and automated parking technology combined allowing transit to move within buildings making connections even more viable. This can provide more flexibility and synergy for multiple overlaps and connections allowing all of the issues to interact; creating more sustainable solutions. Human living spaces and automobile and transit options can now be developed on multiple levels as envisioned by many in the 1970's. Other environmentally sustainable solutions such as green roofs and integrated water and food production solutions can also be integrated within multi-levels. A fully networked sustainable physical environment can be created.

Automated Systems

An automated parking garage in combination with PRT allows all the sustainable possibilities to fully emerge by completely integrating architectural and planning solutions. By utilizing the three-dimensional movement technology of the automated parking facility, new connections are possible for PRT systems and their users further reducing time and distance issues. The three-dimensional capabilities of many types of movement systems of today's automated parking facilities allow for a transfer point of personal rapid transit pods similar to the early train and automobile storage areas. Turntables in combination with lift systems will allow the PRT, especially the rubber wheeled variety, to move in different vertical levels and change in any direction in a minimal amount of space. (**Image 7**) The individual pod of the

PRT system can utilize the existing mechanization of the automated parking garage to change levels and change direction all within the space of a car elevator shaft. This approach allows flexibility in the

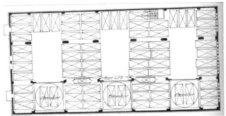

Image 7: "Hill Garage", *American Architect* (April 5, 1928):480.

layout of an entire PRT system especially within existing urban fabric where developable space is limited. By adding transit and parking were needed within existing urban environments on the typical small urban infill lot the use of all the other infrastructure pieces such as water, sewer, electricity and schools allows development to occur more cost-effectively, inherently becoming more sustainable. This combination would also allow space to be conserved by maximizing parking (the automated parking facility is the most efficient way to park automobiles) and minimizing space required for transit system layout flexibility. Therefore, land-use is optimized while solving some of the most difficult issues in integrating a PRT system, changing level and direction, so that the unique benefits of a networked system can be maximized. **(Image 8)**

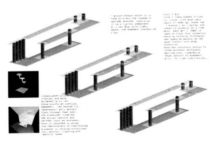

Image 8: Damien Kolash, Montana State University, 2002.

Automated technology in its earliest forms, elevators and turntables, has been around since the inception of parking garages and is a proven technology. One early garage moved vehicles vertically and horizontally mechanically: the August Perret Garage, Rue de Ponthieu, 1905 in Paris. **(Image 9)** In the 1920's many excellent examples of automated facilities existed in the United States. Some of these garages are still in existence today such as the Jewelers' Building in Chicago; although the center space once used for parking is now used for storage it could be retrofitted to park cars again. **(Image 10)** A completely glass tower with a ferris-wheel system constructed in 1932 in Chicago is the early vision of today's' smart tower. In Detroit, early 1930 a fully developed underground system with small beautifully detailed pavilions on the street above was presented to the city council as the downtown waterfront parking solution. **(Image 11)** The 1950's also saw resurgence in automated parking in the United States in cities large and small with the Bowser, Pigeon Hole and many local systems. Currently automated facilities can now be found all over the world many based upon the innovations in the United States in the 1920's and 1950's.

Trevi Park based out of Italy is one system that integrates automated parking and the foundation of a building, placing the automation under the building addressing several sustainable issues all at once. **(Image 12)** Using an automated parking facility will allow a PRT system to run easily at grade and switch to either above or below ground as appropriate, easily and efficiently. Some of these automated systems can be quite small while others can be expanded as needed and where space exists. Since the traditional gas powered automobile is not running or operating while in these garages, the facility does not need to be designed for exhaust and emissions. Therefore the standard open air façade requirements for parking

 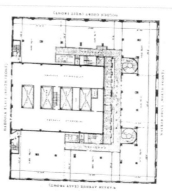

Image 9: "Notes From Paris", Architectural Review 24 (1908).
Image 10: R. Stephen Sennott, "Chicago Architects and the Automobile, 1906-1926", in Jan Jennings, ed., Roadside America: The Automobile in Design and Culture(Ames: Iowa State University for the Society of Commercial Archeology, 1990)166-168.

facilities are not required allowing for more design flexibility. Transit such as PRT can now also be fully integrated along with the automobile into the urban fabric allowing the benefits of the PRT system: shorter distance and time traveled to be fully realized. This in combination with the elimination of ramps as part of the parking garage design allows these new facilities to be minimal in size reducing the block size and to visually integrate seamlessly in the urban fabric or to be designed in small urban infill lots, sliver, unique or odd shaped plots of land where development could typically not have occurred.

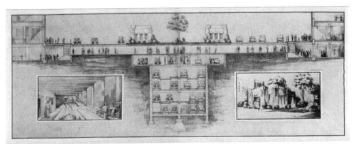

Image 11: Nolan S. Black and Winfred V. Casgrain, *A Plan for the Construction of Underground Mechanical Garages in Downtown Detroit* (Detroit:1930).

By utilizing the automated parking facility in this type of linkage with PRT, a person will be able to park their car and interface with the transit system within a minimal amount of space. Entry at different levels is possible and the driver can enter and move to different levels. All of this transfer and change – shifting of systems – can occur hidden within a building. Automated facilities are currently fully integrated into buildings in Tokyo so allowing PRT into a building is an achievable next step. As well the overlapping use of elevators and other systems for the multiple uses of man and machine will allow the construction of these connection places to be less expensive as a whole. Lobbies for patrons to wait out of the weather for either their automobile or as part of PRT stations will help promote usage and usability and can be easily designed to occur within the transportation facility. Lobbies can now serve multiple purposes and link to other uses. As PRT and the automobile due to their small size become powered by electricity, solar and fuel cell they also will be able to fully operate within typical building design allowing for better access, handicap accessibility and station location.

Image 12: Trevi Park, Courtsey of MitchCo. Inc. – exclusive licensee for Trevipark in the United States www.mitch-co.com.
Image 13: Ove Bjorn, Project Manager, Birch & Krogboe, Consulting Engineers, "Danish Automated Parking", Parking (2005):45-48.

A project in Denmark places automated parking between two housing towers under a green roof. The lobby is a high tech place for community connections as you wait for your car to arrive. **(Image 13)**

Currently, automated parking garages are being constructed again in the United States. Automated parking facilities now can be found in Boston, Hoboken NJ, New York, Washington, D.C. with some of the older automated facilities being renovated such as in Florida. The time is now to understand how they can link with PRT to develop a fully networked transit system that will allow more spatial choices. These facilities are also being imagined linking with bicycle parking and with the zip car strategy. A fully developed multi-modal station that can best provide a multitude of transportation choices for denser walkable environments by using automation. **(Image 14-18)**

Image 14: 123 Baxter Street, AutoMotion, www.automotionparking.com
Image 15: Trevi Park, Courtsey of MitchCo. Inc. – exclusive licensee for Trevipark in the United States www.mitch-co.com.

Image 16: 7 State Circle – Siena Corporation, Bohl Architects, Lasater/Sumpter Design.

 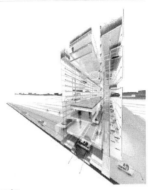

Image 17: ZipCar Dispenser, proposal Moskow Architects
Image 18: Filter Garden – Levin Betts Studio

Other Automated Parking Technologies

All parking facility and PRT linkages can connect to many other new technologies such as pass cards, fully automated computer monitoring technologies and cell phone activation. Finding ways to reduce the use the number of automobile trips as well as time spent seeking a parking space is a key factor for sustainability in transportation. **(1)** Looking at ways that the parking facility can assist with minimizing circulation is a key part of sustainable solutions. Parking management now involves several technological applications. One of these applications is the use of the cell phone for location of parking space and payment to reduce the time spent looking for a parking space. The cell phone as well as other technologies related to reducing time such as Radio Frequency Identification (RFID) and license plate recognition (LPR) could be also be used to allow for an on-demand PRT transit device to arrive just in time at the parking facility, further encouraging the use of transit by reducing travel and parking times, directing people quickly to their final destination. **(2)** Another application is the automated signage to direct parkers to the open parking space, so that the driver does not continue to circle around the parking facility. With a parking space identified and linked to real-time messaging, a PRT device could even meet the arriving user at their parking space further minimizing time connecting from point to point making transit more accessible to many. PRT as a transit vehicle can also more easily access multiple spatial levels as another solution to provide time-efficient transit within centers.

Security issues can also be integrated seamlessly as a PRT device can be "scanned" similar to a person or an automobile and this could occur within the travel route. Any vehicle that would require further scrutiny could be automatically directed to a specific area designed for this purpose.

Energy Use

As environmentally sustainable fuels sources have greater applications the automobile and the PRT system due to their size and scale they can refuel or be fueled by a generating source such as solar panels that power the automated facility or with many other potential energy sources. Environmental issues are then addressed at several levels of architecture, land use and eventually could connect to a national power grid that is "fueled" sustainability. Also, buildings and transportation vehicles can provide and interactive power source and generation between and for each other. Currently operational is a solar powered ramp garage in Northern California; however the technologies could be extended with small intermediate solar-powered electric charging stations so that the parking facility becomes the point of connection and linkages for transportation power sources as well. Energy sources such as electric, solar and fuel cell are being considered as more viable solutions, today, although their research has been on-going for over 40 + years, with the electric vehicle popular on and off since the early 1900's. **(Image 19)** On-street charging stations for electric vehicles are reappearing. They were found on our streets at the turn of the 20th century for the emerging electric car. **(3)**

Image 19: "The Edison Electric Garage, Boston, Mass.," *Horseless Age* 31, no.19 (May 7,1913): 841,842.
Image 20: ULTra Parking Connection,© ULTra Advanced Transport Systems Ltd.

Parking facilities have provided charging stations several times in the past 100 years and we will soon see this again. The electric vehicle is an appropriate solution for commuting needs. In transit applications if the vehicle size decreases these energy sources can be used for powering these devices. PRT can be powered by these more sustainable energy sources. The ULTra vehicle being implemented at Heathrow **(Image 20)** is a low power, electric vehicle, while the City of Santa Cruz is investigating the development of a solar powered PRT vehicle for their cities transportation needs.

When this advanced form of transit, PRT, is placed in combination with parking facilities further sustainable energy interrelationships and solutions can be developed. These new energy sources all have application to buildings as well as transportation energy needs. A recently constructed parking facility, the Fairfield Multi-Modal Transportation Center, Fairfield, California, 2002, designed by Gordon Chong (the firm is now known as Stantec Architecture) **(Image 21)** has solar panels on the façade to assist with energy use of the building. Synergy between building and vehicle providing power sharing between building and vehicle can create a totally linked sustainable system. Creating small strategically placed multi-modal parking facilities can link sustainable power sources for multiple efficient uses of building and machine. PRT is the perfect system to begin to connect all of the possibilities to create more sustainable power and energy sources due to its new paradigm of smaller vehicles and multiple off-line stations.

Image 21: David Wakely, photographer: Santec

Conclusion

Sustainability occurs on many levels in transportation. Energy, time-efficiency, air-quality, efficient land-use, cost-effectiveness, maximization of existing infrastructure and multiple transportation choices for user needs are the key points in defining transportation sustainability. PRT (understood as both purely transit and dual-mode in this paper) linking with parking facilities opens up unique solutions to transportation sustainability. Adapting existing parking structures to accommodate PRT and PRT stations are an initial way to begin implementing PRT therefore allowing its many benefits to be experienced as a preferred mode of transit. PRT in combination with ramp parking can also offer sustainable possibilities for PRT, PRT stations, automobile and other movement interfaces. PRT in conjunction with automated parking can create a seamless transit - parking system for a total sustainable approach to integrating a preferred transit device and its stations into the urban fabric. How these three different parking garage options can link with PRT to provide fully sustainable approaches is one of the keys to successful implementation of PRT systems in multi-modal transit connections.

Sources
(1) Shoup, Donald. *The High Cost of Free Parking*. American Planning Association, 2005.
(2) Rainey, Steve. *High-Security, Paid, Automated Smart Parking Design for a Large Office Park*, ITS World Congress, November, 2005.
(3) McDonald, Shannon. *The Parking Garage – Design and Evolution of a Modern Urban Form*. Urban Land Institute, Washington, D.C., 2007.

The Impact of PRT on Army Base Sustainability

Peter J. Muller, P.E.*

* President, PRT Consulting, Inc. 1340 Deerpath Trail, Ste 200, Franktown, CO 80116; PH 303-532-1855; pmuller@prtconsulting.com

Abstract

This study investigates the ability of a personal rapid transit (PRT) system to enhance sustainability at the Fort Carson Army Post in Colorado Springs, Colorado. Staff and stakeholder values are explored in relation to the implementation of PRT and alternative systems. Trade-offs between level of service, convenience, visual impact, etc. are weighed and the overall desirability of PRT is determined. Stations are located, a preferred alignment is developed and ridership is estimated. The system is modeled to determine guideway capacities, walk, wait and travel times as well as the number of transportation pods (T-Pods) required and its ability to carry sufficient passengers to meet the Post's goal for reducing single occupancy vehicle use. PRT benefits and costs are determined and compared. Recommendations are made regarding possible next steps to be taken toward implementing the PRT system.

Introduction

The Fort Carson community has adopted sustainability goals for the post, which include significant reductions in single occupancy vehicle (SOV) use. At the same time, the Post is experiencing significant growth, which potentially challenges its ability to meet the goals. In addition to encouraging people to share rides, the community is considering numerous alternatives to reduce SOV trips, including providing improved sidewalks and the use of low impact vehicles, such as bicycles and Segways. However, average on-post trip lengths are about 5.6 km (3.5 miles) long, and these alternatives will probably not always be appropriate (especially in times of inclement weather). Bus services on the Post have historically been poorly utilized and thus, traditional transportation options seem very limited in their ability to facilitate achieving the SOV goal.

Personal rapid transit (PRT) is a relatively new form of transit, which utilizes small, automated vehicles travelling on guideways, to transport passengers directly to their destinations, without stopping or transferring. It provides a high level of service more akin to an automobile than a bus. Due to the automation, PRT is relatively inexpensive to operate. However, the infrastructure involved requires considerable up front capital expenditure. PRT can carry significant numbers of people in all kinds of weather and, working with low impact vehicles, could potentially allow the Post to meet its SOV goal. The purpose of this study is to undertake a preliminary investigation of the feasibility of a PRT system, on the Fort Carson Army Post.

Considerable growth is planned at the Post, and this study is based on conditions as they are projected to be at build-out (approximately 2015) – also referred to, herein, as the Planning Year.

Comparison with Other Modes

This section compares PRT with other modes of travel, from the point of view of existing Fort Carson commuters.

Transportation Preferences of Fort Carson Commuters

A Transit & Parking Options Workshop, open to Fort Carson leaders, employees, residents, soldiers and community stakeholders, was held on October 1, 2008 from 0900 to 1500. There were 21 participants, of which one was a soldier living on post. Participants were exposed to descriptions of numerous transportation options. Most of these descriptions were fairly brief, since most options (such as buses) were already familiar to the participants. Where options were typically unfamiliar (such as PRT), a more in-depth description was provided. Participants were given the opportunity to ask questions and take part in discussions. During the workshop, participants responded to a number of questionnaires, the results of which are tabulated and discussed below.

Table 1. Travel Pattern Survey

Travel Pattern Survey Summary			
Survey #	Description	No. of Parking Spaces used	Note
1	PPACG Staff Member	1	Rarely visits Post
2	Civilian	3	POV to/from post then GOV vehicle
3	Civilian	0	Bus only
4	Civilian	0	visits base 1 time per month via GOV
5	Civilian	5	POV to/from post then GOV vehicle
6	Student/Intern @ UCCS	0	N/A - did not consider base visit
7	Civilian	1	POV only
8	Civilian	4	POV only
9	Civilian	5	POV & GOV use
10	Civilian	2	POV to/from post then GOV vehicle
11	Civilian	4	POV only
12	Civilian	1	POV only
13	Civilian	3	POV only
14	Civilian	5	POV to/from post then GOV vehicle
15	Civilian	3	POV only
16	Civilian	3	POV only
17	Civilian	2	POV only
18	Soldier Living On-Post	6	POV only
19	Civilian	2	POV only
20	Civilian	2	POV only
21	Civilian	2	POV only
Average Parking Stall Requirement		2.6	

POV = privately-owned vehicle. GOV = government-owned vehicle

Table 1 categorizes the workshop participants. It also shows that people who drive on the post typically use more than one parking stall. In fact, if the four people, who indicated they rarely travel to the post or use the bus, are eliminated, the results indicate that drivers on the post use more than three parking spaces on average.

Table 2. Travel Preference Survey

> **Travel Preference Survey Question:**
> Please vote on which of the following transportation characteristics are most important to you for Post-related trips. You have a total of 100 votes. You may not use more than 25 votes on any one characteristic. Ordered from most to least important.

Ordered Travel Preference Survey Analysis Results		
Reliable	13.22	Highest Priority
Flexible Departure and Arrival	10.22	
Low Cost	9.50	
Easy to Use	9.22	
Short Walking Distance	7.72	
Short Waiting Time	7.61	
Energy Efficient	6.72	
Short Travel Time	6.39	
Low Emissions	4.56	
No transfers	4.39	
Consistent Travel Time	4.17	
Safe	3.72	
Comfortable	3.44	
Visually Appealing	2.67	
Seated Travel	2.28	
ADA Compliant (disabled persons access)	1.94	
Personally Secure	1.94	
Private	0.28	Lowest Priority
Total	100.00	
Median	4.47	
Mean (Average)	5.56	
Average Deviation	2.91	
Standard Deviation	3.48	

This survey sought to discover which transportation characteristics were most important to participants. Prior to the survey, participants were given the opportunity to modify and/or add to the list of characteristics.

Since the focus of this study was on-Post travel, it was desired to calculate mode preference for on-Post trips only. Since there was insufficient time to have the participants do this, we (PRT Consulting) rated each Mode against each characteristic, shown in Table 2, on a scale of 1 (poor) to 5 (excellent). For example, the mode "bicycle" was rated 5 for low emissions and 2 for safety. Note that we assumed trip lengths of 0.8 to 16 km (1/2 to 10 miles).

Table 3. Mode Preference

Mode Preference	
Mode	Score
Personal Rapid Transit	661 (Best)
Low Impact Vehicle	588
Car	572
Bicycle	540
Walk	532
Jitney	467
Light Rail	467
Monorail	451
Commuter Rail	451
Paratransit	443
Maglev	443
Heavy Rail	435
High Speed Rail	435
Bus Rapid Transit	403
Express / Regional Bus	387
Shuttle Bus	387
Local Bus	346 (Worst)

Each rating was then multiplied by the average number of votes that characteristic received. For example, the bicycle rating of 5 for low emissions was multiplied by the 4.56 average vote for low emissions, for a score of 22.8. The scores for each mode and each characteristic were then added to arrive at the score in Table 3 (540 for bicycle).

The table has been colored to highlight the interesting result, that all small vehicle modes outscored all rail modes, which in turn outscored all bus modes (except paratransit).

In summary, although the attendance was insufficient and too homogenous to provide scientifically valid data, there was fairly clear evidence of preference for small vehicle modes and resistance to bus modes.

PRT Layout

Figure 1 shows the proposed PRT guideway layout and station locations. The layout was developed keeping the following considerations in mind:

- The layout was constrained to those areas of the Post generating the most traffic. For this reason zones 5 and 6 in Figure 1 are not served. Arrows at gates 1, 2, 4 and 20 are indicative of the desire to connect to other transit systems and/or to expand the PRT system off post at some future date.
- Stations were located in such a way that walking distances exceeding ¼ mile were avoided to the extent reasonable.
- Stations were located close to proposed low-impact vehicle share locations where feasible.
- One-way guideways were laid out connecting the stations and avoiding existing structures.
- Guideway directions were determined so as to minimize out-of-the-way travel to the extent reasonably possible. In some instances, connecting loops were inserted that do not serve stations, but serve only to facilitate reversal of direction of travel.

The layout depicted is necessarily approximate. The scope of work in this project did not allow for detailed design and the layout was only developed to the extent necessary to determine system requirements and approximate costs. Prior to implementation, much additional work will be required in order to finalize station and guideway locations and details. It will be desirable to ensure that the guideways and stations fit in well with the existing infrastructure and do not impact historic foot traffic and physical training routes. The guideway layout should be fine tuned, to eliminate sharp curves where possible, and to optimize opportunities to place it (and stations) at grade, to facilitate access and reduce costs. The addition of over-/under-passes at key intersections should be investigated to determine the impacts on capacity, number of T-Pods required and trip times. It is anticipated that this could have a significant positive impact on the circuitousness of PRT trips.

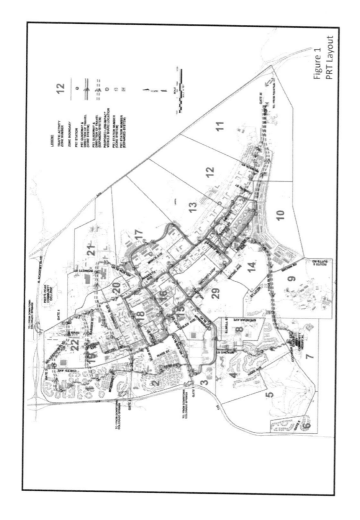

Figure 1
PRT Layout

At this time, it has been assumed that 20% of the guideway and stations will be at grade, and the remainder will be elevated.

PRT stations at entrance gates have been located inside the secure perimeter. It has been assumed that parking lots will be located outside the secure perimeter (but within the Post fencing system). PRT users will then walk through security to enter the station. Mr. Carl Backus (1) has confirmed that this arrangement will alleviate some of the screening burden and reduce the need to enlarge the gate facilities. If the PRT system is ever expanded to travel off-Post, it is envisioned that riders entering the post will exit their T-Pod to be screened and then enter a different T-Pod to continue their journey, thus, greatly reducing the requirement to inspect the T-Pods themselves. Load cells on the vehicles can sense objects left in the vehicles down to 4 kg (10 lbs.), and this capability can be backed up by on-board CCTV, coupled with left luggage detection software. In this way, allowing T-Pods to enter the post should be less risky than allowing automobiles to do so.

Where PRT stations are adjacent to low impact vehicle (LIV) share locations, it is envisioned that the LIVs will be utilized to extend the reach of the PRT system. For example, at station 23, the LIVs could be used by the disabled to get closer to the portion of the hospital they are visiting. Alternatively, they could be used to facilitate access to zones 5 and 6, which are not served by the PRT system. LIV vehicles should be of low speed and/or restricted from road travel, to avoid them being used in place of the PRT system and, thus, potentially causing traffic problems.

Stations 1 through 17, together with their associated 17.1 km (10.7 miles) of guideway, are envisioned as the core PRT system – large enough to have a significant impact, yet about half the size of the expanded system with its 35-station, 36.6 km (22.9 miles) of guideway. It is interesting to note the core system will only carry 18% of the traffic that the expanded system will carry, using the same ridership assumptions. This is because doubling the number of stations effectively quadruples the number of station pairs available and, thus, the utility of the system.

PRT Ridership

The Fort Carson Community has a sustainability goal of reducing non-mission SOV trips by 40%, by the Planning Year. No data is available for this study as to the percentage of mission vs. non-mission trips. All trips have, therefore, been assumed to be non-mission. Since the United States has a very poor record of persuading people to reduce SOV trips by offering conventional alternatives, it has been assumed that the PRT system will have to carry the bulk of the diverted trips, in order for the goal to be met. Since the LIV system is planned for an 8% mode share, the mode share of PRT must be 32%. This mode share is unlikely to be achieved without some type of PRT incentive/automobile disincentive program. Such a program is beyond the scope of this study, but has been assumed to be feasible.

Table 4. Projected PRT Peak-Hour Person Trips

Station	1	2	3	6	8	10	12	14	16	19	21	23	25	27	28	29	30	31	33	35	Totals
1	0	20	9	48	30	13	7	41	13	16	15	57	37	21	5	7	23	3	12	52	430
2	20	0	0	70	25	23	13	48	23	35	0	26	11	9	2	4	31	0	20	20	378
3	9	0	0	2	2	1	1	4	1	5	0	3	4	3	0	0	2	0	0	5	42
6	48	70	4	0	23	0	0	0	0	109	48	25	32	16	24	55	1	1	0	29	485
8	20	25	2	23	0	0	0	0	0	50	18	19	21	15	14	34	0	0	0	40	282
10	13	23	1	0	0	0	11	4	26	20	21	9	13	7	6	15	0	0	0	16	187
12	7	13	1	0	0	11	0	1	11	36	13	4	5	0	2	5	0	0	0	5	114
14	41	48	4	0	0	4	1	0	4	54	42	18	23	9	22	47	22	3	21	26	389
16	13	23	1	0	0	26	11	4	0	20	21	9	13	7	6	15	15	0	0	16	187
19	16	35	5	109	50	20	36	54	20	0	23	55	43	20	15	34	14	1	13	24	587
21	15	0	0	48	18	21	13	42	21	23	0	7	10	7	5	13	26	0	17	16	300
23	57	26	3	44	19	9	4	18	9	55	0	0	25	15	11	35	12	2	10	29	388
25	37	11	4	32	21	13	5	23	13	43	10	25	0	14	11	25	14	4	21	23	349
27	21	9	3	16	15	7	0	9	7	20	7	15	14	0	16	18	5	4	15	20	223
28	5	2	0	24	14	6	2	22	6	15	5	11	11	16	0	13	7	4	24	12	202
29	7	4	0	55	34	15	5	47	15	34	13	35	25	18	13	0	13	9	40	22	405
30	23	31	2	1	0	0	0	22	0	14	26	12	14	5	18	38	0	3	19	7	235
31	3	5	0	1	0	0	0	3	0	3	0	2	4	4	4	9	3	0	2	5	50
33	12	20	2	0	0	0	0	21	0	13	17	10	21	15	12	40	19	2	0	25	234
35	38	20	5	29	29	16	5	26	16	24	16	29	23	20	12	22	17	5	25	0	377
Totals	405	383	44	502	282	187	114	389	187	589	300	369	349	223	206	430	209	42	240	393	

Total Peak Hour PRT Person Trips	5,844

Table 4 shows projected peak-hour, station-to-station, person trips the PRT system would need to accommodate. This demand matrix was derived from the inter-zonal trip demand matrix, provided by Jacobs Consultancy, showing projected average daily person trips between each transportation analysis zone (TAZ) for the Planning Year. To account for the proportion expected to use PRT, the trips were factored by 32%. This resulted in a daily projected PRT ridership of 53,500. Multiplying by 365 provided the expected annual ridership of 19.5MM.

The daily person trips were then multiplied by 10% plus a 10% contingency factor, in order to develop the matrix of peak hour trips shown in Table 4. Ten percent of average daily trips is a commonly used factor (confirmed by Jacobs Consultancy) to determine peak hour trips. Since no data was available regarding the peak hour directional split, one was not applied, but a 10% contingency factor was added. This is to say that the peak hour traffic was assumed to be equal in both directions. This probably underestimates the flows in the vicinity of the gates. However, the flows in the interior of the network are larger and more likely to be equal in both directions.

PRT System Requirements

In order to meet the demands outlined in this report, the PRT system will need to meet the following requirements:

- 29.3 km (18.3 miles) of elevated one-way guideway (excluding station access guideways)
- 7.4 km (4.6 miles) of at-grade guideway (excluding station access guideways)
- 28 elevated stations
- 7 at-grade stations
- 800 T-Pods, each capable of transporting at least three adults and their luggage, and operating at 25mph, with headways (time between T-Pods) as low as three seconds
- Capable of accommodating a 32% mode share
    - 19.5 million annual passengers
    - 53,500 average daily passengers
    - 6,000 peak hour passengers

In order to determine if the PRT layout depicted in Figure 1 could accommodate 6,000 peak hour passengers, the system was simulated using NETSIMMOD, a proprietary PRT simulation program developed by PRT Consulting. It was found that these passengers can be accommodated at a T-Pod average occupancy rate of 2.0 and a minimum operating headway (time between T-pods) of 3 seconds. The average occupancy rate of 2.0 has been assumed, based on the expectation that some ride sharing can be encouraged and will occur – particularly during the peak hour. If riders are charged for PRT use, this can be done on a per-vehicle rather than per-person basis, thus, encouraging ride sharing. In addition (for example), in the evening peak, many trips will have a gate station as a common destination, and riders waiting for a T-Pod are likely to offer to share rides. An off-peak occupancy rate of 1.1 has been assumed, yielding an average occupancy rate of 1.37 (2.0 x 30% + 1.1 x 70%).

The following four tables show NETSIMMOD results from simulations run with differing values for T-Pod occupancy, minimum headway and total number of T-Pods. As can be seen, there are a number of different ways that a PRT system can achieve satisfactory results (average wait time under one minute, less than 10% waiting more than three minutes and in-vehicle delays under 30 seconds). The configuration shown in Table 5 is the one that has been adopted for the purposes of this report. It shows that 700 active T-Pods are required in order to provide satisfactory service levels. In order to allow for contingencies and maintenance needs it has been assumed that a total of 800 T-Pods will be needed.

Table 5. PRT Simulation with 3 second headway, 2.0 occupancy and 700 T-Pods

Time (min)	Pax Processed	Ave Wait Secs	Max Wait Secs	% Waitin g > 1 min	% Waitin g > 3 min	No. of T-Pod Trips	Pax Km/ Veh KM	Ave Delay Secs	No. of Inline Stations	No. of Offline Stations	Total Guideway Length	No. of T-Pods	Max T-Pod Occ	Ave T-Pod Occ	Min Head way
0-15	1430	7	264	10	2	503	0.00	0	0	35	41145	700	4	2	3.00
15-30	1578	21	600	26	8	1128	1.60	4	Wave-Offs			Pax KM	Veh KM		
30-45	1567	23	369	29	10	1116	1.52	4	Occ.	Unocc.		34,082	21,925		
45-60	1478	18	423	23	8	1125	1.58	4	0	0					
60-75	1561	14	612	22	2	667	1.51	6							
15-75	6184	19	612	25	7	4036	1.55	5	Max Delay =	34					

The first column of each table shows five quarter-hour (15 minute) time intervals and one, one-hour (60 minute) time interval. The first fifteen minutes (row labeled 0-15) is used for the simulation to settle down. The following four fifteen-minute intervals are summarized in the last row (labeled 15-75). The first column shows the number of people processed. The second column shows the average waiting time in seconds. The third column shows the maximum time anyone had to wait. The system analyzed here needs some optimization, since maximum wait times should not exceed five minutes (300 seconds). The next two columns show the percent of people waiting more than one and three minutes respectively. The No. of T-Pod trips is the total number, including empty vehicle movement. Pax Km/Veh Km is the ratio of passenger kilometers travelled to vehicle kilometers travelled. The average delay and maximum delay are in-vehicle delay times. These account for delays a T-Pod may have in leaving or entering a station.

Table 6. PRT Simulation with 3 second headway, 1.5 occupancy and 800 T-Pods

Time (min)	Pax Processed	Ave Wait Secs	Max Wait Secs	% Waitin g > 1 min	% Waitin g > 3 min	No. of T-Pod Trips	Pax Km/ Veh KM	Ave Delay Secs	No. of Inline Stations	No. of Offline Stations	Total Guideway Length	No. of T-Pods	Max T-Pod Occ	Ave T-Pod Occ	Min Head way
0-15	1493	15	273	17	2	1065	0.00	0	0	35	41145	800	4	1.5	3.00
15-30	1475	33	360	32	10	1108	1.28	6	Wave-Offs			Pax KM	Veh KM		
30-45	1551	37	384	39	13	1128	1.39	5	Occ.	Unocc.		34,369	25,765		
45-60	1545	50	636	39	23	1217	1.29	5	0	0					
60-75	1594	47	738	41	16	1182	1.38	6							
15-75	6165	42	738	38	15	4635	1.33	6	Max Delay =	174					

Table 6 shows that reducing the average occupancy (of occupied vehicles) requires approximately one hundred more T-Pods, to reach a similar level of service, to that indicated in Table 5.

Table 7. PRT Simulation with 2 second headway, 2.0 occupancy and 700 T-Pods

NETSIMMOD - Data Summary														
Time (min)	Pax Processed	Ave Wait Secs	Max Wait Secs	% Waitin g>1 min	% Waitin g>3 min	No. of T-Pod Trips	Pax Km/ Veh KM	Ave Delay Secs	No. of Inline Stations	No. of Offline Stations	Total Guideway Length	No. of T-Pods	Max T-Pod Occ	Ave T-Pod Occ	Min Head way
0-15	1368	9	244	16	1	1101	0.00	0	0	35	41145	700	4	2	2.00
15-30	1515	16	246	23	2	1210	1.28	3	Wave-Offs			Pax KM	Veh KM		
30-45	1607	18	626	28	4	1312	1.26	3	Occ.	Unocc.		35,012	26,143		
45-60	1562	10	738	14	0	1206	1.44	3	0	0					
60-75	1576	13	224	18	1	1319	1.38	3							
15-75	6260	14	738	21	2	5047	1.34	3	Max Delay =	72					

Table 7 indicates that reducing the headway to two seconds provides no improvement over Table 5, which had identical parameters, except for a three-second headway. This is an indication that the guideways are not overloaded.

Table 8. PRT Simulation with 2 second headway, 1.5 occupancy and 800 T-Pods

NETSIMMOD - Data Summary														
Time (min)	Pax Processed	Ave Wait Secs	Max Wait Secs	% Waitin g>1 min	% Waitin g>3 min	No. of T-Pod Trips	Pax Km/ Veh KM	Ave Delay Secs	No. of Inline Stations	No. of Offline Stations	Total Guideway Length	No. of T-Pods	Max T-Pod Occ	Ave T-Pod Occ	Min Head way
0-15	1394	15	222	18	1	1133	0.00	0	0	35	41145	800	4	1.5	2.00
15-30	1576	19	374	20	1	1253	1.23	3	Wave-Offs			Pax KM	Veh KM		
30-45	1538	20	294	26	1	1224	1.27	4	Occ.	Unocc.		35,742	27,978		
45-60	1592	34	512	34	8	1233	1.34	3	0	11					
60-75	1585	33	476	35	9	1327	1.27	3							
15-75	6291	27	512	28	5	5037	1.28	3	Max Delay =	107					

Similarly, Table 8 indicates that reducing the headway to two seconds provides no improvement over Table 6, which had identical parameters, except for a three-second headway.

Benefit/Cost Analysis

This section compares the benefits of a PRT system with the costs, in order to determine economic feasibility. PRT and new park & ride facility capital costs were summed and then reduced by road and parking expansion costs that will no longer be needed. The net capital costs were then annualized over a forty-year life at an interest rate of 6%. PRT O&M costs were increased by new park and ride O&M costs estimated at 5% of the capital costs per year. PRT costs were adjusted after communication with two PRT vendors.

Savings in road and parking lot maintenance were estimated at 5% of the capital costs per year. Surveys and calculations were made to estimate congestion and delay reduction costs (2) due to the PRT system during normal and adverse weather conditions (3). Savings in automobile costs were estimated (4). Lifecycle emissions, including emissions during petroleum extraction and refining, vehicle manufacturing and maintenance, as well as roadway construction and maintenance were estimated and the associated costs determined (5). Muller (6) found that PRT is approximately 100 times safer than other modes but a factor of ten was used in estimating annual savings in accident costs (7). Fare box revenues were estimated based on a $2.00 per trip fare (similar to Denver light rail fares). Since government employees receive a $115 per month transit allowance this should not be a hardship for soldiers and other government employees.

The quantifiable costs and benefits are summarized below in millions (MM) of dollars.

Costs:
- Annualized net capital costs — $15,42MM
- Annual net O&M costs — $26.12MM
- Total annual costs — $41.54MM

Benefits (savings):
- Savings in annual road and parking maintenance costs — $9.37MM
- Savings in travel time — $48.65MM
- Savings in automobile costs — $31.69MM
- Savings due to emission reductions — $5.78MM
- Savings in accident costs — $21.61MM
- Fare box revenues — $39.05MM

Total annual savings (benefits) — $156.15MM

The benefit/cost ratio is 3.76. This indicates a significant benefit (personal, societal and/or governmental) and implies that large changes would need to be made in the data, analysis and/or assumptions used for this proposed PRT system, not to be feasible from a benefit/cost point of view.

Conclusions and Recommendations

Based on the preliminary study undertaken, a PRT system at Fort Carson appears to be feasible and has a highly favorable benefit/cost ratio. Most remarkably, it appears that the potential fare-box revenue could not only cover the operating costs, but also the majority of the annualized capital costs. This is unheard of in conventional public transit, where fares are typically subsidized, just to cover operating costs.

To put this project in perspective, it is compared to the recently-funded Salt Lake City Mid-Jordan Light Rail Extension (8) and the Dulles Rail Project (9) and in Table 9 below.

Table 9.

	Mid-Jordan LRT Extension	Dulles Rail Project	Fort Carson PRT Project
Miles of track	11 (two-way)	23 (two-way)	23 (one-way)
Stations	9	11	35
Daily passengers	9,500	60,000	53,500
Capital cost	$428,300,000	$5,200,000,000	$529,420,000

Clearly this appears to be a very viable transit project. However, this initial study was of limited scope and did not address all issues, nor was the work undertaken of sufficient depth to provide a fully credible result. In addition, while the project is economically viable, funding and financing mechanisms need to be established before it can proceed.

While funding and financing are key hurdles to be overcome, it appears that these may not be insurmountable obstacles. The PRT project should obviate the need for some $23MM presently-planned road expansion projects, and these funds could potentially be used to seed the project leaving $15.42MM in annualized net capital costs. If the annual maintenance costs of $26.12MM are reduced by the savings in annual road and parking maintenance costs of $9.37MM, the total annual net operating and capital costs are $32.17MM ($15.42 + 26.12 − 9.37MM). To cover this cost, each of the 19.5MM annual passengers would need to pay a fare of $1.65. This fare-box revenue could be used to finance the project, but fares by themselves are typically insufficient for bonding of conventional transit projects.

It appears that this project will pay for itself in deferred capital and operating costs, for other projects no longer needed and in revenues from fares. However, mechanisms for utilizing the savings and revenues to finance the project will have to be found.

This study has shown that a PRT system could bring significant benefits to the Soldiers and people living and/or working on the Post. When monetized, these benefits far outweigh the system's costs. A PRT system would go a long way towards allowing the Post to meet its transportation-related sustainability goals. The sprawling nature of the present development on the Post is such as to not be conducive to a PRT system. The positive results of this study are, thus, somewhat surprising and indicate a potential for PRT to have beneficial transport and sustainability impacts in other military or civilian developments of similar type.

Bibliography

1. Meeting with Carl Backus, Chief, Physical security Branch and Richard Orphan, Traffic Engineer, Planning Division, 1/9/2009
2. Gannett Fleming, Fort Carson, Colorado, Comprehensive Transportation Study (2008 Update)
3. The Weather Warehouse, Colorado Springs Municipal Airport 1/1/1988 − 1/1/2008 http://weather-warehouse.com/
4. American Automobile Association, Cost of Operating a New Vehicle Rises in 2008, 5/5/2008
5. Litman, Todd, Transportation Cost and Benefit Analysis, Victoria Transport Policy Institute, January, 2009.
6. Muller, Peter J., Personal Rapid Transit Safety and Security. TRB Paper No. 07-0907, 2007.
7. Cambridge Systematics, Inc., Crashes vs. Congestion − What's the Cost to Society?, prepared for AAA 3/5/2008.
8. CE News, Salt Lake Transit Project Gets Federal Boost, February, 2009
9. The Washington Post, U.S. Transportation Chief Backs Dulles Rail Project 1/8/2009

Ride Sharing in Personal Rapid Transit Capacity Planning

John Lees-Miller[1], John Hammersley[2] and Nick Davenport[2]

[1]Department of Engineering Mathematics, University of Bristol, Queen's Building, University Walk, Clifton, Bristol BS8 1TR. Email: enjdlm@bristol.ac.uk

[2]Advanced Transport Systems Ltd, Unit B3, Ashville Park, Short Way, Thornbury, Bristol BS35 3UU. Email: johnhammersley@atsltd.co.uk; ndavenport@atsltd.co.uk

ABSTRACT

Personal Rapid Transit (PRT) systems are designed so that passengers usually travel together only by choice, but strangers may choose to share a vehicle at peak times, when the system is near capacity. By predicting whether and to what extent this *ride sharing* will occur, PRT planners can better estimate the impact on system capacity and passenger experience. This paper develops a model for ride sharing based on queueing theory and applies it to explain the relationships between vehicle occupancy, passenger queue length and passenger waiting time. The effects of multiple destinations, passengers who are unwilling to share and passengers arriving in preformed parties are considered. A case study is provided to show how the model can be applied to a simple point-to-point system; in this case study it appears possible to reduce the size of the vehicle fleet by at least 30%, while still maintaining a high level of service for passengers during peak times.

1. INTRODUCTION

A Personal Rapid Transit (PRT) system provides on-demand, non-stop transportation using compact, computer-guided vehicles running on a dedicated network of guideways. In normal operation, each vehicle carries an individual passenger or a small party traveling together by choice; each party (an individual is a party of one) travels directly from their origin to their destination, without sharing with other parties, stopping or changing vehicles. However, during peak times, the number of vehicles required to provide one vehicle per party may be prohibitively large. In this paper, we consider *ride sharing*, in which several parties may choose to share a vehicle.

Previous work indicates that ride sharing can greatly reduce the number of vehicles needed to provide an acceptable level of service during peak times. Johnson (2005) reports that peak capacity for a given fleet size is roughly doubled, using a model with a single origin station and several equally likely destinations. He also discusses the passenger experience in the origin station and describes a station management strategy that facilitates ride sharing. Johnson's ride sharing model does not explicitly represent the passenger arrival process; instead, immediately after a passenger is served, a new passenger arrives to replace him, thus maintaining a queue of constant length. While this is analytically convenient, it is difficult to justify, and it limits the utility of the model for PRT planning, because the passenger arrival process is a crucial input in the planning process. Andréasson (2005) gives a good overview of the operational issues created by ride sharing, including the implications for passenger safety and security. He reports a similar increase in capacity in a full system simulation for a single case study. However, the paper does not explore how these results may be generalized to other systems. Also, both authors assume that all passengers who have the opportunity to share will choose to do so, which is a potentially misleading assumption.

In this paper, we develop improved models for ride sharing in simple networks, discuss aspects of PRT system design and operation in the context of these models, and show how these models can be used in capacity planning. Section 2 explains a ride sharing model based on queueing theory, and section 3 shows how to use this model to explain the effect of ride sharing on system capacity; it also compares our results to those in the literature. A discussion of the effects of passengers who are unwilling to share follows in section 4, and section 5 explores the effects of larger non-separable parties on system capacity. Section 6 is a case study that shows how the models in this paper can be applied to a simple point-to-point system; it also deals briefly with the questions of how to operate stations to facilitate ride sharing and how to account for demand that changes with time.

2. A QUEUEING THEORY MODEL FOR RIDE SHARING

Consider a system with one origin and N destinations, where all passengers are traveling from the origin to one of the N destinations. When $N = 1$, this models a system of two stations or regions with dominant tidal demand from one to the other, like the point-to-point system studied in section 6. When $N > 1$, the model might describe traffic from a transit hub to several buildings, for example.

Parties arrive at the origin station bound for destination i according to a Poisson process with rate λ_i, in parties / hour. Assuming these N arrival processes are independent, the aggregate arrival process is also a Poisson process, with rate $\lambda = \lambda_1 + \ldots + \lambda_N$. Upon arriving at the origin station, passengers queue in first-in-first-out order, each waiting for a vehicle to serve them. There are s vehicles in the fleet, each of which can carry up to C parties with the same destination. Any vehicle can serve any destination, but it serves only one destination on a given trip; when a vehicle becomes available, the first party in the queue determines its destination, and up to $C - 1$ other waiting parties with the same destination can board. The vehicle then takes d hours to serve the group and return to the origin station; these service times could vary between destinations, but for simplicity we fix them all at d. Note that a vehicle cannot leave the origin when empty; it must wait for at least one party to board. The following approximations are implicit in this model; we revisit some of them later on in the sections indicated.

A1. The service time d is approximately deterministic because it is dominated by the vehicle round trip transit time, from the origin to the destination and back to the origin; the true transit time also includes stochastic terms for passenger loading and unloading, and for delays due to network congestion, but these are less important when the origin and destination are reasonably far apart.

A2. While the capacity of a vehicle is a constant number of passengers, the number of *parties* it can carry depends on the number of passengers per party, which is stochastic. For simplicity, we scale the mean passenger demand and the vehicle capacity into parties; if each vehicle seats 4 passengers, and we expect less than 1.33 passengers per party, we set $C = 3$ parties and scale the demand λ appropriately. This is only approximately correct (see section 5).

A3. The system capacity is limited by the number of vehicles available. Another limiting factor that we do not consider is the station throughput at the origin; this is mainly a function of the number of berths in the station, so we effectively assume that the origin station is large.

A4. The total party arrival rate λ is constant over time (but see section 6).

A5. Once the first party in the queue has determined a vehicle's destination, parties with the same destination can share the vehicle, regardless of their position in the queue (but see section 6).

A6. All parties who can share will choose to do so (but see sections 4 and 6).

More formally, our model is known as an $M^N/D^C/s$ queueing system, in the notation of Cromie and Chaudhry (1976) and Huang (2001), which is based on the standard Kendall notation. The M^N refers to the Markovian (Poisson) arrival process with N destinations. The D^C refers to the deterministic service times and *bulk service* rule, where each vehicle has capacity C. The s denotes the number of servers; that is, we treat each vehicle as a server.

To our knowledge, there are no useful analytical results for the performance measures of the $M^N/D^C/s$ queueing system, in the literature. Cromie and Chaudhry (1976) give useful analytical results for many performance measures of the $M^1/M^C/s$ queueing system, in which service times are Markovian, rather than deterministic. While there is some variation in the service times, which we have neglected, using a Markovian service model introduces far more variation than is desirable; this is why we have not chosen an $M^N/M^C/s$ queueing system as the basis for our analysis. Tijms (2006) gives useful approximations for the $M^1/D^1/s$ system, but ride sharing is not allowed when $C = 1$. Even these analytical results are only suitable for computer calculation; we use them to validate the statistical properties of our simulations when $N = 1$ and $C = 1$. Huang et al. (2001) derive analytical results for an $M^N/M^C/s$ queuing system, in the context of semiconductor manufacturing, but again they assume Markovian services, and they use a 'largest batch first' service discipline that is not appropriate for our application. The value of the $M^N/D^C/s$ model is as a theoretically sound starting point for further extensions. We rely on Monte Carlo simulation to obtain quantitative data on our models, but we note that these models are well-suited to computer implementation, so this is not an onerous limitation. In all of our figures, each point is the mean of ten runs of one million seconds each, unless otherwise noted.

3. SYSTEM CAPACITY WITH RIDE SHARING

We now apply our model to explore the effects of ride sharing on *system capacity*, which is the largest number of parties that the system can serve per hour. When the system is saturated, vehicles become available for service at rate $\mu = s/d$ vehicles per hour, and all vehicles operate at their full capacity, C, so the system capacity is μC parties per hour. That is, if the party arrival rate λ remains constant (assumption A4) at or above μC, the number of waiting parties grows without bound. So, for fixed fleet size s and service time d, increasing the vehicle capacity C results in a proportional increase in system capacity. Figure 1 shows this effect; when $C = 1$, no ride sharing is allowed, and the queue grows without bound as the arrival rate λ exceeds 110 parties/h. For $C = 2$, divergence is delayed until λ exceeds 220 parties/h. This increase in capacity is explained by an increase in mean vehicle occupancy, which approaches the vehicle capacity ($C = 2$), as λ exceeds 220 parties/h.

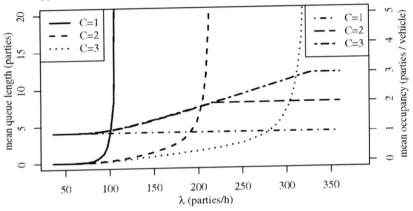

Figure 1: Mean queue length and mean occupancy for fixed service time and fleet size, with increasing party arrival rate ($d = 660$s, $s = 20$ vehicles).

This increase in mean vehicle occupancy requires an increase in mean queue length. When a vehicle becomes available, only those parties currently waiting in the queue can share with one another. If there are fewer than C parties (with the same destination) in the queue, then the vehicle makes that trip at less than full occupancy. The queue length fluctuates because of randomness in the arrival process, but high mean occupancy requires, on average, a standing queue. Moreover, as a consequence of Little's Law (Little 1961) for queueing systems, the mean party waiting time is directly proportional to the mean queue length; so, using larger vehicles increases system capacity at the cost of increased passenger waiting time. The degree to which ride sharing can increase capacity in practice thus depends on how much additional waiting time the passengers will accept; we return to this subject in section 6.

Next, we consider systems with more than one destination and compare these results with the existing results in the literature (Johnson 2005). For simplicity, we assume that the demand is split evenly among the N destinations. Then, in a queue of a given number of parties, the number of parties that are bound for any particular destination is inversely proportional to N. Only parties with the same destination can share a vehicle, so for larger N, a longer queue is needed to achieve a given increase in the mean vehicle occupancy, and hence the system capacity. This suggests that ride sharing is most effective when the number of destinations is small.

To quantify this, and for comparison with Johnson's results, we refer to Figure 2, which shows a linear relationship between mean passenger waiting time and the number of destinations. Johnson also finds a linear relationship between mean waiting time and the number of destinations, but for a different ride sharing model. In Johnson's model, the arrival process is chosen so that the queue length is held constant at $N+1$ parties, in order to make the model more tractable. The mean waiting time is then $(N+1)/(2\mu)$, in our notation; that is, the constant of proportionality is fixed at $1/(2\mu)$. Figure 2 shows that the constant of proportionality varies with the total arrival rate. In this sense, Johnson's results also hold in our model, for a limited number of arrival rates. It is also worth remarking that passenger waiting time increases considerably as the number of destinations grows; when $\lambda = 180$ parties/h and $N=1$, passengers wait 0.4 minutes on average, but when N increases to 24, as in Johnson's paper, this increases to 4.8 minutes. This indicates that ride sharing is less helpful for such a large number of destinations.

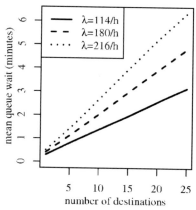 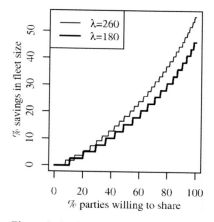

Figure 2: Mean waiting time for fixed service rate and several total arrival rates, with increasing number of destinations ($C = 3$, $d = 660$s, $s = 20$).

Figure 3: Savings in fleet size are sensitive to the percentage of parties that are willing to share ($C = 3$, $d = 660$s, $N = 1$).

4. PASSENGER WILLINGNESS TO RIDE SHARE

All ride sharing models that we are aware of (Johnson 2005; Andréasson 2005) allow parties to choose whether to ride share with other parties. These models also assume that all parties are willing to share (assumption A6), which is potentially misleading. There are many factors that can influence whether a party is willing to share; here, we restrict our analysis to waiting time, monetary incentives, and peer pressure.

Another major factor is the station design; while this is largely below the resolution of our model, we return to it briefly in section 6.

If passengers are rational, and they act to minimize their remaining waiting time, no sharing will occur. This is because the first party in the queue must consent to sharing their vehicle, something which gives them no waiting time benefit; once they have selected their destination, they can either choose to share, in which case they incur a small extra wait due to other passengers loading, or not to share, in which case they leave as soon as the vehicle arrives. Thus, although parties further back in the queue can usually reduce their remaining waiting time by sharing, the passenger at the front of the queue has no incentive (in terms of waiting time) to allow others to share his vehicle.

There are, however, two mitigating factors. Firstly, we speculate that there is considerable peer pressure to allow sharing when in a crowded station; taking a private vehicle might be frowned upon by those left waiting in the queue. This effect can only be quantified by experiment. Secondly, the operator can adjust the fare policy to offer a monetary incentive for sharing. Suggestions include charging by vehicle rather than by person (Andréasson 2005), or giving a discount to those who are willing to share (Andréasson 2005; Johnson 2005). A more thorough analysis of such fare collection policies is required, but it is beyond the scope of this paper. We also note that some systems (in airports, for example) are operated without fares; in these systems, peer pressure is the only incentive for sharing.

While further experiments and analysis are needed to properly answer these social engineering questions, our model can be modified to provide some sensitivity analysis. We consider the effect of varying a fixed probability w that a party is willing to share; so far, w has been 100%. This fixed probability is a fairly crude approximation, because it assumes that a party's decision on whether or not to share is entirely intrinsic; in reality, it may depend in a complex way on the actions of other parties around them. For example, parties may see that a vehicle is filling up and become less willing to share, further preventing high occupancies. However, this assumption provides a reasonable starting point.

Figure 3 shows the effect of w on the number of vehicles needed to ensure that 90% of parties wait less than 60s (see also section 6). For example, when $\lambda = 180$ parties/hr and $w = 100\%$, the number of vehicles required is reduced by 46% (from 39 to 21, in the particular system under study). When w drops to 80%, the required fleet size is reduced by only 30% (to 27). We note that a small change in w when w is near 100% can significantly affect the required fleet size; that is, system capacity is quite sensitive to w. The main reason is that the probability of n parties sharing is w^n, so achieving high vehicle occupancy $(1 < n < C)$ requires a disproportionately longer queue as w decreases.

5. THE EFFECT OF PARTY SIZES ON RIDE SHARING

We have so far assumed that a vehicle can always carry up to three parties (assumption A2). In reality, party sizes will vary stochastically, allowing a possible conflict between the number of passengers arriving in a new party and the number of remaining empty seats in a vehicle. In this case, the arriving party will have to decide on whether to split up or stay together. The distribution of party sizes differs considerably between applications. For example, many of the parties in a theme park will be families, and each family would require their own vehicle; ride sharing would be less effective in this case. In most applications that the authors have considered, however, the vast majority of parties will be individuals or pairs. We now explore several relaxations of assumption A2 to assess its validity.

We consider a model in which parties arrive according to a Poisson process, but, each time a party arrives, X passengers with the same destination join the queue; here, X is a random variable taking positive integer values. This is known as a compound Poisson process (Woodward 1994). The vehicle capacity in this model is defined to be S passengers, rather than C parties. The distribution of X would be based on the actual group size data for the application under study, but here we use a parameterized distribution. For simplicity, we still assume that party sizes cannot exceed vehicle capacity (no party has more than S passengers). We also note that in many applications, parties arrive by automobile, and so the party size is limited by the capacity of a typical automobile. These considerations lead us to define X by a binomial distribution with

$$\Pr(X = x) = \binom{S-1}{x-1} p^{x-1}(1-p)^{S-x}, \quad x = 1, 2, \ldots, S$$

where $p = (G - 1)/(S - 1)$ and G is the mean party size. This means that a party consists of at least one passenger, accompanied by up to $S - 1$ additional passengers; each additional passenger occurs with probability p. In general, the group size distribution could vary between destinations, but we ignore this for the sake of simplicity. The distribution of X when $G = 1.33$ is computed in Table 1.

X	1	2	3	4
$\Pr(X = x)$	0.705	0.261	0.032	0.001
$\Pr(X \leq x)$	0.705	0.966	0.999	1.000

Table 1: Distribution of group size X when mean group size $G = 1.33$.

The cumulative density indicates a 97% chance of party size one or two, and our simulations indicate that this does not have a significant impact on the number of vehicles required to provide satisfactory service; assumption A2 is a reasonable approximation when $G \leq 1.33$. This is the case for most applications. For mean party sizes up to $G = 2$, results are mainly the same, but when $G = 3$ there is a 70% chance of a party with size 2 or 3, and assumption A2 significantly overestimates the potential for ride sharing.

6. POINT-TO-POINT PRT SYSTEM CASE STUDY

We now apply our model to a simple but useful PRT system that connects two locations, where we assume that there is one station in each location and that the network layout and the peak demands are given. Our objective is to determine how many vehicles are needed in order to provide an acceptable level of service. The level of service is defined in terms of the 90th percentile of the party waiting time distribution; for example, service might be acceptable when 90% of parties arriving in the peak period wait less than 1 minute before boarding a vehicle. For each combination of peak demand and fleet size, the peak period is simulated 1000 times to build an accurate estimate of the waiting time distribution; we then choose the smallest fleet size that provides an acceptable level of service for all of the expected peak demands.

We have so far assumed that the passenger demand λ is constant (A4), but this is not usually true in peak periods. The party arrival rate will usually rise to a maximum and then fall off. It is straightforward to extend our model to capture this. We use the representative demand profiles shown in Figure 4, which were generated from Gaussian curves with "standard deviations" of 15 minutes for the AM peak and 30 minutes for the PM peak. The AM peak is 2 hours long, and is sharper and higher than the PM peak, which is 3 hours long. The simulator records waiting times for all passengers arriving in peak hours, and it terminates upon serving the last passenger that arrived during the peak. Waiting times for passengers who arrive after the peak are discarded; waiting times from the first two hours are also discarded, to reduce the importance of the simulator's initial conditions (all vehicles begin at the origin, ready to serve passengers).

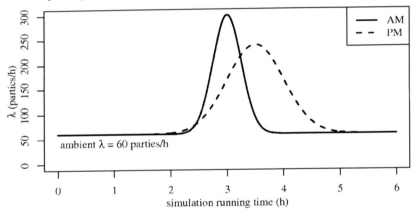

Figure 4: Demand used for AM peak and PM peak simulations.

We have also assumed that parties anywhere in the queue can share with one another (assumption A5). Whether this can be achieved in practice depends on how the stations operate; our assumptions about this are as follows. Each station contains a fixed number of *berths*, at which parties can load into or unload from vehicles. Each berth has a destination selection panel, with which a party tells the system where they are traveling to. This layout is typical of stations in the ULTra PRT system, developed by Advanced Transport Systems Ltd.; it differs from the station layout in (Johnson 2005), which separates destination selection from berths, but the following discussion suggests that our layout can also facilitate ride sharing.

At low intensity, there will usually be some empty vehicles parked in the berths, waiting for passengers to arrive. Ride sharing is unlikely at low intensity, because parties will arrive, choose a berth, select their destination and then depart immediately on a waiting empty vehicle. However, at high demand there will usually be a queue of parties waiting for vehicles (section 2). We assume that the party at the head of the queue will go to a free berth, select their destination and wait there, while the other parties wait in first-in-first-out order. When a party selects their destination, they are asked whether they want to share (section 4); if they choose to share, their destination is displayed on a screen above their berth. Other parties with the same destination can then "jump the queue" to share a vehicle with that party.

It is unlikely that the station process outlined above will be perfectly efficient (assumption A5). The apparent complexity of the human factors involved suggests that more work, including experimental work, is required in this area. For now, we examine what happens when parties can only communicate with their immediate neighbors in the queue; this assumption is intended to provide a lower bound on the likely level of interaction between parties in a station. When the "neighbors only?" column in Table 2 is "Y," the parties can only share with their neighbors; otherwise, assumption A5 is in effect.

To fix the remaining parameters, we set the vehicle capacity at $C = 3$ parties (see assumption A2 and section 5) and the vehicle round trip time at $d = 660s$ (ten minutes travel plus one minute for passenger loading and unloading; see assumption A1). Table 2 shows the predicted fleet sizes for several ride sharing scenarios. The "% willing to share" column corresponds to the probability of a party sharing, as defined in section 4. We consider two possible definitions of acceptable service, one where 90% of parties wait less than 1 minute, and another where 90% of parties wait less than 3 minutes.

peak profile	% willing to share	neighbors only?	vehicles needed for "90% wait < ..."	
			60s	180s
AM	0		53	47
AM	60	Y	44	36
AM	60	N	40	32
AM	80	Y	37	28
AM	80	N	34	25
AM	100		26	18
PM	0		45	41
PM	60	Y	37	31
PM	60	N	34	27
PM	80	Y	31	24
PM	80	N	29	21
PM	100		22	15

Table 2: Fleet sizes for case study system under varying ride sharing assumptions.

First, the AM peak consistently requires more vehicles than the PM peak, so the AM peak determines the fleet size. At the "90% wait < 60s" service level, the system requires 53 vehicles if no ride sharing is allowed, but only 26 vehicles under the most optimistic ride sharing assumptions; this is a 51% reduction, which is in line with other results in the literature (section 1). If a lower service level is acceptable, the savings can be greater; at the "90% wait < 180s" service level, and under the most optimistic ride sharing assumptions, the fleet size is reduced by 62%, from 47 vehicles to 18 vehicles. This is because longer waiting times imply longer queue lengths, which in turn allow increase vehicle occupancy, as discussed in section 3.

When not all parties are willing to share, or the communication between parties in the station is more limited, the savings due to ride sharing are reduced, but still significant. Assuming that 80% of parties are willing to share, and that parties are limited to sharing with their neighbors, the fleet size required to provide the higher service level is reduced by 30%, from 53 vehicles to 37 vehicles. The fleet size required to meet the lower service level is reduced by 40%, from 47 vehicles to 28 vehicles. This is still a substantial reduction, but, as noted in section 4, these results are quite sensitive to the fraction of parties that are willing to ride share; when this drops to 60%, the corresponding reductions in fleet size are 17% and 23%.

The number of extra vehicles required because of the "neighbors only" restriction is fairly small (on the order of 10%) in the system under study, because there is only one destination and most parties are willing to share. Its effect is larger when there are more destinations; if there are two equally likely destinations, a party with a given number of neighbors is only half as likely to find a suitable party to share with. Our model indicates that for a similar system with two destinations and the AM peak demand split evenly between them, 45 vehicles are required to provide "90% wait < 60s" when 80% of parties are willing to share; this is a 15% reduction from the number required when there is no ride sharing at all. When there are multiple destinations, the station signage and layout become much more important.

7. CONCLUSIONS

The aim of this paper was to establish a suitable model to analyze the effects ride sharing has on PRT system performance, and examine how station design and passenger behavior factors should be taken into account. To this end, we developed a model for ride sharing based on queueing theory, and although the model requires a number of assumptions (see section 2), we believe it is a sound basis for analysis, and it provides an alternative to anything found already in the literature. This model was then used to explain the relationship between occupancy and queue length in the presence of ride sharing, and to demonstrate the effect increasing the number of destinations has on these relationships, comparing our results with those in the current literature.

A crucial issue seemingly ignored in previous studies is the *willingness* of passengers to rideshare; in both Johnson (2005) and Andréasson (2005) it is assumed that all parties are perfectly willing to share. As discussed in section 4, if all parties behave rationally and seek to minimize their waiting time, no ridesharing will occur as it is the decision of the party at the head of the queue whether to share or not, and they get no benefit from doing so. Whilst incentives such as peer pressure and monetary savings may increase the likelihood of ride sharing occurring, as the effect of unwillingness to share on the beneficial effects of ridesharing is quite pronounced (see figure 3), one must take this issue into account in any analysis.

A factor which appears to have a much smaller effect is the arrival party size; although larger, non-separable parties reduce the mean vehicle occupancy, this reduction is only significant when the mean party size approaches three. Thus under our assumption of less than 1.33 passengers per party (assumption A2), this effect is negligible.

In the case study of section 6, our models were applied to a point-to-point system to determine the required fleet size to provide an acceptable level of service. In order to more realistically approximate peak period behavior, we dropped the assumption of a constant demand (A4) and instead used the two profiles shown in figure 4, representing AM and PM peaks. The simulation results presented in table 2 reveal that it is the sharper and higher AM peak which determines the fleet size, and under most optimistic ride sharing assumptions, we find a 51% reduction in the number of vehicles required at the "90% wait < 60s" service level, consistent with the findings in other literature (section 1).

What our results also show, however, is the reduction in savings one obtains if some passengers are unwilling to share, or the station isn't properly designed to promote ridesharing. At the same service level, but only assuming 80% of parties are willing to share, and that parties are limited to sharing with their neighbors, the reduction in fleet size drops to 30% (from 51%), and if the willingness is further reduced to 60%, the saving on vehicles is only 17%. Generating an environment which encourages passengers to rideshare at busy times is thus very important for it to be effective in allowing for smaller fleet sizes.

Facilitating the passenger's ability to rideshare also plays a crucial role, as the final analysis of section 6 demonstrated; for a station with two equally likely destinations, a willingness to share of 80%, and neighbors only interactions, the fleet size was only reduced by 15% (rather than 30% in the single destination case)

Thus the optimistic projections of a 50% reduction in fleet size requirements due to ride sharing need to be tempered by the observations that such a figure makes potentially unrealistic assumptions about passenger behavior and station design. In order to achieve a benefit anywhere close to this figure when there are multiple destinations, station design (signage and layout) needs to be carefully considered so as to both facilitate and provide sufficient incentives for ride sharing in PRT.

ACKNOWLEDGEMENTS
For useful discussions and feedback, the authors wish to thank Prof. Martin Lowson and Phil Smith of Advanced Transport Systems Ltd, and Dr. R. E. Wilson of Bristol University Department of Engineering Mathematics.

REFERENCES

Andréasson, I. (2005). "Ride-sharing on prt." *Automated People Movers*.

Cromie, M. V. and Chaudhry, M. L. (1976). "Analytically explicit results for the queueing system M/Mx/C with charts and tables for certain measures of efficiency." *Operational Research Quarterly (1970-1977)*, 27(3), 733-745.

Huang, M.-G., Chang, P.-L., and Chou, Y.-C. (2001). "Analytic approximations for multiserver batch-service workstations with multiple process recipes in semiconductor wafer fabrication." *Semiconductor Manufacturing, IEEE Transactions on*, 14(4), 395-405.

Johnson, R. E. (2005). "Doubling personal rapid transit capacity with ridesharing." *Transit: Intermodal Transfer Facilities, Rail, Commuter Rail, Light Rail, and Major Actvity Center Circulation Systems*, 1930, 107-112.

Little, J. D. C. (1961). "A proof for the queuing formula: $L = \lambda W$." *Operations Research*, 9(3), 383-387.

Tijms, H. (2006). "New and old results for the M/D/c queue." *AEU – International Journal of Electronics and Communications*, 60(2), 125-130.

Woodward, M. E. (1994). "Modelling with discrete-time queues." *IEEE Computer Society Press, Loughborough*.

WIRELESS COMMUNICATION BASED COMPUTER SIMULATOR TO ASSESS THE OPERATIONAL SCENARIOS FOR THE PRT SYSTEMS

Jun-Ho Lee* and Yong-Kyu Kim**

* Senior Researcher, Korea Railroad Research Institute, 360-1, Woram-Dong, Gyeonggi-Do, 437-757, Korea ; PH +82-31-460-5040; jhlee77@krri.re.kr
** Principal Researcher, Korea Railroad Research Institute, 360-1, Woram-Dong, Gyeonggi-Do, 437-757, Korea ; PH +82-31-460-5434; ygkim1@krri.re.kr

Abstract

In this paper we deal with a computer simulator that can be used to assess the operational scenarios for the PRT (Personal Rapid Transit) systems. The computer simulator is consists of central control module, virtual vehicle module, and graphical user interface that are implemented by the commercial embedded processor boards operated in the real time operating system. Communication networks between processor boards to transmit the control command and the vehicle status information are realized by using wireless communication network. The virtual control scenarios for the vehicles that is pre- designed are coded into the control processor board for the central control module and the on-board virtual vehicle control module. The experimental results performed in the proposed simulator present the effectiveness of the proposed evaluation simulator.

Introduction

The control algorithm for PRT is different from the conventional train control method such as ATS(Automatic Train Supervision), ATC(Automatic Train Control), ATP(Automatic Train Protection), ATO(Automatic Train Operation). The conventional control systems for trains are based on the track circuit to detect the train position, but PRT control scheme should realize a safe vehicle control in conditions that the guideways are interconnected in a network configuration without track circuit to detect the vehicle position and that vehicles are operated in non-stop from origin to destination. These require a novel control strategy for PRT system [1]-[5].
In order to construct the vehicle operational control scheme it is necessary for the PRT system to employ a specific communication system that makes it possible to transmit control command and vehicle status information between wayside facilities and the on-board vehicle computer because PRT system does not use the conventional rail system. The wireless communication method which is very popular technology during the last ten years may be one of the methods that can be employed for the control of the PRT vehicles, if the reliability of the wireless communication is guaranteed.

Since the fundamental concept of the PRT system is to make it possible for the vehicle to go to its final destination without stopping and with very short headways, the vehicle control scheme plays a very important role in avoiding collisions between vehicles [6]-[8]. In order to control the vehicles effectively, several elements are necessary. Among them the most important elements are: the status information of the vehicles in front and in rear, vehicle dynamics, and the speed profiles or brake curves to control the vehicle speed.

The speed profile is produced by the central control computer or by the vehicle on-board computer based on the state information of the vehicles in front and in rear. In order to develop the vehicle control algorithm that determines the system performance, it is necessary to use an effective simulator and an evaluation tool to test the designed controller [9][10].

In this paper we focus on the design of the operational control scheme providing the avoidance of the impact between the vehicles when they are operated in some operational speed, which employs VME Bus type PowerPC process module and monitoring devices.

Equations for Brake Curves

When vehicle speed is controlled by a completely automated system such as PRT system the speed control equipment for vehicles is one of the most important parts in the overall PRT control systems. In order to achieve the collision avoidance performance each vehicle should follow its speed pattern produced by the central control system or by the vehicle on-board computer system.

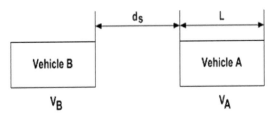

Figure. 1. Relative speed between two vehicles

Figuire 1 considers the relative speed properties between two vehicles. As seen in the Figure 1 if the vehicle A (the vehicle in front) reduces the vehicle speed, the vehicle B (the vehicle in rear) should also reduce the speed to keep the safety distance d_s. In this case the initial speed of the vehicle B should be reduced to the final speed of the vehicle B. It is possible to employ Eq. (1) to produce the speed pattern to reach the final speed of the vehicle B with a deceleration to maintain the safe distance.

$$v_{Bf} = \sqrt{2a(D_B - d_{Bp}) + v_{Bi}^2} \tag{1}$$

Equation (1) means that if the initial speed of the vehicle B, v_{Bi}, the instantaneous vehicle position d_{Bp}, the block distance or the brick wall safety distance D_B, and the acceleration or deceleration a, are known, it is possible to calculate the final speed of the vehicle B, v_{Bf}. Generally the vehicle speed is a function time, however Eq. (1) indicates the speed versus distance which represents the vehicle speed pattern or the vehicle brake curve.

Eq. (1) does not consider the brake reaction time of the vehicle B, which means the delay time to activate the brake system of the vehicle B from the moment that the vehicle A has activated its brake system. By inclusion of the delay time for the brake reaction t_{Br} the Eq. (1) is modified as

$$v_{Bf} = \sqrt{2a(D_B - d_{Bp} - v_B t_{Br}) + v_{Bi}^2} \qquad (2)$$

In order to accurate analysis of the speed patterns it is necessary to include the mobile characteristics of the PRT vehicle in the eq. (2) such as the dynamic properties of the vehicle or the friction force between the guideway surface and the traction component (rubber tire or steel wheel etc.). However in this paper we do not consider the dynamic properties of the vehicle or the friction force because the scope of the paper is to design the fundamental simulation apparatus to evaluate designed operational control algorithm rather than to design a physical vehicle. The fundamental idea for the simulation apparatus to evaluate the designed operational control algorithm is not changed even if eq. (2) is modified to include the dynamic characteristics of the vehicle.

Virtual Operational Scenarios

Table 1. Speed transitions

Speed transition	Distance /step	Total distance	Initial speed	Final speed
1	100 m		0 kmh	40kmh
2	150 m	260 m	40 kmh	40 kmh
3	140 m	400 m	40 kmh	30 kmh
4	260 m	760 m	30 kmh	30 kmh
5	240 m	1000 m	30 kmh	60 kmh
6	500 m	1500 m	60 kmh	60 kmh
7	100 m	1600 m	60 kmh	40 kmh
8	160 m	1760 m	40 kmh	40 kmh
9	140 m	1900 m	40 kmh	30 kmh
10	360 m	2260 m	30 kmh	30 kmh
11	240 m	2500 m	30 kmh	60 kmh
12	200 m	2700 m	60 kmh	60 kmh
13	200 m	2900 m	60 kmh	30 kmh
14	100 m	3000 m	30 kmh	0 kmh

In order to verify whether the proposed configuration of the apparatus is proper or not, it is necessary to design a control algorithm to be tested in the proposed system. The control algorithm for the test is divided into two parts. One is for normal mode shown in Table 1. and the other is for an emergency mode. In normal mode fourteen virtual speed transitions are set for the 3 km guideway. The final speed limits in each step are set arbitrarily. For an emergency mode shown in Figure 2 both vehicles assume that there is no activation of the emergency brake for either vehicle running on the guideway at a constant speed. However, once the vehicle in front activates the emergency brake, the vehicle in rear should activate its emergency brake as soon as it recognizes the activation of the emergency brake of the vehicle in front. Then the vehicle in rear should stop while maintaining the safe distance.

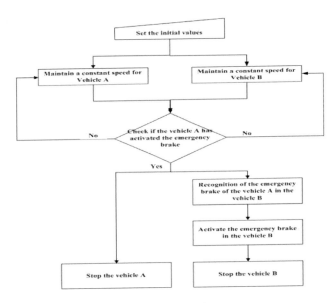

Figure 2. Task flow for emergency mode

Apparatus Configuration

In this section a simulation apparatus that makes it possible to test and evaluate the designed virtual operational algorithm is presented. The configuration of the simulation apparatus is composed of the central control module, the virtual vehicle module, and graphical user interface. The central control module collects the information from the virtual vehicle module that includes the vehicle operational status and speed for the four different virtual vehicles. It sends the parameter information to each vehicle for the calculation of the speed pattern in the virtual vehicle module. We employ a MPC7410 microprocessor-based VME bus processor module of Motorola Inc.. The wireless communication modules shown in Figure 3 are used to transfer the vehicle status and the control information between the central control module and the virtual vehicle module Figure 4 and Figure 5 show the conceptual configurations and the real hardware configurations of the simulation apparatus. As seen in Fig. 4 the microprocessor (MVME 5100) is provided by the VMEbus rack. A laptop computer that shows graphical user interface(GUI) is connected to the MVME 5100 microprocessor by way of Ethernet Lan hub. The four embedded wireless communication modules which are installed in the virtual vehicles communicate with the MVNE 5100 microprocessor (central control module) by way of the wireless communication. Figure 5 is the hardware configuration of the apparatus.

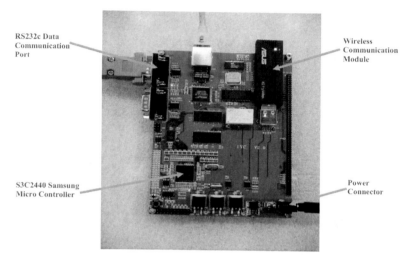

Figure 3 Embedded wireless communication modules including on-board vehicle micro computer

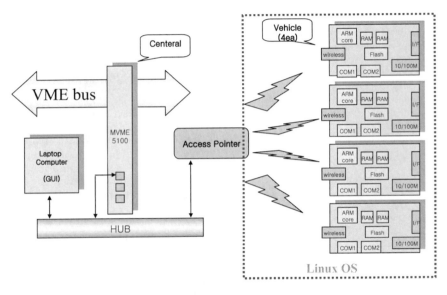

Figure 4 Conceptual configurations

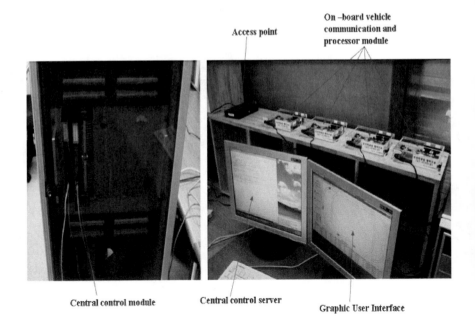

Figure 5 Hardware configuration

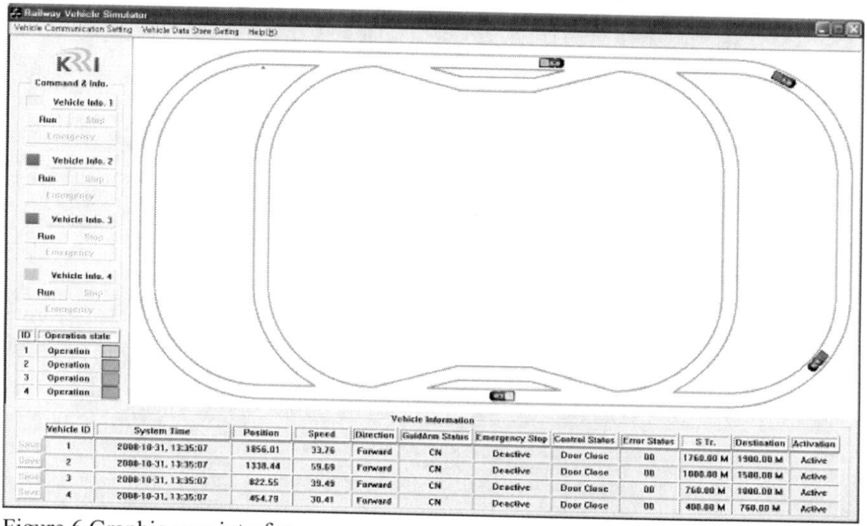

Figure 6 Graphic user interfac

Figure 6 represents the graphic user interface. This figures shows the four vehicles that are operated on the guideway based on the normal mode operational scenario. The vehicle status and the control information are transferred between the central control module, virtual vehicle module and GUI. In the lower side of the figure there are information boxes indicating the vehicle status and the control information for each vehicle. The information for the vehicle operational status is shown in the left-hand side of the figure.

Simulations

The simulation results of the MPC7410 microprocessor for the normal mode and for the emergency mode are shown in Figure 7 - Figure 9. In Figure 7, fourteen speed transitions are presented which are predetermined as the test operational scenarios for the normal mode (see Table 1.). In the figure the vehicle in front (dashed line) departed 200m earlier than that of the vehicle in rear (solid line). Each vehicle tracks the predetermined speed transitions very well, which means that the proposed simulation apparatus can be used as an effective evaluation tool of the vehicle operational algorithm. Figure 8 and Figure 9 show the simulation results for avoiding the impact between vehicles when the vehicle in front activates the emergency brake. In both figures the vehicle in front (dashed line) activates the emergency brake 1500m from the origin (dashed vertical line) and will be stopped. On the contrary the vehicles in rear (solid line) recognize the activation of the emergency brake of the vehicle in front with some delay but no matter where they recognize the activation of the emergency brake of the vehicle in front they follow the speed patterns to be stopped while maintaining the safe distance.

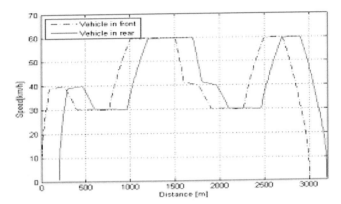

Fig. 7 Calculation results for the normal mode

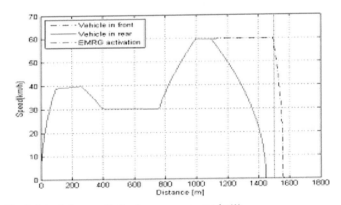

Fig. 8 Calculation results for the emergency mode (1)

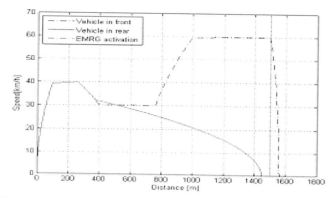

Fig. 9 Calculation results for the emergency mode (2)

Conclusions

First, in this paper we have introduced a test algorithm to control a vehicle on a guideway of 3 km in length. The test algorithm is composed of the normal mode that has fourteen speed transitions and the emergency mode to test the impact avoidance algorithm between vehicles. Speed patterns for the speed transitions were provided by the virtual vehicle module that receives the vehicle control information from the central control module.

Second, we have shown a hardware configuration for the assessment of the designed operational control algorithm. The processor that has been employed by the central control module and the virtual vehicle module is a commercial off-the-shelf processor. This has the advantage that the processor used for testing can be the same processor that is applied to the real system to control the real vehicle, with minor changes for the implementation of the control algorithm.

Finally, the operational control algorithms for PRT that have been reported up to now were focused on the computer simulation of vehicles, of system operations, and of line management in the overall control hierarchy point of view. However this paper proposes an apparatus which makes it possible to directly evaluate the characteristics of the vehicle operations on the guideway using real hardware. Further, this real hardware can use the same processor and operational control algorithms being designed for a real system. In this sense the apparatus proposed in this paper can reduce the time for the development, implementation and evaluation of the operational control algorithm for PRT.

References

[1]. J. E. Anderson, *Transit Systems Theory*, Lexington Books: 1978
[2]. Jack H. Irving, *Fundamentals of Personal Rapid Transit*, Lexington Books: 1978
[3]. Markus Theodor Szillat, "A Low-level PRT Microsimulation", *Ph. D. dissertation, University of Bristol, April 2001.*

[4]. Duncan Mackinnon, "High Capacity Personal Rapid Transit System Developments", *IEEE Transactions on Vehicular Technology, Vol. VT-24, No. 1, pp. 8-14, 1975*
[5]. J.E. Anderson, "Control of Personal Rapid Transit", *Telektronikk 1, 2003*
[6]. Ollie Mikosza, Wayne D. Cottrell, "MISTER and other New-Generation Personal Rapid Transit Technology", *Transportation Research Board, 2007*
[7]. Wayne D. Cottrell, Ollie Mikosza, "New-Generation Personal Rapid Transit Technology: Overview and Comparison", *Transportation Research Board, 2008*
[8]. Carnegie, J.A. and P.S.Hoffman, "Viability of Personal Rapid Transit in New Jersey", *New Jersey Dept. of Transportation, Division of Research and New Jersey Transit, Feb. 2007.*
[9]. Jun-Ho Lee, Ducko Shin, Yong-Kyu Kim, "A Study on the Headway of the Personal Rapid Transit System", *Journal of the Korean Society for the Railway, Vol. 8, No. 6, pp. 586-591, 2005.*
[10]. Jun-Ho Lee, Kyung-Ho Shin, Jea-Ho Lee, Yong-Kyu Kim, "A Study on the Construction of a Control System for the Evaluation of the Speed Tracking Performance of the Personal Rapid Transit System", *Journal of the Korean Society for the Railway, Vol. 9, No. 4, pp. 449-454, 2006.*

EXTENDING PRT CAPABILITIES

Prof. Ingmar J. Andreasson*

* Director, KTH Centre for Traffic Research and LogistikCentrum AB.

Teknikringen 72, SE-100 44 Stockholm Sweden, Ph +46 705 877724;
ingmar@logistikcentrum.se

Abstract

Personal Rapid Transit (PRT) offers direct, on-demand travel in automated vehicles seating 3-6 passengers on exclusive right-of-way. Commercially available systems now offer speeds up to 45 kph at headways from 3 seconds.

With 3-second headways, a typical load of 1.5 passengers and 30 % empty vehicles, the link capacity will be 1200 passengers per hour (one direction).

This paper explores ways to extend the capabilities of PRT with respect to capacity and speed. Strategies have been developed and verified with the generic simulation software PRTsim.

Conventional PRT may not provide the required capacity

One sceptic (Vuchic 2007) claims that Personal Rapid Transit (PRT) "in suburban areas … is economically infeasible, and on major arterials … cannot provide the required capacity". The present paper addresses the second part of his claim by extending the traditional PRT concept.

Safety approved and commercially available PRT systems now offer speeds of 40-45 kph at headways of 3-4 seconds. With 3 seconds headway, a typical load of 1.5 passengers and 30 % empty vehicles, the link capacity would be 1200 passengers per hour in one direction. This is sometimes insufficient.

We have explored ways to extend PRT capacity with respect to both capacity and speed. Operational strategies have been developed and verified with our generic simulation software PRTsim.

Network vs corridor systems

As opposed to line-haul transit, PRT offers network-wide transport without stops or transfers. Vehicles take the fastest route to each passenger's destination. If one link should be overloaded then vehicles will avoid that link when there is an alternative route. The concept of capacity applied to PRT should refer to the total capacity of alternative routes for each relation. Therefore corridor or link capacity is not the dimensioning factor for PRT. We shall still discuss ways to increase link capacity of PRT while remembering that the network capacity will be the sum of one or more link capacities.

Ride-sharing

The most efficient way to improve passenger capacity is to increase the load of each vehicle. In contrast to scheduled services PRT vehicles operate only when there is a demand. Since operation is on demand, PRT vehicles are made small, seating 3-6 passengers. In pure on-demand service the average load will be 1.1-1,5 passengers.

Passenger trials performed by BAA indicated that passengers with the same destination spontaneously share vehicles without being told to do so. Ridesharing can be encouraged if destinations are displayed over standing vehicles.

Fig 1. Destinations signs over vehicles in stations encourage ridesharing.

Simulations have shown that ridesharing is efficient and worthwhile only in stations where many trips start and where passengers show up in bunches as is the case in transfer stations from scheduled services, especially from trains or subways during peak hours. In such situations it is possible to fill vehicles with passengers sharing the same destinations without anyone waiting. Instead, average waiting can be reduced by shorter queuing for vehicles.

The Fornebu area in Oslo is a newly developed area served from a nearby commuter train station. In our study Andréasson (2005) we demonstrated that the application of ridesharing would increase the average vehicle load from 1.5 to 3.1 passengers with the same short waiting (0.9 min). As a consequence the required fleet was reduced from 610 to 285 vehicles. On the critical link leading out from the train station there were no empty vehicles and most vehicles could be filled to capacity. The resulting link flow was 4500 passengers per hour with 4-passenger vehicles and 3-second headways.

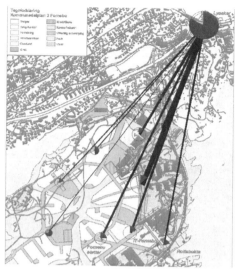

Fig. 2. PRT demand pattern during the morning peak in Fornebu.

More elaborate ridesharing strategies with more than one destination for each vehicle did not bring any further improvement.

Vehicle capacity

Without ridesharing it makes little sense to build larger PRT vehicles than 3 or 4 passengers. With the introduction of ridesharing at peak load from transfer stations, there is prospect of higher passenger loads. A vehicle seating 4 in comfort may be modified to seat 6 children or even 6 adults in less comfort when needed.

Train transfer stations

The normal PRT station is designed for randomly arriving passengers. The station can be made small since most passengers depart immediately so that only few people are waiting and then only for a short time. Small stations can be placed in city streets.

Transfers from scheduled services with large units such as commuter trains create a challenge for systems based on small vehicles. People do not prefer to travel in large bunches but they are forced to do so by transit planners in order that money can be saved on driver wages. Now that the large train units are here, we need to find ways to cope with surges of sudden demand for transfers from trains to PRT.

Train stations have passenger platforms as long as the longest train. A PRT station at a train station can take advantage of long platforms to find space for a long PRT station. A PRT station platform as long as a typical train platform may accommodate around 65 PRT vehicles holding up to 250 passengers.

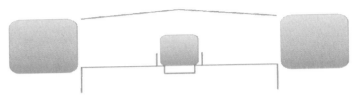

Fig. 3. Cross platform transfers between train and PRT.

We assume that train tickets are valid on PRT as well. Then no handling of PRT tickets is needed at the train platform. The system will collect statistics on demand to each destination and show destination signs over each PRT vehicle to encourage ridesharing. If someone enters the wrong vehicle then some passengers may have to suffer an extra stop on the way to their destination.

The station can be filled up with PRT vehicles (empty or with passengers taking the train) when a train is anticipated. Clearing all vehicles from the PRT station when the train has departed may take 3-4 minutes. If more than say 250 passengers transfer from the train then another platoon of waiting PRT vehicles is moved to the platform.

Platooning of empties

Moving vehicles are separated by a minimum safe distance, depending on speed. This is a requirement to safeguard passenger safety. The spacing between two empty vehicles can be shorter without jeopardising passenger safety. It may be up to the operator to balance risks of hardware against increased system capacity.

If empty vehicles can be spaced closer together, it makes sense to try to group several empty vehicles together in platoons. Platoons of empty vehicles can be created from stations and/or by choice of routes, by appropriate priorities in merges and by allowing empty vehicles to catch up on each other.

Typically about 30 % of all running vehicles are empty. If they can be grouped together and closely spaced then link capacity can be increased by 15-25 %.

High-speed links

Initial PRT networks will probably be planned for local circulation within limited distances. If PRT networks are expanded to long distances then 40-45 kph will be too slow to be accepted. So how about high-speed PRT?

Two undesirable consequences come together with high speed. One is air resistance, which grows with speed squared. The other - worse - is reduced capacity following from increased safe time headway between vehicles, proportional to speed.

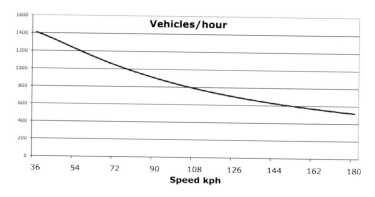

Fig. 4. Line capacity is reduced at high speeds due to increased safe headway.

Train formation

The natural way to increase capacity and at the same time reduce air resistance is to couple vehicles to form trains. Safe distance is only required between trains.

The French Aramis system in the 1970:ies was designed to form trains dynamically en route. Although this can be done, it is not clear if it will be considered safe by the regulating authorities. Before a connecting vehicle has reached the one in front, it has to pass an unsafe area (closer than the safe distance). In any case it is safe to form trains during standstill in stations.

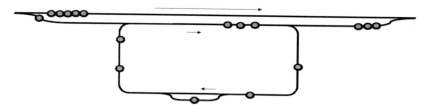

Fig. 5. Formation of PRT trains before entering a high-speed link.

A local PRT network can be connected to a high-speed link via a transfer terminal. Passengers remain in their vehicle while vehicles get connected into trains, before entering the passing track.

Splitting up trains

One particular feature of PRT is vehicle switching on passive guide-ways. Vehicle switching makes it possible to switch out a vehicle from a train at the passage of a track diverge. In this way trains can be divided as necessary en route, at least as long as one or more vehicles brake out from the front or from the rear of the train.

Breaking out from the middle of a train would create a gap, which is smaller than the safe distance. It would be desirable to arrange vehicles in the train so that separations can be made from the front and rear of the train. However rearranging vehicles is a difficult operation requiring both space and time.

Until break-outs from the middle of a train has been safety approved, the length of trains is limited by the order in which vehicles are connected and the destinations of individual vehicles.

We have so far limited the implementation of train formation to pairs of vehicles.

Pair-coupling

Trains of two vehicles can always be separated, by switching apart at speed. Pairs are easy to form in stations as long as destinations do not matter. Running vehicles in pairs will almost double the line capacity as long as the pairs are kept together, as they typically may be during about half of the trip. The effect is almost a factor of two near departure stations. In some networks, such as connecting a suburb to a city centre there may be a long stretch where the pairs can be kept together increasing the line capacity on that stretch. This is often the part where higher speeds are desirable and where the capacity would otherwise be a bottleneck.

We have applied pair-coupling in stations, at merge points when queues are formed and at points of speed increases where queues may form due to a drop of capacity.

Implementation and effects

We have implemented the features discussed in this paper in our generic simulation system PRTsim. The same features can, with limited amendments, be incorporated into the control systems of commercially available PRT systems.

The effects on link capacity of the various improvements depend on network and demand patterns but are in typical cases estimated to be:

Feature	Capacity improvement factor
Ride-sharing	1.5 – 2.1
Platooning of empties	1.15 - 1.25
Pair-coupling	1.5 - 1.9

Combinations

Obviously ride-sharing can be combined with the other features. Platooning of empties will have less additional effect if vehicles are already pair-coupled. However empty vehicles can be dynamically platooned whereas pair-coupling is only applied in stations and pairs are broken up successively along the route. Combining all features it is reasonable to expect a capacity increase by a factor of 3.

On the main link leading out from a large transfer station there will be no empty vehicles during the peak. With pairs of 6-seater vehicles departing every 3 seconds the theoretic capacity would be 14 400 passengers per hour. During the peak, waiting passengers can be expected to fill up most vehicles if destinations are displayed over each vehicle. Provided that passengers board the right vehicles they still get to their destination without timetables and without stopping en route.

Comparison with LRT

The new LRTs for Stockholm accommodate 213 passengers (most of them standing). They run at 10 minutes headway but plans are to introduce 7.5 min headway. It is possible to operate LRT down to 3 minutes headway and LRT vehicles can be coupled in pairs provided that all stations have been made large enough. That gives a theoretic LRT capacity of 8 520 passengers per hour per direction.

Conclusions

We have discussed several ways to increase the capacity of "conventional" PRT meaning individual trips in individual vehicles. Without giving up the traditional PRT qualities of direct non-stop travel on demand it is possible to offer capacities similar to capacities of LRT systems.

At least where LRT is an option we claim that PRT can provide the required capacity. It may not be "conventional" PRT during peak demand but it very much looks like it. Further the available commercial PRT systems can with small software amendments incorporate the required features.

In comparison with LRT these PRT systems offer practically no waiting and about half the travel times. And they cost less to install and a lot less to operate.

References

Vukan Vuchic, Urban Transit Systems and Technology, Wiley 2007

Ingmar Andréasson, Ride-sharing on PRT. Proceedings of the ASCE Automated People Mover Conference, Orlando 2005.

Open-Guideway Personal Rapid Transit Station Options

Peter J. Muller, P.E.*

*President, PRT Consulting, Inc., 1340 Deerpath Trail, Ste 200, Franktown, CO 80116
www.prtconsulting.com

Abstract

Open guideway personal rapid transit (PRT) systems are inherently more flexible than captive-bogey systems or elevated systems and, thus, lend themselves to an almost infinite variety of station configurations. This paper explores alternative station layouts for open-guideway PRT systems. The station layouts studied include configurations resulting from consideration of various combinations of such variables as, in-line station; off-line station; single bay; multiple bay; in-line bay; off-line bay; elevated; at grade; below grade; one-way guideway; two-way (shuttle) guideway; access from one side of transportation pod (T-Pod); access from both sides of T-Pod; in building; attached to building; elaborate; simplistic. The wide variety of stations presented provides potential solutions for PRT stations in many different applications.

Introduction and Basic Philosophy

Captive bogey PRT systems, such as those being developed by Vectus and Skyweb Express have their wheels captured in the guideway and show little variation in station design. The stations are always off-line, and the bays are always arranged in line with each other. This lack of variation probably results from the intended relative high capacity of these systems and their inability to accommodate tight radii. Open guideway systems such as those being developed by ULTra and 2getthere on the other hand have rubber tires running on flat surfaces (similar to automobiles on roads), are steerable, can accommodate tight radii and are probably better suited to handling low capacity situations. The flexibility of open guideway PRT systems invites a wide variety of station design, meeting a wide range of capacity requirements and customer/passenger needs.

This paper discusses various station configurations suitable for a range of applications. Drawings are provided, depicting some preferred layouts. The drawings are mostly not to scale and are focused on the stations themselves, so required items, such as guideway safety fencing/railing and adequate acceleration/deceleration lengths, are often not shown or incorrectly depicted. This paper is focused on layout and operational considerations, and architectural aspects are not addressed. All of the stations shown could be rendered appealing, through appropriate architectural means.

Two important factors drive the philosophy behind the station designs, shown herein. These are, the desire to keep stations simple with low costs, and the desire to keep platforms as close as reasonable, to the elevation of the users.

Transfers and mode changes are the Achilles heel of transit. The worst involve climbing steps, while lifting luggage, combined with unknown transportation arrival and travel times. The best involve walk/roll on/off, such as, transitioning onto or off a moving sidewalk. The closer PRT station design comes to achieving the latter, the better.

Stations that are at a different level to the general pedestrian level suffer from two problems. They are difficult to find, and they require vertical circulation means that add to capital and maintenance costs and reduce safety. The elevated people mover, in Concourse A at Detroit Metropolitan Wayne County Airport, can easily be missed, altogether, by passengers unfamiliar with the airport. The Washington Metro's underground stations are difficult to spot and are served by elevators and escalators, 20% of which are usually out of service.

Station Basics

Stations designed for use in the USA should meet the requirements of the Americans with Disabilities Act. At-grade station platforms should include wheelchair ramps, down or up, to the surrounding pedestrian grade. Elevated or underground stations should include elevators.

Since common PRT design includes very short wait time (often less than one minute), the need to provide an enclosure, or even a roof for passenger comfort, is minimal. However, a roof will often be advisable, to protect parked T-Pods and station equipment from the weather and sunlight-induced heat loads, in particular. If such a roof is provided, it would make sense to extend it over the passenger platform too.

Station doors are considered unnecessary from a safety standpoint – particularly for open guideway systems having no third rail. They may be desirable for preventing conditioned air from escaping down the guideway. People and animals should be constrained from accessing at-grade guideways by safety fencing/railing. The fencing/railing should be far enough from the vehicle path to avoid potential pinch points. This is particularly important in a station where an arm, extending over a railing, could be pinched by a slow-moving T-Pod. The fencing/railing can have an opening opposite the door position of a parked T-Pod. Floor texturing, floor color, signage and cctv monitoring has been 100% successfully used, for over 30 years, to prevent accidents on the Morgantown PRT system (which has a third rail), as depicted in Figure 1.

Figure 1. Morgantown PRT station. Note the lack of station doors.

In-Line Stations

In-line stations, on a main guideway, are contrary to the basic PRT philosophy, but might be appropriate in very low capacity situations. A more common use, of an in-line station, could be where one or more stations are on a loop, off the main guideway, that is necessary but has low demand. This whole loop could then be treated like a long station bypass.

Off-Line Stations

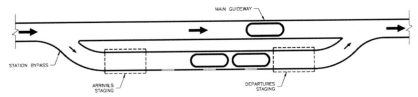

Figure 2. Off-line station.

A typical PRT off-line station layout is shown in Figure 2. The station is on a bypass guideway separate to the main guideway. Depending on the control system, station bypass guideways may include areas set aside to stage arriving and/or departing T-Pods. T-Pod bays/access points are indicated on this illustration and a number of others by gaps in the side wall.

The station allows T-Pods, on the main guideway, to continue on their way, without slowing for station operations by other T-Pods. The station bypass guideway shown is diagrammatic. The bypass guideway is used for acceleration and deceleration, to and from main guideway speed, and its length has to be designed accordingly.

In-Line Station Bays

The station bays, in Figure 2, are in line with each other. This has the advantage of simplicity, but the disadvantage of a delayed T-Pod blocking following vehicles. This is the typical configuration for captive-bogey system stations. When there are numerous bays in this configuration, the T-pods usually leave in platoons. This means that the dwell (in-station) time of each T-Pod is often dictated by the dwell time of the slowest T-pod, in the station in front of that T-Pod. Another drawback of this layout is that a problem in the station, in peak periods, can fairly quickly lead to the bays and the arrival staging area filling up. At this point, the station has to stop accepting T-Pods and, any T-Pods destined to it, must be either held in their departure stations, sent to an intermediate staging destination, or waived off (made to go on past the station exit and loop around for another attempt). Whether to choose an intermediate staging destination, or a waive-off, is typically dependent on the type of control system being used.

Off-Line Station Bays

Off-line station bays allow each T-Pod to function almost entirely independently of the others (they must, of course, avoid bumping into each other). ULTra has developed a saw-tooth bay arrangement (depicted in Figure 3) that maximizes the use of space. It does require that T-Pods back out into the station bypass guideway, but controlling this is only slightly more complex than controlling the merging that must take place, in the layout depicted in Figure 4. It has been argued that the saw-tooth arrangement reduces capacity significantly, if there are numerous bays. However, it is trivial to show that it can have similar capacity to the layout in Figure 2, since a platoon of T-Pods could all back out and depart simultaneously, taking approximately an extra six seconds (for the entire platoon) than had they departed without having to back up.

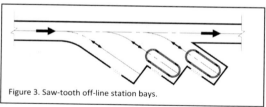

Figure 3. Saw-tooth off-line station bays.

Saw-tooth station bays need to provide sufficient platform to facilitate access to the vehicle doors. To accomplish this, it seems optimal to arrange the bays at approximately 35° to the bypass guideway direction.

The layout shown in Figure 4, results in a longer but slightly narrower station, which may better fit some locations. This layout could be shortened if the T-pods were to back up a little, immediately upon arrival.

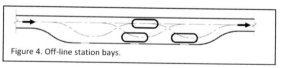

Figure 4. Off-line station bays.

Both the Figure 3 and the Figure 4 layouts violate basic PRT principle, by either having a backing up or stopped T-pod on the guideway, or by having crossing paths. Since these situations take place at very low speeds, this is not thought to be a safety or reliability issue.

Stations In or Attached to Buildings

Unless there is a need for interior transportation, PRT stations should not be located inside a building. Bringing a guideway into a building introduces issues related to weather proofing the building, tracking of rain and snow, etc., and should not be done, unless the system is intended to provide both interior and exterior transportation. The simplest solution is to use the building exterior wall as the boundary between the station platform and the T-Pod, as

Figure 5. Building station.

depicted in Figure 5. This low-capacity station makes use of the building vertical circulation systems and would require station doors. The capacity of a station, attached to an existing building, could be increased by using the layout in Figure 4, or for new construction, the Figure 3 layout could even be accommodated.

Figure 6 shows a parking garage, where it is desired to provide a high level of service, by having PRT stations on every floor. The guideways encircle the building and are interconnected by up- and down-ramps. In this way, the PRT system itself provides the building's primary vertical circulation.

Figure 6. Parking garage stations and guideways.

Airports are facilities that require internal transportation systems. The station and guideways, shown in Figure 7, were designed to fit inside Concourse B at Denver International Airport. The footprint of this station takes up the same length and less width than the existing moving sidewalks.

Another airport application is providing stations within terminal buildings. Figure 8 shows how this could be accomplished, at Denver International Airport, with minimal changes to the building. Although the guideway and station in the foreground look quite large, a better sense of scale can be achieved, by observing the return guideway and columns in the background.

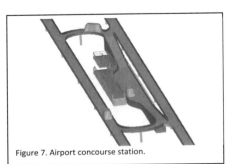

Figure 7. Airport concourse station.

Figure 8. Airport terminal station.

Urban Stations

Elevated Guideway

Figure 9. Urban elevated guideway station.

Figure 10. Close up – urban elevated guideway station.

As stated previously, it is recommended that stations be at grade whenever possible, even if the guideway is elevated. One problem with doing this is that, bringing the guideway all the way down to grade takes up a significant amount of space. A compromise arrangement is shown in Figures 7, 9 and 10, where the guideway is brought down to a station platform, raised about four feet (1.2m) above grade. This compromise results in the station taking the space of approximately 8 parking stalls. This will not be a problem, if the PRT system reduces the need for parking. The station is provided with stairs and a wheelchair ramp for pedestrian and handicap access. This arrangement requires slightly longer station bypass guideways, since gravity opposes both acceleration and deceleration.

Below-Grade Guideway

While stations for below-grade guideways can be provided completely underground, as for a subway, it is recommended that they be brought close to grade, for the same reasons mentioned before. Figure 11 depicts a station, adjacent to a three-lane road, with the platform depressed about 1.2 m (four feet). The station, plus sidewalk, fits in the width previously assigned to parking and sidewalk, plus about 1 m (three feet). Approximately six parking stalls must be sacrificed to the station, plus an area about 43 m (140 feet) by 1 m (3 feet) on the building side of the station. Greater length (and, possibly, width) would be needed for additional bays.

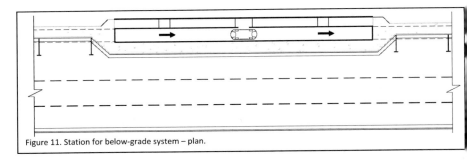

Figure 11. Station for below-grade system – plan.

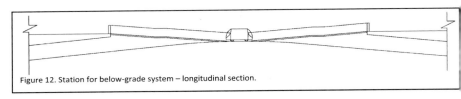

Figure 12. Station for below-grade system – longitudinal section.

Figure 12 shows a longitudinal section through the station, while Figure 13 shows a cross section. Note that the walls protecting pedestrians from falling into the station are about 1.2 m (four feet) high. The sloping pedestrian ramps, leading down to the station, are about 1.2 m (four feet) wide. The 1.2 m (four foot) wall, on the station side, slopes down with the ramp to maintain its

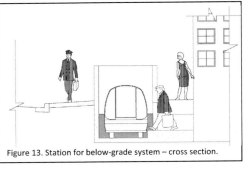

Figure 13. Station for below-grade system – cross section.

height and avoid a claustrophobic effect. If this station needs additional bays to accommodate additional demand, the pedestrian ramps will probably need to be widened.

Some applications may require a walkway, adjacent to an underground tunnel. However, this is not thought to be necessary for the station bypass guideway, since it should be easily possible to allow emergency pedestrian use of the bypass guideway, while simultaneously ensuring no T-Pod usage.

In this arrangement, gravity aids both acceleration and deceleration. Since the PRT system should reduce parking needs, the additional space requirements of this station are minimal.

At Grade Parking Lot Stations

Large airports often have large surface parking lots that are inefficiently served by shuttle buses. These parking lots are often for long-term parking and, thus, do not have a high trip demand. The problem with serving them economically with a PRT system is that the overhead guideways are expensive, and, if the stations are also elevated, there is the added cost and maintenance hassle of elevators. Heathrow has partially solved this problem, by bringing the overhead guideways down to grade in the parking lot. Since their two parking lot stations are at the end of their respective branch guideways, the vehicles have to be turned around, and this is accomplished in the way depicted in Figure 14. An alternative way of accomplishing this is depicted in Figure 15. This takes up less area, requires less pavement and slightly reduces trip length. Note that the upper right T-Pod is in a location where it could be temporarily staged, allowing others to pass.

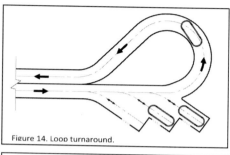

Figure 14. Loop turnaround.

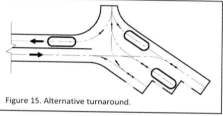

Figure 15. Alternative turnaround.

Figure 16 depicts a large parking lot served by a PRT system that is entirely at grade. The one-way main guideway traverses three sides of the perimeter of the parking lot. Automobile access is from the fourth (top) side and so does not need to cross the guideway. The inset, on the bottom of Figure 16, shows a typical off-line perimeter station. However, walking distances would be too great, if a parking lot of this size were to be only served by perimeter stations. The figure depicts two-way guideways serving internal stations. The left central inset shows a terminal turnaround station. This station has room for one station bay and one staged T-Pod. An alternative one-bay turnaround is shown in Figure 18. The right central inset, in Figure 16, shows a mid-block station. This station adds to the complexity, since it would require doors on both sides of the T-Pod. Note that the guideways would be fenced to prevent people or vehicles from crossing them.

Figure 17 shows design details of the center aisle guideway and hammerhead turnaround. One problem with the trough-shaped guideway depicted is that it will trap blown sand or snow. For this reason, the side walls are designed as beams on posts, thus, letting the wind blow through the walls at the level of the guideway surface. Other mitigation measures are available but beyond the scope of this paper.

Some might argue that two-way guideways are contrary to accepted PRT practice. From a safety standpoint, it would certainly be wise to operate them at slow speeds, and acceleration/deceleration sections (not shown) would have to be provided, where they join the main guideway. From a capacity standpoint, two one-way guideways should be provided, if modeling shows wait times are excessive with one two-way guideway. In this event, a turnaround, such as depicted in Figure 15, can be provided.

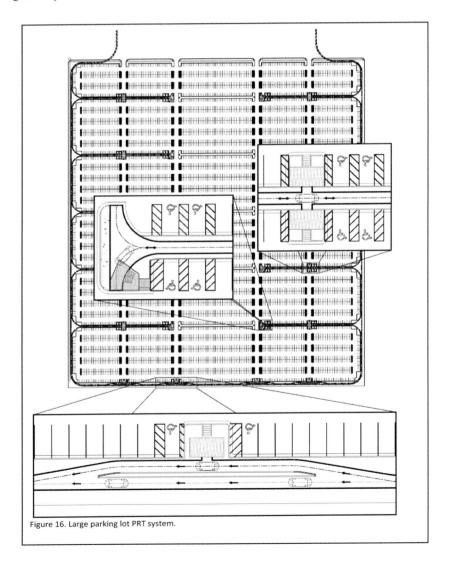

Figure 16. Large parking lot PRT system.

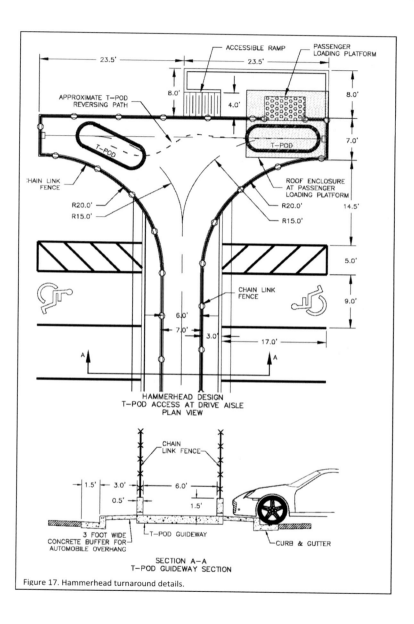

Figure 17. Hammerhead turnaround details.

Conclusions and Recommendations

It is clear that numerous station configurations are possible for open-guideway PRT systems. This is particularly true when considering low capacity applications. PRT is unlikely to be viable in low capacity situations, unless every effort is made to keep the infrastructure as simple and economical as possible. This paper is intended to begin the dialogue as to how best to accomplish this. The concepts depicted can certainly be improved upon and will hopefully stimulate others to develop different concepts.

Safety and reliability are paramount – especially in early PRT systems. Some of the concepts depicted in this paper violate basic PRT principles and require low operating speeds and special control and safety system adaptations, in order to operate satisfactorily.

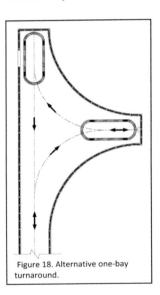

Figure 18. Alternative one-bay turnaround.

INTRODUCING PRT TO THE SUSTAINABLE CITY

Robbert Lohmann* and Luca Guala**

* Commercial Director, 2getthere, Proostwetering 16a, 3543 AE Utrecht, the Netherlands; PH +31 30 2387203; robbert@2getthere.eu
** Luca Guala, Area Manager, Systematica SpA, Via Marengo 34 – Cagliari 09131 Italy; PH +39 070 275939 guala@systematica.net

Abstract

The zero carbon, zero emission city of the future will require a high-level-of-service passenger transit system to accommodate the trips that in cities are typically performed by automobile. Mass transit, or group transit, is badly suited for this purpose as it can't replicate the service as supplied by the automobile. Also from the energy use point of view, with the exception of peak hours, mass transit is extremely inefficient as it requires that large vehicles travel nearly empty to respect a schedule. The sustainable city requires an on-demand, door-to-door, personal, zero emissions, energy efficient transport service that can be obtained by means of automated, electric powered taxis.

The sustainable city will employ Personal Rapid Transit, the solution that provides on-demand, private transit directly from origin to destination. The city will feature a network of guide-ways with a large station density ensuring short walking distances (maximum 150 meters). The stations are off-line, since the vehicles will make no intermediate stops, the guide-ways are located at grade, while the pedestrian level is elevated to create a new street level (the podium). The guide-way allows for multiple lanes, incorporating acceleration and deceleration lanes to allow vehicles to speed up and slow down for stations away from the main through-lane.

The entire network has been modeled using both static (macro-scale) and dynamic (micro-scale) simulation software. The simulation has been extended to pedestrian traffic and the interaction and mode split between walking and use of PRT has been modeled as well. The extensive use of simulation models is a fundamental step to the assessment of a novel transport system such as the PRT. The modeling allowed a precise assessment of the traffic volume in all branches and nodes, and the determination of parameters of exercise such as headway, trip time, wait time, energy use etc.

The network will also accommodate the movement of a variety of freight and waste services. As the PRT replaces automobiles, dedicated (automated) vehicles are required to replace delivery vans and trucks. The freight and waste vehicles will feature similar driving characteristics (acceleration, deceleration and top speed) to ensure mixing of traffic on the network is not made more complex. A generic freight vehicle will accommodated different types of loads; creating flexibility in the operations. Most loads will be transported in standardized containers, adopting an existing standard. It is essential that the freight system takes into account the supply chain to and from the city, ensuring seamless connections while taking into account liability issues.

Introduction

How would you build a city if you could start from scratch? Would a city look different form the cities of today? How would you accommodate the accessibility? With sustainability in the back of your mind, would you still allow access to cars? If not, how would you accommodate mobility of people and goods? Would you be able to with today's technology? Today's Concepts? Or do we need to introduce a new transit concept to allow the city of the future to be build differently, taking into account our natural environment – changing the focus to sustainability without it being at the expense of accessibility and comfort? A 'dream'? No, certainly not: a vision for the future, yes. And being realized now!

The City of the Future Today

The city of the future is carbon neutral, zero waste; a sustainable dwelling place acting as an example for future urban developments. The city features green buildings, waste management and reusage and natural energy taken from the sun or the wind. A sustainable city can't feature fossil fueled cars; it can't even feature cars at all! A truly sustainable city ensures accessibility but not at the expense of space or living comfort.

For the city of the future, a modern and reliable system of transport is needed to replace the private car. It relies entirely on the energy produced within the city from renewable and carbon-free sources, be free from congestion and significantly safer than any transport system based on private cars.

Personal Rapid Transit (PRT), an automated taxi-like service concept, has the qualities to provide the mobility desired, meeting the requirements of the sustainable city, without having to compromise on any other aspect of the development of the dwelling. It features the car's privacy guaranteed by the fact that only the individual, or group of individuals that board a vehicle at the first starting station will occupy it: each vehicle, once a person or group has boarded it and planned the route, will not stop until the chosen destination has been reached. PRT is a combination of the characteristics of the personal automobile, the advantages of public transportation (congestion, parking) and clean technologies to ensure a sustainable transit system.

PRT vehicles run on electricity, with a significantly lower energy consumption than other means of transport. The level of energy saving is significant also compared to mass-transit systems as the vehicles only run on-demand, so they never run empty, with the exception of the vehicles that are automatically routed to pick a passenger, and their ride is uninterrupted, so they do not have to expend extra energy to accelerate after an intermediate stop.

When compared to proper public transport systems, the PRT may have a lower capacity, since a public transport vehicle can increase its occupancy during peak hour. From the user's point of view, compared with public transport, the PRT offers better comfort, lower wait time, higher travel speed, no need to plan routes or transfer from one vehicle to another. From the community point of view the PRT offers very low energy consumption, high reliability and safety, non intrusive infrastructures and the possibility to build a thick network, capable to cover an urban area thoroughly and requiring very short walking distances from and to any point.

The PRT system functions as a local area network, connecting the locations within its network, and a feeder system to both other means of public transit as well as parking locations where access to more traditional private transit systems is provided.

Considering PRT

Personal Rapid Transit is selected as one of the transit options for the city of the future on the bases of several distinct characteristics in comparison to other options such as cars, taxis and public transport. Summarized the advantages are:
1. Shared usage: one PRT car can perform the task of 30 to 40 private cars.
2. Through automation congestion on the network is avoided through dynamic rerouting.
3. Automation leads to predictability, creating safety by avoiding human error.
4. The minimal footprint through a reduced guideway width and not requiring parking ensures only 13% of the surface is dedicated to transport (1/3 of the surface required for a traditional city).
5. PRT provides direct travel and on-demand service, ensuring trips are quicker, seamless and energy consumption is less.
6. Off-line stations warrants the level of service is not reduced if the number of stations is increased. The density of stations in the urban area is limited only by the space available and the cost.
7. PRT guarantees the privacy of the passengers; users can allow other passengers with the same destination to board the PRT vehicle with them, but only at their choice.
9. At off peak times the level of service increases as typically a car will be waiting at the station already.

Although PRT has significant advantages, there are several aspects that need to be addressed to be able to properly configure the system for the city of the future.

One clear aspect needing to be addressed in the accessibility of the stations. Where cars (and bikes) provide door-to-door transit (if parking is available at both origin and destination), the best effort for PRT requires a network with a high station density. Within extreme climates the maximum acceptable walking distances are relatively small (100meters or 1,5 minutes), ensuring the transit system remains attractive to use. This does impact the costs of the network significantly, as stations are not located at grade.

The PRT system will be public transportation. As a result it is not possible to leave objects in the car and the wear-and-tear faced is associated with public transit rather than personal ownership (where people tend to be more careful with personal possessions).

Traditional transportation system by the sheer size of the vehicles provide better capacity during peak hours, allowing the seats and standing places to be used to and over the maximum. However uncomfortable, this contributes to increasing the capacity of the line. PRT still allows for private usage, but ride sharing could be encouraged. The psychology is comparable to airports of larger cities with a shortage of taxis available; people will resort to ride sharing rather than waiting longer being able to travel by theirself.

The capacity of a lane for manual vehicles is based on a headway of 2 seconds or less, although at times, through human error, this will result in accidents paralyzing the system and its capacity. As PRT needs to comply with the current legislations imposing brickwall stop requirement, a headway of 2 seconds is not yet achievable. PRT's lane capacity might be lower, but when relating it to the space consumed, it is actually much better (as the required lanes are smaller).

Based on these considerations PRT is determined to be a useful supplement to the transit network of the city of the future. It supplements public transit (an LRT and metro line guaranteeing external connections), slow traffic (bikes, pedestrians and segways) and car traffic (at the city perimeter).

PRT Blueprint

Mobility, and accessibility in particular, is an important element for people in the selection of their housing or place of work. Hence the transit system in the city of the future is an integral part of the urban planning. The network needs to be planned to provide the required capacity, while also minimizing its footprint to ensure space can be used for value adding (money making) activities.

Urban Planning

To be able to ensure the throughput of any transit system, avoiding the congestion on 'normal' roads and leaving the space at grade for other activities (such as walking), systems require a dedicated, grade-separated infrastructure (guideway). For Personal Rapid Transit the popular choice is an elevated infrastructure, a result of the costs of underground installation and working within existing spatial planning in build-up areas. In the city of the future, as a green field development, these drawbacks were less constraining.

After analyzing the impact of an elevated network of PRT guideways on the dense built fabric of the city, especially considering the required thickness of the network (in order to minimize the walking distances and optimize the accessibility) and its visual impact, alternative possibilities were researched. The analysis clearly showed that a raised pedestrian level with an 'undercroft' created at-grade (a basement at grade level), would allow to exploit the entire available road surface, without disturbing the image of the city and minimizes the extreme weather impact contributing to the energy efficiency of the system; although at the sacrifice of the view of passengers during the trip.

This solution is not new, although, clearly, it has never been implemented with PRT. The township of Louvain la neuve in Belgium, and the district of la Défense in Paris, France, are built like this but in those cases, it's car and truck traffic and parking which take place under the elevated pedestrian free circulation space.

This concept also allows quick and direct access to any location in the city for special vehicles (emergency, maintenance, exceptional freight), providing the infrastructure of the system also allows access to these types of vehicles. The running surface hence has to be flat and free of obstacles, while featuring a bearing capacity to support a large freight vehicle for a width of at least 3.5 m.

Network Design

The transit system is one of many elements of the city of the future, which means the characteristics of its network design are influenced by all of the other elements. In the city of the future, the PRT network needs to take into account that:
- Stations need to be featured near main attractors of traffic;
- Stations need to be spaced such that the walking distance is minimized;
- The exact location for a station is based on the space available at each location;
- The corridors must follow the boundaries of the plots in which the city is divided
- The PRT running surface must be accessible to other vehicles in case of need;
- PRT tracks can't cross each other (no intersections);
- PRT tracks must preferably be one way;
- The junctions of the PRT network allow only merge/diverge maneuvers;
- maneuver lanes are required along most of the network (as acceleration and deceleration on through lines constrains the network capacity).

The complexity in the design is matching the architectural needs of the city with the attraction of traffic to the characteristics of the PRT system. In order to optimize the transport network and its efficiency in serving the needs of the city of the future, a model was designed to allow a dynamic interaction between the use of the land in the city of the future and the PRT network. Use of the land and network were defined in successive iterations in order to reach a satisfactory distribution of functions and a good network with the least number of stations and lines that allow to prevent congestion in any foreseeable situation. The first iteration was done on the initial land use and population data; a first PRT network was tested and the feedback used to modify the land use pattern. Several iterations were performed until a satisfactory combined solution of land use and transport was obtained.

In the final layout, the main attractors were positioned along the "spine", that runs diagonally through the city. This contains the LRT line as well as the main PRT connections. A Business District expands in the other direction, along which the roads are straighter and therefore better connections can be provided. A Light Industry area is positioned mostly along the edge of the city, so that it can receive its prime matter and deliver its finished goods with the least disturbance to the inner road network. The network was tested for robustness with the flow of passengers assumed for the two morning peak hours (7AM to 7AM) and a final network and station layout were defined

In the creation of the PRT network in the undercroft, the creation of a grid compound of outer and inner loops and transversal connections entering and exiting the spine proofed most effective. This design process follows the layout of the pedestrian-level inner roads , which are twistier in the SE-NW direction and straighter in the SW-NE direction. Therefore, the fastest and most direct PRT connections were designed in SW-NE direction. Designing the network is an iterative process during which several different networks are designed and verified against the geometrical constraints, architectural and environmental requirements.

All the transport levels were specified with their main features, using several geometric and functional attributes, and properly connecting the LRT line, Metro train line, PRT system and the pedestrian network. All the transport systems are connected to the pedestrian network and are connected each other using the pedestrian network (i.e. walking to the PRT stop or LRT station); the PRT lines are all one way except the connection to the external car parks.

The design of the network also required an analysis of the connections between the two separate built districts which make up the city of the future. Since these connections could become a bottleneck for the whole network, the main aim was to design as many connections as possible between these districts to let the PRT flows spread across them toward their destinations instead of having all the traffic between the two districts over one or few links.

Without reference in literature or field data, all hypothesis made have a significant level of uncertainty, even though they have been defined as accurately as possible. The results of the simulation have therefore been considered "safe" only when they yielded a flow of less than 60% of the saturation flow in every branch and node.

The least number of roads were used to connect all the PRT stations required to guarantee the accessibility level needed. The resulting transport system is composed of:
- Walk network: ~104.7km
- PRT network: ~45.0km, all one way except ~10.0km

- LRT network: ~40.1km (including connection with parking lots)
- Metro train network: ~6.7km
- 103 PRT stops (including external parking lots),
- 154 merge/diverge nodes
- 6 LRT stops inside the city

In parallel with the design of the route network, the locations of the PRT stations was analysed, based on the requirement to ensure the maximum walk distance from any point to the nearest PRT stop is no greater than 150 meter. The design constraints are basically generated by a compromise between opposite goals: achieving maximum coverage and not having an excessive number of stops, which leads to an uniform land coverage.

Since the roads of the city are arranged in a square angle grid layout, the walk distance from the PRT stops has been evaluated along a right-angled path and not along a radius; The theoretical scheme that arises from these considerations is a triangular, or staggered grid of stations. This layout allows the minimum overlap between areas of influence and the minimum number of stations to cover the whole city area. The actual network follows the plot division and is not as regular as the theoretical scheme, so some areas of influence have a higher overlap and the number of stops is slightly greater. Punctual attractors (LRT stations, main office building, hotels, shopping areas, Places of Worship, Schools, etc.) were considered as fixed points in the station scheme.

About 1/3 of the whole area has 2 PRT stations within a distance of 150m. The best coverage has been obtained on the busiest areas (BD, spine, special buildings, etc.). If a uniform distribution of population is assumed over the territory, then at least 50% of PRT users will have a station less then 100m away, and often they will have more than one station within 100m.

Only some very small areas fall beyond the 150m walking distance goal from the nearest station; part of these are included in the "green finger" linear parks that cross the city, other small areas lie within the light industry built up area. Since the population is not uniform, but more concentrated in the areas where also the PRT stations have also been located, it can be assumed that more than 50% of the PRT users will walk less then 100m to the nearest station. This result is consistent with the goal of providing a near-door-to-door transport service throughout the built area.

Transit System Modelling

The estimation of demand is the process that produces the number of trips from each zone of the area analyzed to every other zone. To be used in a traffic model, these trips must be arranged in a matrix format (matrix estimation process). The final result of the process was a matrix of 283 zones based on the plot subdivision of the built area of The city of the future, with about 62,400 trips over 2 hours (7AM-9AM).

The analysis performed is based on:
- The attraction and generation rates for each land use, as well as the inbound/outbound traffic rates and the AM traffic rates (when trips rates are referred to the whole day) are those found on literature Trip Generation Manuals (TG).
- It has been assumed that 70% of the commuters that reach the city of the future for work/study purpose will travel in the peak 2 hours between 7AM and 9AM. this assumption is purposely cautionary.
- Trip rates represent the number of trips generated and attracted for each unit of reference and for each land use category. The values in the TG manuals are referred to different types of units of measurement: areas (typically, 100 sqm GFA in TG Manuals), number of students for University and School, dwelling units for Housing, keys for Hotels, and employees for Light Industry.
- The zoning system represents the 280 plots inside the two built districts, plus three external zones, one zone by mode used by the commuter to reach the city, that is Light Rail Train, Regional Metro Train line underground, Car+Bus+Hov (those using the external Roads and car parks).
- For each land use the inbound/ outbound ratio were taken from the TG manuals. As an example, for residents this is equal to 83% outbound, 17% inbound.
- Every land use and activity inside a single plot was examined individually and its specific data were used in the trip generation process, then the trips generated and attracted were handled as uniformly spread over the whole plot.
- Total trips generated and attracted by each plot were calculated multiplying the trip rates by the value of the relevant parameter per each land use per plot (for example, area, population, no. of students etc.).
- For what concerns the University, only the employees and 10% of the students were considered in the trips generation process, as potential inbound travellers, because 90% of the students were considered as resident inside the University quarters (and not travelling towards the city during peak hour).
- Since trip rates from the manuals are expressed in vehicle trips, a vehicular occupation coefficient equal to 1.2 was used to convert vehicular trips in passenger trips where appropriate. This is equal to the average occupancy of cars and is cautionary because the occupation coefficient will probably be closer to 1.5 if a HOV promotion strategy is implemented (Single Origin, Multiple Destinations - SOMD).
- Trips attracted by each plot have been estimated allocating the number of commuters entering in the city on the base of the "power of attraction" of each plot compared to the others. "Power of attraction" has been calculated applying attraction trip rates to each plot on the base of the different types of activities localized in it.
- Trips generated by each plot have been distributed among destination on the base of their "power of attraction". For each type of trips a set of possible destinations have been selected. For example, it was assumed that destinations allowed for commuters are the work and leisure locations but not the places of residence.
- The mode split between the external transport means, and the occupancy of the vehicles have been defined in a trip generation scenario.

- It has been assumed that car and HOV passengers will board the PRT vehicles with the same occupancy as the car they traveled into, while bus passengers will fill up the PRT vehicles to almost their capacity. Other travelers (LRT users and residents) will board the PRT cars with an average occupancy of 1.2 passengers per car. This too is a cautionary assumption, since it can be assumed that LRT users will accept higher occupancy levels of the PRT vehicles (SOMD).

The biggest generators are three zones that represent the virtual points of origin of the trips by private transport, LRT and Metro. Since the plots are all mixed use areas, the relation between land use and the number of trips generated is not as clearly visible but, as a general rule, the residential area is a shallow but vast generator. The main attractors during the peak hours are all the offices and business areas where people usually work. The residential area has a low attraction level, during the morning peak hour, mainly related to the community activities.

The traffic model based on the multi-level network and the matrix obtained from the demand analysis was drawn, employed an equilibrium iterative model to evaluate the traffic flow on each link. Specific BPR flow curves were used to fit the sharp PRT network capacity drop off.

For what concerns the commuters who use the light rail, no assumption is made about their distribution over the rail way stations inside the city of the future: they are assumed to be free to choose the best railway station (the nearest) to reach their destination. This also applies for commuters using cars, buses and HOV: no assumption is made about their distribution over the external car parks and their use of the LRT to enter the city: they are assumed to be free to choose the most suitable car park to reach their destination and to choose between PRT and LRT where available. The "best path choice" and the "path-finding" algorithms handled these processes (with the parking lot capacity serving as a constraint). The capacity of the network was defined based on a PRT system which allows a 3 seconds headway to the vehicles. This is a baseline number, and not a precise estimation of the performance of any specific PRT system.

The network resulted to be almost everywhere within capacity although over 60% of capacity in about half of the network. The model also showed some locations in the network close to 100% capacity, which requires further analysis to determine how capacity can be increased; a shorter headway being one of the options. This means the PRT system implemented will need to accommodate for a reduction of the headway once the build out of the city and the increase of demand on the network show this to be required.

PRT (and FRT) Systems

Besides the PRT system the network will also need to accommodate the FRT (Freight Rapid Transit) system to allow for the delivery of goods and the removal of waste. For deliveries the FRT system will operate during the same period as the passenger vehicles, although avoiding the absolute peaks, while for transport of waste will be scheduled for the low demand hours of the passenger transit system.

In normal operations the priority is the passenger transport. The graph below is an indication of the time interval assigned to the different vehicles. The PRT system will be operational 24/7. During the peak hours of the system, no other system will be operational in the undercroft. The undercroft will be accessible at all times for emergency vehicles. During the night it is also possible for exceptional loads (using manually controlled vehicles) to be transported. Exceptional loads in other hours of the day should be avoided, but if it can't be avoided these would be scheduled before and after the lunch period.

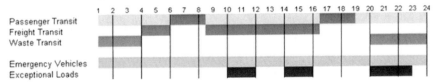

As the PRT and FRT vehicles operate at the same time intervals and on the same infrastructure, their driving characteristics should be similar to avoid disruptions of the system and network. This means that characteristics such as the maximum speed, turning radius, acceleration and deceleration capabilities and profiles should be equal. This is most easily achieved by means of a shared platform. The build-up on the platform and the body would differ considerably. The shared platform greatly facilitates the planning issues of the supervisory control system.

Supervisory Control System

Although vehicles are the most visible aspect of a PRT system, the (supervisory) controls are the most critical factor for the success of the system. Personal Rapid Transit is all about network (and vehicle) controls.

The supervisory system TOMS (Transit Operations Monitoring and Supervision) will co-ordinate the different vehicles and the scheduling, but will also be expandable towards the future to allow for expansions and/or phased introduction. The supervisor should allow for 'easy' expansion, allowing the network the be extended and the fleet size to be increased. Hence the control system will feature a distributed network architecture, making it more flexible and allowing for easy expansion while also ensuring the system's robustness. Local control and autonomous decision making are essential to the success of the network and the application.

The extensive experience of 2getthere with automated vehicle systems results in a 4th generation supervisory control system based on object-oriented Holonics software architecture. The objects (holons) within the framework are identifiable, self-organizing units that both comprise subordinate parts and constitute part of a large system. The interactions between these objects (each object influences and is influenced by both subordinate and superior objects) enhance the ability to respond locally while maintaining a global goal – ensuring the system is flexible, robust and scalable. The strength of the architecture is that it enables the construction of a (very) complex system that ensures efficient use of resources, is highly resilient to disturbances (both internal and external), and is adaptable to changes in the environment in which it operates. The supervisory software is platform independent.

An important element of the supervisory control system is the newly developed Graphical User Interface, allowing the operators of the system to monitor the process real-time in either 2D or 3D, selecting the track sections, stations or individual vehicles of interest at any time. The GUI is developed in-house, incorporating previous applications experience.

Vehicle Guidance

The transit system is a free ranging system, operating on rubber tyres. The most notable shared characteristics vehicles are the controls and the obstacle detection system.

The vehicle control software operates based on the patented FROG (Free Ranging On Grid) technology; creating intelligent vehicles that can operate in any environment. The on-board computer controls the vehicle based on electronic maps (route planning). While driving, the vehicles measure distance and direction traveled by counting the number of wheel revolutions and measuring the steering angle (odometry). External reference points (magnets embedded in the road surface) are used to correct possible small inaccuracies in reference to the planned route (calibration). The system has continuous longitudinal and lateral position calculations, ensuring external influences, such as wind, are automatically corrected. The passive reference points merely serve to improve the accuracy even further. The patented Magnet Measurement System has tested and proven up to speeds of 100km/ph, showing only marginal deviations from the planned path. For PRT applications the maximum speed will be restricted to 40km/hr in light of the applicable safety regulations.

The navigation system is not dependent on any physical infrastructure. It does not require any physical guidance (rail) or guiding infrastructure elements (curbs and/or walls), ensuring complete liberty in design. Dependency on infrastructure elements is avoided as it entails a higher vulnerability and increased inspection and maintenance costs of the infrastructure. The guidance system is inherently safe and can not be 'derailed' by placement of additional magnets taking the vehicle of the planned path.

Although the vehicles operate on a dedicated guideway, each vehicle features an advanced obstacle detection system. The obstacle detection system functions as a safety measure both related to people in the immediate environment (in the unlikely case that there are any), as well as in respect to other vehicles operating in the same environment. The obstacle detection system (ODS) uses sensors to scan the base-hull and extended hull in front of the vehicle. The base-hull (the vehicle envelope plus a small safety margin) is specific for each track section and calculated based its' characteristics (straight, curve, etc.). The extended hull is wider and incorporates space adjacent to the base hull where objects might be present that could move into the base hull. The vehicle would slow down and ultimately stop for obstacles (objects present in the base hull), while it would only slow down to pass at a lower speed for objects in the extended hull. In case the object would move into the base hull from the extended hull, thus becoming an obstacle, the lower speed ensures that it is still possible to stop in time. Both hulls are divided into a large number of detection cells. Each detection cell is scanned several times per second. The measurements of the scans are repeated and based on them a probability factor of an object being present (in that detection cell) is determined. The probability factor is among others determined based on object size (presence in multiple adjacent detection cells) and its' presence in continuous repeating scans. Based on the object probability number the reaction requirement is computed, which in its' turn is translated to an adapted speed profile taking into account the obstacle. The new speed profile takes into account acceleration, deceleration and jerk constraints as defined for the application.

PRT Vehicle

2getthere's PRT vehicle has been developed in co-operation with Zagato (design) and Duvedec (realization). Based on stringent customer requirements and reviews the vehicle design was developed in several design cycles. It reflects the appeal and characteristics of the personal car, while being resistant to the wear and tear associated with public transit. The design is in-line with the design of the city of the future, appealing to energy efficiency and innovation.

The exterior is compact with optimal interior space. Ease of access is provided by large, automatically operated, sliding doors. Although stations might be featured on both sides of the tack, the vehicle will only feature sliding doors on one side. Access to the vehicle will be ensured from the platform in adaptation of the platform design. The opening width of the doors allows for wheelchair or pram access from the platform which is flush with the vehicle floor.

The cabin is spacious and light. The height of the cabin is such that passengers will not be able to stand during transit and passengers will be notified to seat their children. Large, heat reflective, glass surfaces give good all round vision and add to the security feeling of the passengers. The cabin is well illuminated when driving at night. It accommodates 4 to 6 passengers (maximum 4 adults, 2 children). The seating is configured in the form of two benches, placed opposite of each other and located over the wheels. The benches feature two seats sunken-in each. The space in between the seats accommodates a child. Seating is comfortable with space clearly exceeding normal public transport standards.

The vehicle's user interface consists of an information screen and interfaces for the vehicle activation, intercom, doors, medical assistance and emergency stop. Ease of use and ample travel information are important prerequisites for People Mover systems. Therefore layout and design of the controls are inviting and simple in such a way that passengers intuitively find their way. This process is supported by a display for guidance and feedback. It also provides further travel information. The display can be upgraded for commercial messages.

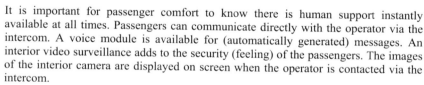

It is important for passenger comfort to know there is human support instantly available at all times. Passengers can communicate directly with the operator via the intercom. A voice module is available for (automatically generated) messages. An interior video surveillance adds to the security (feeling) of the passengers. The images of the interior camera are displayed on screen when the operator is contacted via the intercom.

In the application actually two types of PRT vehicles are featured: standard and VIP. The VIP vehicle can be used by indentified individuals only and feature several luxurious features not present in the standard vehicle (such as leather seating and privacy glass).

FRT (Freight Rapid Transit)

To optimize the operations and ensure maximum flexibility, while minimizing the fleet size, a common platform was developed for both freight and waste transit. The platform is a flatbed vehicle capable of carrying containers. The type of container determines whether the vehicles are being used for freight (multiple variants such as refrigerated, valuables, flammable/dangerous) or waste.

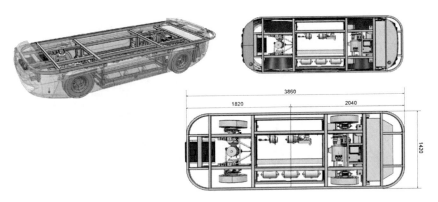

The containers are designed specifically, color-coded based on the purpose they serve:
- Blue: freight containers
- White (reefers): refrigerated containers
- Black: waste containers
- Yellow: valuables (including key-pad with security code to be able to open the container but also to release clamp mechanism of PRT vehicle)

The containers allow for 2 pallets per container. Roll containers would also fit, although the size of the roll container platform would have to be adapted specifically to the width of the containers. With the interior dimensions being approximately 1050mm, the base should be 500mm wide (allowing for two roll containers to be placed adjacent to each other). All the containers would feature fork-pockets at the base, allowing them to be handled manually by means of a fork-lift truck. At the same time the pockets could also be used of automated handling (e.g. by a miniature stacking-crane).

Furthermore the pockets would allow for securitization of the freight during transit. The vehicles will feature automatic clamps which will lock the container after it is loaded onto the vehicle. These clamps will be automatically released once the container reaches its destination, with the exception of the containers carrying valuables – in this case it will be required to enter a security code first before the clamps will be released from the vehicle. Entering the security code would either be done through the interface on the container (same panel as required for opening the doors to these containers, but a different code), or via a terminal of the supervisory control system.

By using the color-coded containers and flatbed vehicles, the vehicles can serve any kind of transport without any risk to hygiene. The vehicles will feature a load handling mechanism similar to airplane-freight. This would mean that the pallets, once the clamps are released, can be pushed off by hand.

Energy Management Concept

The vehicles are powered by an on-board energy source. As there are many parallels with other applications (e.g. cars) the development of these type of energy sources is rapidly progressing.

In light of the various application characteristics, including the climate and the sustainable nature of the project, it is essential the (battery) technology applied ensures the best fit. Opportunity charging using regular car batteries is an option, but is certainly not the most sophisticated or sustainable solution. The city of the future application will feature Lithium-Phosphate batteries, which, although considerably more expensive, offer the same capacity at a lower weight in comparison to lead-acid batteries. To optimize the life-cycle of the battery deep-charging (with a full charge within 2 hours) is preferred over opportunity charging, although this is possible. The range of the vehicles is approximately 60 kilometers per charge.

A battery management system is available to ensure all cells are discharged equally and an indicator for the remaining energy is provided. The chargers are actually located inside the vehicles based on the requirement to be able to recharge at the stations and the fact that direct current (DC) energy is provided. This allows for the sensitive process of charging to be checked immediately. Otherwise each possible parking location would also need to be connected to a dedicated charger.

Headway and Throughput

The minimum headway (distance) between vehicles is a safety factor, which also influences the maximum capacity of the infrastructure. The headway will largely depend on the requirement for a brick-wall stop. When a brick-wall stop requirement is imposed, the CyberCab PRT system will feature a 3 second headway. NOTE: this is calculated according to the requirements of safety certification procedures, with the leading vehicle braking at the technical maximum, with the trailing vehicle decelerating at the (fail-safe) minimum emergency deceleration speed while taking into account sensory, communication and activation time delays. When the brick-wall stop requirement would be lifted, a shorter headway would be possible.

The capacity of a track lane is determined by the length of a vehicle moving block (headway + vehicle length). At lower speeds the vehicle moving block is longer (as each section of the track is occupied longer by a vehicle). At lower speeds, the throughput will decrease. Hence the infrastructure design (especially the curves) will determine the maximum throughput of each track-section. The throughput will be as high as applicable for the lowest speed in that section. The curve radii are vital in establishing the throughput of a track section) are thus vital in determining the potential throughput per hour; which is why close co-operation with the design of the network is required to ensure the PRT system is able to make true on its capacity to the largest degree possible.

Infrastructure

For a network with a high-station density and a large number of vehicles, acceleration and deceleration lanes become a pre-requisite. As a metaphor please use the resemblance with highways. When vehicles would need to slow down on the main guide way before being able to turn into a curve or into a station, all the trailing traffic would need to slow down as well; seriously impacting the capacity of the network.

For this reason the sustainable city will feature multiple lanes, consisting of one or more highways and one or more deceleration lanes. From this perspective a lack of physical separation between lanes is a great advantage as it allows the system to switch lanes at any point along the trajectory. In addition the lack of physical guidance ensures the infrastructure costs are minimized (both in construction and maintenance).

Surface

The infrastructure surface of the guideway is an important factor often neglected; it impacts comfort (noise, vibration) and the passenger experience very directly! Although the weight of both the PRT and FRT vehicles is limited, the consistency in driving (maximum normal lateral deviation of 1cm) ensures rutting is a serious issue. A concrete infrastructure would solve this issue, but the longitudinal evenness (or roughness) could be a point of concern, especially as it directly impacts the ride comfort. In addition concrete provides less comfort and more noise hindrance, as well as it is more difficult and expensive to maintain.

From these perspectives there is a preference for asphalt, which, however, is more suspect to rutting. An asphalt surfaced pavement for the underlying infrastructure will provide a smoother surface. Furthermore in case of incidental surface defects it will be easier and more economical to provide local resurfacing. With modern asphalt mixes and polymer modification of bitumen it is possible to provide tough and rut resistant asphalt pavements.

A surface water drainage collection system is required to drain tracks and all non-permeable areas within the APM structures. The surface drainage design includes the design of the gravity sewers, road gullies and connection to a main drainage connection point at the property boundary. For the magnets to be embedded in the road surface it is important to take into account a 10 centimeter clearance underneath the magnet (15 centimeters below surface level) for the steel reinforced grid. This to avoid any potential disturbance of the magnetic field to be measured by the vehicles sensors.

Stations

All stations feature all the necessary amenities and security measures to ensure the comfort and safety of the passengers. The stations are ADA compliant ensuring easy access for less able bodied passengers. 2getthere opts for a station with angled berths (avoiding the gas-station queuing problem), with the design allowing for independent entry and exit of all berths (optimizing capacity) while ensuring passenger do not need to cross tracks (minimizing costs) and remaining intuitive in usage (minimizing the signage required).

Please note that the operations at angled berth stations are more complex as a result of the maneuvering required; vehicles reversing out of berths need to be coordinated with vehicles coming into the station. To enable these operations the distributed network architecture of the supervisory system is important as the maneuvering can be controlled locally rather than centrally. It should also be taken into account that reversing out of the berths does increase the driving time slightly. However, the average trip time is reduced as delays will not build up for all vehicles. The independent exit and entry ensures the system is more robust, less vulnerable to disturbances and single point of failures.

Amenities that are a requirement are passenger information consoles and a PA-system. The station lay-out should make usage intuitive and easy to use for all passengers – both young and old. Despite the fact that there is minimum waiting time, and is most cases a vehicle will be present in the station already, it is still important to provide passenger information. The information concerns the operations of the system, especially if there are exceptions, but also e.g. if there is no vehicle present the possibility to indicate a transport request and the waiting time for the vehicle to arrive. The transport consoles at the stations will also feature a possibility to contact the operator via an intercom. The PA-system will allow general messages to be broadcasted at all stations simultaneously.

From the perspective of security each station needs to feature the possibility to contact the operator (via intercom) and have camera surveillance. The transportation request modules at the stations all feature an intercom facility. The operator can be contacted directly. Once the operator is contacted the images of the CCTV cameras will be displayed on screen immediately; in this way the supervisor can actually observe the surroundings of the person contacting him. Each station will feature camera surveillance. The cameras cover all areas of the station (avoiding blind spots). Stations will feature multiple cameras with overlap, to avoid a non-functional camera creating blind-spots.

Wrap Up

The city of the future will allow for door-to-door transportation, but not by car. Personal Rapid Transit can become a part of the transport mix offered to residents and workers of the city. The most prolific advantages are the savings in energy, its' environmentally friendly nature and the huge reduction of the space required for transit systems – allowing this space to be used for other purposes.

PRT is a concept that can only now be realized with the developments of numerous technologies contributing to it. More and more applications will be realized in future years; in niche markets first, but in complete city applications thereafter. On what term will become evident based on the success of the first applications…

The Need for High Capacity PRT Standardization

Raymond MacDonald

President, 21st CENTURY PRT Inc., 40 Palatine #108, Irvine, CA 92612
Phone: 949-379-6861 E-mail:raymaciain@yahoo.com

Abstract

The development of PRT technology is now proceeding into the proliferation phase. Numerous companies are developing PRT systems that vary profoundly from the original concepts. While this process is both natural and desirable, it is also disruptive to the effort to commercialize PRT on a sustainable worldwide market basis. Of particular concern is the division of the technology into Low Capacity PRT which has adopted the APM Standard Operating Criteria and High Capacity PRT which rejects that APM Criteria. The author considers that LCPRT is ultimately not financially viable, and urges that PRT be standardized according to the HCPRT Criteria, even if used for initially low capacity applications. The argument often heard, that it is too early to standardize PRT is gainsaid by the development of computer aided design which can telescope the rational evaluation of numerous technologies in a short time framework.

Introduction

After a hiatus of many years during which PRT development was limited to a few companies all of which were starved for funding, there is now a revival of interest in this technology. This revival has taken the form of development projects in England with the ULTRA System, in Korea with the VECTUS System, in UAE with the 2GETTHERE system and in the USA with the Taxi 2000 System. A few other systems are at a lesser stage of development elsewhere. The author sees an urgent need to standardize these efforts.

Within the Transport Engineering Community there has been an effort to link PRT with Automated People Mover (APM) technology. The most vivid evidence of this is the presence of PRT Promoters at a series of APM Conferences such as this. Partly this happens because the PRT Community is too small to hold its own conferences and presumably because they feel warmer within the context of a larger, although still small, group.

This sharing of the same bed appears to be leading to a number of disturbing results, not the least of which is the re-classification of PRT as an APM Technology. Why is this disturbing? The purpose of this paper is to explain the dangers inherent in this association and the need for standardization of High Capacity PRT (HCPRT).

The PRT concept is very different from that of APM which is essentially a group travel concept that relies on large vehicle or trains of vehicles operating in a Line

Haul manner, that is to say following a fixed route and stopping at a number of stations with a mix of seated and standing passengers. This operating mode leads inexorably to large heavy vehicles that require a heavy guideway and on-line stations at substantial spacings to permit adequate average speeds. The APM concept characteristically falls into the Line Haul category and it becomes virtually indistinguishable from Light Rail Transit (LRT) in its operating mode. It is only by virtue of grade separated guideways that APM is able to offer automated operation and where LRT systems can be similarly grade separated, they too can be automated.

In most countries railway systems are operated according to rules or conventions that require the headway between vehicles or trains to be long enough to permit a clear stopping distance plus a margin of safety between succeeding trains. This is termed the "Brick Wall Stopping Criteria". Existing APM Design criteria have adopted this standard and it appears that the promoters of many PRT systems such as ULTRA, VECTUS and 2GETTHERE have also adopted it. These systems are therefore examples of Low Capacity PRT (LCPRT).

While this appears to make good political sense in countries that espouse railway operating practice and that will countenance no other, it makes no sense for PRT to adopt this criteria. The future of PRT depends upon it being FINANCIALLY FEASIBLE for implementation by cities and private owners. There is no future for LCPRT which is just another APM subject to the tyrrany of FTA Funding Regulations due to its lack of capacity and hence its lack of Financial Feasibility. The only form of PRT which has any chance to be Financially Feasible and hence MARKETABLE without subsidization is HCPRT. This will allow PRT to break out of the constraints and shackles that Public Transit is subject to in the USA and the rest of the World. As an example we now see the transport planning community happily moving their efforts towards streetcars that average 9mph.

Standardization Topics

This paper therefore calls for the total divorce of PRT from the APM Operating Criteria and the adoption of a HCPRT Standard. The reasons for this divorce are simple and logical:

(1)   **Headway**

The use of Brick Wall Stopping Criteria makes no sense for LCPRT in practical financial terms since it limits minimum headway to about 2.50 or 3.00 seconds giving a maximum line capacity of about 1200 vehicles/hour. Therefore the primary goal of PRT Standardization should be to abandon the Brick Wall Stopping Criteria. We already see the automobile system operating sometimes at fractional second headways and our highways would appear to be almost empty if we operated at 65 mph with average gaps of 240 to 290 feet between vehicles. Most of us present appreciate that automated operation is inherently safer than manual operation, therefore we need to move into the modern age and use the power of the technology available to us. All PRT systems should be standardized to be capable of operation in HCPRT mode, whether their initial implementation requires this or not.

(2) Capacity

Since true PRT operates to carry individual passengers or groups non-stop from origin to destination, the vehicle occupancy will be similar to automobile average occupancy and the line capacity of Low Capacity PRT will thus be limited to round 1,500 passengers/hour at best. Since all PRT networks involve merging lines, it follows that many or most lines in a given network will only carry some fraction of the maximum line capacity. The economics of a transit system limited to a few hundred passenger/hour on the average line are dismal and it is unlikely that such systems can be financially viable. On the other hand HCPRT operating at headways of 0.50 seconds or less offers maximum line capacity of about 6,000 vehicles/hour with capacity approaching 7,500 passengers/hour, which is comparable to many LRT lines. If we do not adopt standards that lead to rational financial criteria when designing PRT then there can be little hope for this concept to succeed as a heavily subsidized underperforming system.

(3) Control

The key to developing financially viable HCPRT is of course the development of control systems that are absolute safe and reliable. This can only be done by means of computer controlled operating systems that involve multiple computers that are rigorously failure monitored. This is the operating concept of most high tech aeronautics systems and it allows a vehicle experiencing any system failure to be directed to a maintenance depot long before the failure of one control leads to a vehicle breakdown. It follows that multiple computer controls with failure monitoring should be the standard for HCPRT

(4) Emergency Stopping

The Brick Wall Stop is largely a mis-nomer since there are extremely few instances where a vehicle running on a guideway experiences a sudden collision with an obstacle on the guideway that is immovable. In those cases where such a sudden stop has been experienced, the failure was the lack of a suitable detection system. Two cases of Brick Wall stops come to mind. The collision of the Transrapid Mag-lev with a maintenance vehicle on the Emsland Test Track and the collision of the German ICE with an overpass at Escheide. In both cases the lack of adequate failure warning systems resulted in disaster. HCPRT must have comprehensive warning and control systems as a Standard.

Those present who take comfort from the thought of operation at 2.50 second headways, or 15 second or 90 seconds are obviously not understanding the reality behind such collisions. The requirement is for a standard of absolute control and failure monitoring. Conventional Line Haul Transit systems may indeed require operating headway including safe stopping distance plus a safety margin since they generally have no passenger safety equipment on board and they are not designed to survive collision beyond the dubious provision of buffing strength. Any automobile designer worth his/her salt will inform you that the last thing a vehicle wants is buffing strength. The standard passive safety equipment required for HCPRT vehicles includes on-board passenger restraints such as seat belts, low velocity air-bags, adequate throw distances, crushable front and rear body structures and shock absorbing buffers. In addition HCPRT vehicles should be equipped with active collision avoidance systems such as infra-red detectors for preceding vehicles such as are now available on automatic cruise control systems. Another standard should be emergency braking systems that are capable of applying deceleration rates that are well in excess of the rates attainable by friction braking. Deceleration rates of 1.0g emergency to 2.0g extreme emergency should be possible with fail safe mechanisms that lock the vehicle to the guideway in the event of a power failure. Since a failed HCPRT vehicle will generally roll to a halt if its power fails, it should be easy for the following vehicles to match that gentle deceleration rate without emergency braking and in the event where a vehicle operating at full speed comes upon a stopped vehicle the detection system and the emergency brakes should be able to avoid a collision completely.

(5) **Accident Resistance**

PRT vehicles have substantial advantages over APM vehicles when passenger safety is considered both in terms of accident damage resistance and personal safety. This factor alone calls for a different design criteria and standards for PRT compared to Line Haul transit systems such as APM and LRT.

(a) PRT passengers are all seated and can therefore be protected by seat belts and low velocity air-bags whereas standing APM passengers can not be afforded any protection at all and protection for seated passengers is minimal. Seating arrangements in which passengers face each other are fundamentally unsafe due to the lack of protection for passengers thrown forward and the lack of crush protection for the passengers facing rearwards.

(b) PRT vehicles should be locked to the guideway so that derailment is impossible even under the worst conditions of wind, seismic or collision forces. Few APMs have this capability and no rail systems.

(c) PRT vehicles can be made fire proof and the vehicle running gear and propulsion system can be solid state so that any overheating can easily be detected. Fire suppressant systems can be fitted to the propulsion systems of each vehicle.

(d) PRT vehicles should be capable of high rates of emergency deceleration that would be unacceptable in a Line Haul Transit vehicle with standing passengers or unprotected seated passengers.

(e) PRT vehicles can be made crashworthy using similar design techniques to those in the auto industry. PRT vehicles are not subject to the wide range of forces from all directions that an unrestrained automobile may experience so the PRT vehicle structural design can be more effective. The crash protection requirements for APM and LRT vehicles on the other hand relies on buffing strength which permits no energy absorption at all thus subjecting passenger to the full deceleration forces.

(f) The number of passengers involved in any disastrous event on a PRT system will by definition be small. The worst condition would be where some very heavy object fell onto the guideway. The first vehicle to encounter such an object may in all probability crash into it, but succeeding vehicles would be able to stop in time either to avoid or limit the collision force. Only a small number of passengers would be injured. Seismic damage to a PRT guideway is another possible accident mode. PRT guideways should be made flexible and due to their light weight and steel construction this is quite easily accomplished. A seismic monitoring system would allow the PRT system to be stopped quickly during an event so that even with a disruption of the guideway the amount of damage or injury would be strictly limited.

(g) A breakdown of the operating system is another area in which PRT standards differ from APM and LRT. Passenger evacuation onto a walkway is required in many Line Haul systems. This requirement stems from a variety of sources. Line Haul systems carry large numbers of passengers in a single vehicle or train, thus any incident affects a large number of people whereas on a PRT system only the passengers in a single vehicle may be affected. There is no need to evacuate a PRT vehicle since there are a number of strategies available to handle such emergencies such as the provision of battery power on-board, pushing strategies for failed vehicles and so on.

(h) Walkways beside the track have become mandatory for mass transit systems whose passenger loads per train are very large. Many APM systems also incorporate walkways between the tracks Some PRT designers have attempted to show emergency passenger evacuation from a stalled vehicle onto a channel shaped guideway, but this requires passengers to crawl around the sides of the vehicle and is essentially impractical. It is actually far preferable for passengers in a stalled vehicle to wait until the vehicle can be moved by pushing or other means.

(6) Financial Viability

It is in the area of Financial Viability that the need for PRT Standardization will be most keenly required. Those of you who have studied the history of transportation development will realize that there has been a proliferation of ideas and technical concepts in every transport field. There have been some 4,600 automobile companies since the late 19th century with every imaginable mechanical device incorporated into their designs. The plethora of streetcar manufacturers in the early 20th century was causing a major problem for cities as the suppliers went out of business due to competition from buses and automobiles and spare parts and replacements became impossible to find. The standardization of streetcar design following the Presidents Conference Committee led to the PCC cars that partially resolved the problem. Now the automobile industry has been condensed down to a few companies turning out essentially similar products. The aircraft industry has seen a similar drastic consolidation and convergence of design. In urban transit the USA has completely abandoned the field with the exception of a few bus manufacturers and one streetcar manufacturer operating under license.

It appears that the PRT industry will follow the same path of proliferation before there is a cull of the less competent designs. Why is it necessary for us to repeat this painful and financially dismal process? Some of you will say that it is necessary to test all of the possible ideas before we achieve a superior system. The theme of this paper is that with modern design simulation equipment and with an agreed consensus we can complete the trial, test and refinement process without going through the laborious trial and error process that costs so much and takes so long. Here is what a HCPRT Design Specification might look like.

(7) Standardization Criteria for High Capacity PRT (HCPRT)

The importance of PRT Standardization is the urgent need to move the technology away from the dead hand of the FTA which has opposed all technical development in the Public Transit Industry since 1975. Certainly the APM Manufacturers sit smugly with their tiny market for subsidized systems and develop nothing that is remotely cost effective, or competitive with the automobile in an urban setting.

The following Standardization Criteria are designed to establish an independent role for HCPRT that will break the shackles of the so called Transit Industry which is currently little or no industry at all in the USA.

The proposed criteria are aimed at a pure version of PRT that offers genuine non-stop transport for single individuals or single passenger groups traveling from origin to destination. Those of you who still espouse 4 and 6 seat vehicles in the fond hope that this will increase ridership should take note of the sales figures for seven passenger SUVs that carry on average 1.08 passengers per trip with a 5,000 lb vehicle achieving 12mpg.

Those of you who still cling to the idea that Group Rapid Transit (or a PRT operated in that mode) is a practical option should work through the operational dynamics of such a concept in a reasonably large application in order to understand its limitations.

Proposed HCPRT Criteria

Operating Criteria

Maximum Speed = 80 mph = 117.33 ft/sec
Maximum Acceleration = $0.25g = 8.05 ft/sec^2$
Maximum Emergency Deceleration = $1.0g = 32.20 ft/sec^2$
Maximum Service Deceleration = $0.25g = 8.05 ft/sec^2$
Maximum Jerk Accel. = $0.25g/sec = 8.05 ft/sec^3$
Minimum Headway @ 80mph = 117.33ft/sec = 0.25 sec
@ 60mph = 88.00 ft/sec = 0.33 sec
@ 40mph = 58.66 ft/sec = 0.37 sec
@ 20mph = 29.33 ft/sec = 0.50 sec

Maximum Gradient = 15%
Normal Design Maximum = 10%
Maximum Guideway Superelevation = 15%
Maximum Uncompensated Lateral Acceleration = 0.20g = 6.44 ft/sec^2

Vehicles

Vehicle Dimensions	Length = 8.00 ft Max.
	Width = 5.50 ft Max
	Height Above Guideway = 4.80 ft
Vehicle Weight	1,000 lbs empty
Maximum Loaded Weight	1,750 lbs
Passenger Capacity	Normal 3 Seated Forward Facing
With Jump Seats	Seated – 3 Forward Facing + 2 Rearward Facing
Wheel Chair Capacity	1 with adequate room to turn inside the vehicle plus at least one seated attendant
Vehicle Propulsion –	For All-Weather Applications – Linear Motors
	For Warm Weather Applications Linear Motors or Rotary
Vehicle Braking –	Dynamic for Linear Motors
	Dynamic and Friction type for Rotary Motors
Emergency Braking -	Dynamic + Mechanical to Guideway
Vehicle Amenities –	Air Conditioning, Electric Heating, Public Address System Speaker Phone, CCTV Monitor, Tilt-Up Seats, Seat Belts, Low Velocity Air Bags, Wheel Chair Restraint, Floor Level with Platform Automatic Door Opening, Int. Lighting, Radio/TV
	Chassis Fire Protection
Vehicle Control System –	Autonomous Vehicle Control, Redundant Control by On-Vehicle Computers with Failure Monitoring and Automatic Return to Depot Control.
	Vehicle Follower Separation Sensing System
	Central Monitoring and Command Control
	Guideway Zone Controls
	Station Zone Controls

Guideways

Lightweight Materials	Steel, Carbon Fiber, Other Composites Including Light Weight Concrete.
Pre-Fabricated	In sections that can be spliced at points of inflection
Structural Support	Continuous over columns
Switches	Open Type with no moving parts

Emergency Walkways	Not required in normal circumstances where access from ground level is practical
	Required in tunnels beside the guideway
	Required on river bridges between guideways
Maximum Gradient	10% Normal – 15% Absolute Maximum
Minimum Radius	In-Service Guideway – 178 ft at 20mph with balanced super-elevation
	In-Service Guideway -77 ft at 20mph with full super-elevation and max. unbalance
	In-Service Guideway – 134 ft at 20mph with no super-elevation
	Depot Guideway – 15 ft at 5mph with no super-elevation
Transition Spirals	$Ls = 0.367 \times Se^0 \times V$
	Where Ls = spiral length in Ft.
	Where V = curve design speed in mph
	Where Se^0 = Actual super-elevation in degrees
	Where Su^0 = Equivalent unbalanced super-elevation in degrees
Superelevation	Max. Balanced Superelevation = 15%
	Max. Unbalanced Lateral Accel. 0.20G (6.46ft/sec^2)
Maximum Roll Rate	4° per second
Vertical Curves –Min. Length	Parabolic
	$Lvc = AV^2/60$ on crest curve
	$Lvc = AV^2/75$ on sag curve
	Where $A = G2 - G1$ = Algebraic difference in gradients in percent
	Where $G1$ = percent grade of approaching tangent
	Where $G2$ = percent grade of departing tangent
	V = design speed in mph
Maximum Vertical Acceleration	0.05g = 1.61 ft/sec^2 on sag curve
	0.04g = 1.29 ft/sec^2 on hog curve

Power Supply

Electric Power Supply	450 to 750 vDC power rail
Power Supply Protection	Incorporated into guideway cover
EMF Shielding -	Interior Sources - Copper sheet within guideway cover (or lead)
-	Exterior Sources – Copper sheet within guideway cover (or lead)
Noise Shielding	Insulated guideway cover – allow 70dbA at 5 ft from guideway

Heat Shielding	Guideway cover protection of structure from direct sunlight
Vibration Damping	Dampers fitted at column supports
	Non-harmonic design of guideway structure

Stations

Platform Loading Berth - Station Loading Berths –	Length = 10.00 ft Maximum Parallel to Platform (No Saw-Tooth Berths) (No Reverse-Out Berths)
Platform Loading Procedure	Queue at each loading berth Single sequential queue
Platform Screen Doors –	Automatic - Essential safety and security feature
ADA Accessible -	Elevators for All Passengers
Access -	Stairs for Normal and Emergency Use
-	Escalators as Required
Station Facilities	Enclosed Platforms - Heated and Air Conditioned as required Fare Collection Equipment – automated dispensers Destination Code transfer at Boarding Berth Boarding Berth Separator Stanchions Security CCTV
Maps and Directories	Dynamic System Maps Dynamic Station Area Maps Dynamic Address Directory and Route Direction Service Public Address System Two Way Phone to Control Center
Station Structure	Modular Pre-Fabricated Berth Units to permit expansion. Free Standing and Integral Stations

Maintenance & Storage Depots

Maintenance Shop Building	Fully Enclosed
Diagnostic equipment	Fully automated
Maintenance Equipment	Fully automated wherever possible
Spare Parts Depot	Automated wherever possible
Washing and Cleaning Equipment	Fully automated
Storage Depot Building	Fully enclosed
Vehicle Storage & Retrieval System	Fully automated

Ancillary Equipment

Guideway Maintenance Vehicles	Guideway inspection vehicle
	Guideway cleaning vehicle
Guideway Maintenance Equipment	Vacuum cleaning and brushing
	Alignment verification and adjustment system
	Power supply rail cleaning & inspection
	Structural verification system
Rescue Vehicles	Manually Operated Towing Vehicle CNG Powered
Passenger Rescue Vehicles	Manually Operated Passenger Evacuation Vehicle– CNG Powered

This specification is offered as an example of a goal for HCPRT Standardization. It is intended to be aggressive in the sense that HCPRT must be competitive with the automobile if it is to create a world market. The author welcomes critique in whatever form since it is only by interaction of opinions that Standardization can come about.

VECTUS PRT CONCEPT AND TEST TRACK EXPERIENCE

Jörgen Gustafsson[1] and Svante Lennartsson[2]

[1] Chief Technical Officer, Vectus Ltd, Husargatan 2, S-752 37 Uppsala, Sweden; jorgen.gustafsson@vectusprt.com
[2] VP Engineering, Vectus Ltd, Husargatan 2, S-752 37 Uppsala, Sweden; svante.lennartsson@vectusprt.com

Abstract

PRT, Personal Rapid Transit, is a transportation concept with small vehicles on a low-weight structure offering a high level of service to its passengers. Vectus, with operations in Korea and Sweden, is at the forefront of PRT technology. Vectus has built a test track in Sweden where the system functionality and performance has been verified. Authority approval has been obtained including a complete safety case. The test track concept and the key features of the Vectus system are described along with the current operational experience. This includes two full winter seasons, proving the system's capability to cope with various ice and snow conditions. Test activities are now in the final stages of reliability and endurance testing, verifying the long term operational aspects of the system.

Introduction

PRT has been discussed as a concept for decades, and extensive research and various investigations have been done to determine its potential as a transportation system for tomorrow. Vectus is one of the few systems that has been built and that has experienced full testing. Vectus is able to fully demonstrate a PRT concept (see Figure 1) with the following key features:

- Flexible mechanical guidance system with:
- Vehicles that are captive on the track
- Vehicles that are mechanically guided along the track
- Vehicles that are mechanically steered through switches
- Headway down to 3 seconds, fulfilling brick-wall stop requirement, also in winter conditions
- Top speed about 45 km/h
- Asynchronous control system, maintaining highest performance under all operating conditions

- Dynamic moving-block vehicle separation maximizing vehicle flow and allowing speed restrictions without capacity reductions
- Distributed, reliable control with dynamic routing capable of virtually unlimited expansion
- Full winter operability with linear motor technology
- Complete and approved safety case
- Full RAM and LCC analysis

Vectus has built and operates a test track in Uppsala, Sweden. It has 400 meters (m) of track, three vehicles and one station. The test track construction started in the spring of 2006. In December the first trial runs were made with a chassis to verify the propulsion system and basic controls. The first complete vehicle arrived in the summer of 2007. Since then, tests and verifications have been made, step by step, with increasingly more complex functionality to cover all aspects envisaged to be needed in a commercial application. A thorough design process has been applied, covering safety, reliability etc., not only for the test track itself, but also for a generic PRT system. The test track has been approved by the Swedish Rail Agency for operation with passengers.

Figure 1. First complete vehicle in full operation, summer, 2007

Vectus Test Track

The test track built by Vectus in Uppsala, Sweden is a key component of Vectus PRT development. Prior to the test track being built, various studies had been conducted over several years. One important aspect of these studies was the overall control and the logistics solutions of a real system, which were studied using advanced simulation tools. Based on these studies, the key parameters for a commercially viable system were identified. From these system characteristics, the different sub-systems and key technologies were selected. Again, for selected areas, more studies were conducted in test rigs, simulators, scaled models, etc. Most of these studies were conducted in England and Sweden. It is noteworthy that Vectus PRT development did not originate from a technical concept for e.g. track and vehicle, but rather from a rigorous top-down process. Development began using the complete system as the main focus with special considerations for safety and control aspects. Systems and components were then selected to suit these criteria.

Having these steps accomplished, it was time to build the first full-scale system in 2005. Selection of a suitable site was important; a site where there was availability of good communications, possibility for testing in winter conditions, internationally recognized authorities for approval, and complementary university and industrial structures. The site selection process finally ended with the choice of Uppsala, just north of Stockholm, Sweden.

The test track configuration (see Figure 2) was carefully chosen to be as small as possible but still with the capability of proving all aspects of the full scale system. The total length of the track is about 400m. The outer loop allows speeds of up to 12.5 m/s (45 km/h). The track in the curves is cambered. The inner loop has a station with two platform positions. The tracks entering and exiting the station are fairly long to allow merge operations at full speed. Acceleration leaving the station is 2 m/s^2. Maximum retardation using the independent emergency brake is about 5 m/s^2. The headway can be as short as 3 seconds. The test track has been designed, built and commissioned with a group of Swedish and British companies as the key subcontractors.

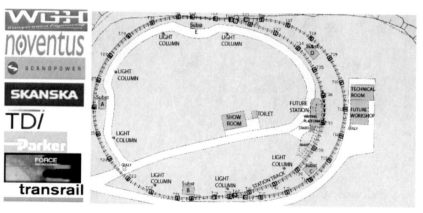

Figure 2. Top view of the test track.

The safety process has been an integral part of the test track design process. Authority approval has also been on-going from the very beginning with the Swedish Railway

Authority, with the same set-up and general requirements as are applied when e.g., building a new metro (subway).

The detailed planning for the selected site in Uppsala started in late autumn, 2005. In April 2006 the ground-breaking ceremony was held, and then the first track sections were delivered in August. Running tests with only a chassis vehicle started in December, 2006 and the first run around the outer loop was accomplished just before Christmas. In the spring of 2007 the inner loop and the station were finalized. The first complete vehicle arrived on-site in the summer of 2007 and the second vehicle arrived shortly afterwards. The system was displayed for a larger audience at the PRT seminars held in Uppsala in the autumn of 2007.

Finalization of the commissioning, safety case, and various testing led to an approval of the safety case in December, 2007, approval for test runs with visitors in March, 2008, and full approval with multiple vehicles and third party passengers in September, 2008.

The test track is a complete and genuine representation of a full scale PRT-system, in particular the function of the control system. The vehicles have been given a realistic design for the look and feel of a real application, whereas the track and parts of the electrical installations have been selected for ease of testing and quick installation rather than from an aesthetic point of view (see Figure 3).

Figure 3. Test track with three vehicles and sample 2-berth station.

The main purpose of the test track is to verify the technical design, and particularly the control aspects. It also makes it possible to carry out a complete safety case including verification and validation on a real application. Another key aspect has been the authority process itself, to identify the "real" requirements for all aspects of a system. There has been ample thought given to various passenger aspects, such as access for the disabled and guidance for the visually impaired. There is an example of a ticketing system, and complete passenger interfaces in the vehicles with displays, intercom, etc.

Another objective is to determine long-term performance. Detailed modeling of the reliability and availability and life-cycle cost were part of the design process, which requires verification. The test track also provides a proving ground for continued development, both for Vectus internal R&D as well as being able to test client-specific modifications at an early stage.

The Vehicle

The vehicle for the test track was designed to give a realistic impression of form, shape, and function. It has low weight and high rigidity. It is vandal resistant, easy to clean, maintain, and repair with no sharp corners or edges. The cabin's interior design is determined by safety requirements and complies with European disability discrimination and accessibility legislation. The layout of seating, grab handles and control devices follows recommended ergonomic practices for mass transit vehicles.

Level access is provided between the platform and cabin floor (see Figure 4). There is a pair of externally sliding bi-parting doors on each side of the cabin. The doors and door aperture curve around into the roof as close to the center line of the vehicle as possible in order to create the greatest possible headroom for passengers when entering the vehicle.

Figure 4. Vehicle access from the station platform.

The interior of the cabin is quite spacious (see Figure 5). The individual seat units are sufficiently wide to provide generous shoulder room for up to four passengers. The seats are fitted with integral 3-point seat belts and headrests and the seat base is angled to help restrain passengers during rapid periods of acceleration and deceleration. An attachment system for a wheelchair and baby or child seat (e.g.

between the two adult seats) can be provided. Interior styling will be developed according to customer specifications.

LED reading lights are available for each passenger which can be adjusted individually with control buttons on the control panel box located between each pair of seats. The control panel box incorporates illuminated disability-compliant push-buttons taken from public transport standard components. The panel also incorporates the vehicle's door control button and alarm call button that activates communication with the system operator. An LCD display/touch screen can also be incorporated for passenger information and entertainment system. The interior temperature is thermostatically controlled with an HVAC (heating, ventilation and air conditioning) unit which can also be adjusted by the passenger.

Figure 5. Vehicle interior with display panel between individual seats and roof-mounted ventilation unit.

A "FeONIC, Whispering Windows" sound system, using advanced "magnetostrictive" technology, provides passenger audio. This system effectively uses the whole interior body shell as a loudspeaker.

The chassis's wheel arrangement and emergency brake system have been designed to meet operational and safety requirements with a minimum of equipment and with as low a building height as possible (see Figure 6). Vectus uses hard tires running on steel track to minimize rolling resistance, giving low energy consumption and less sensitivity to vehicle weight.

Vectus employs switching wheels in the chassis for directional control in junctions. One of the safety advantages of mechanical vehicle-based switching is that direction selection can be made well in advance, before the junction, allowing time for the activation mechanisms to be secured in position and locked. There is also positive mechanical guidance and locking between vehicle and track throughout the whole junction. There are no moving parts at all in the track.

Effective braking in all conditions is a key function to provide a system with short headway and still maintain a brick-wall stop requirement. Specially-developed spring-applied caliper brakes acting directly on the surfaces of the guide rail produce a constant brake force under all track conditions.

Figure 6. First chassis tested for clearance on a section of track.

Track

The Vectus design philosophy has been to develop a simple rail and guidance system independent of the support structure. This allows the architect or system designer the freedom to develop the type of support structure most suitable for the application. The track design at the test track was selected primarily for testing purposes. The guidance system (in blue, see Figure 7) is mounted on a steel support tube as the main structural component. Only the top portion is required (blue) when being installed inside a building or in a tunnel, giving a low overall height requirement. A commercial design may look quite different, while still having the same guidance concept.

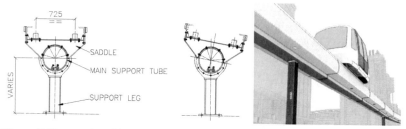

Figure 7. Test track guidance system in blue on steel tube support structure. To the right above is a possible future design of a track.

The Propulsion System

In order to run with a short headway, and with minimal margins for unwarranted safety brake applications, precise and consistent performance of the drive system is a vital component. Linear induction motor (LIM) drives have often been the choice in various PRT concepts. This is also the case in the Vectus system. Vectus has built its own control system to be able to handle all types of propulsion, and for the test track the choice fell to an in-track LIM propulsion solution.

The main reason for choosing an in-track LIM propulsion system was the capability to handle snow and icy conditions without any performance degradation. Linear motors give the same thrust regardless of the available friction between the wheel and the rail, both in acceleration as well as in braking.

Another advantage is the elimination of a current collection system which improves reliability. This is also a safety improvement, further reducing the risk for electrocution by having all electrical installations "safe-to-touch" by means of conventional cables.

Figure 8. Linear motors in the acceleration ramp out from the station in Swedish winter weather.

In the ramp leaving the station, motors are spaced twice as close as normal to provide the thrust required for the vehicles to accelerate. At the test track, the blue buildings (see Figure 8) contain the electrical equipment and the propulsion inverters. Cables are routed from the building to each LIM. In a commercial application this will be integrated into the track using specially-built, small-size inverters fully utilizing the short duty cycle for each LIM.

The Control System

The control system is the most important system for a commercial PRT application. Vectus has performed rigorous analysis of the requirements of the total system together with leading experts in the fields of logistics and transportation optimization. In the early stages, Vectus developed their own PRT simulator and tested various types of large PRT networks to verify functionalities and performance of different control schemes.

The results from the analysis can be divided into four groups of requirements.

Distributed and scalable control.

A distributed system means that the control is carried out locally for a limited part of the system. If there is a fault, it only effects a small part of the system. The rest of the system will continue to work. With the distributed system there is no increase in the load for each individual control segment when the system is expanded.

Asynchronous Control.
With asynchronous control the flow of vehicles is handled as they travel along their path to their destinations. Merging of vehicles is managed as required on a local basis. Occasionally there may be a need to slow down to facilitate merging in switches; there may even be short queues along the route at times. Travel time may be prolonged by a few seconds, but the overall capacity of the system is maintained, which is essential to the overall ability to transport passengers during periods of high system loads (rush hours, events, etc.).

Vehicle Spacing – Dynamic Moving Block.
A moving-block system is superior to any fixed-block vehicle protection system, even if the fixed blocks are very short. The moving-block system continuously updates each vehicle with information on the position of the vehicle in front of it. With this information, each vehicle can run with the shortest allowed vehicle spacing based on the worst case brake performance. With a dynamic moving block, the distance between the vehicles can also be varied depending on the speed of the vehicle (or rather the actual stopping distance depending on the speed). At lower speeds the vehicles run closer to each other; at higher speeds the distance is increased. This allows queuing on the track if required (see Figure 9) without backing up the system for long distances (the vehicles will stop almost bumper to bumper instead of one in each fixed block). Most importantly, it enables varying speed along the track, e.g. in curves, without impacting the flow of the vehicles in that track section, hence overall capacity of the system remains.

Optimal Control.
The above systems are the building blocks in providing safety, as well as adequate capacity. Then it is a matter of optimizing the logistics of the vehicles. Vectus has an adaptive-control which learns from travel patterns of traffic from previous days. This can be manually altered in the event of e.g., delays in a train arriving at one station, special events generating large crowds, etc. Effective empty vehicle management is another key feature, reducing the mileage of empty vehicles.

Figure 9. With the dynamic moving block control, vehicles can run with shorter distance between them at lower speeds.

Safety

The overall safety process in Vectus' PRT project follows the standard EN 50126/IEC62278 "Railway applications – specification and demonstration of reliability, availability, maintainability and safety (RAMS)". This is a standard that is well-implemented in the European Union and defines a process to support the identification of factors which influence the RAMS of railway systems. This safety work has been an integral part of Vectus' development from the beginning.

A number of safety targets were formulated and are as follows:

- The safety level for the PRT system shall be better than the safety levels applicable for competing transport systems such as buses, trams and taxis.
- No single failures in the PRT system shall lead to an accident with serious consequences.
- The PRT system shall be designed so that passengers feel safe and secure during the whole journey, including their time in the station areas.
- The PRT system shall be built without any serious accidents during the construction phase.
- There shall be no accidents with human injuries during the test phase of the system.

For the overall PRT system the following risk acceptance criteria was agreed upon with The Swedish Rail Agency to be met by Vectus (and PRT in general):

- Maximum 0.3 fatalities per billion person kilometres for passengers in PRT system.
- A fatality risk of maximum $1 \cdot 10^{-6}$ per year for the most exposed third person.
- For each of the sub-systems, there must be documentation that there are no single failures that can cause a severe accident.

Safety work has involved various activities such as establishing safety plans and performing safety analysis, FMECAs (Failure Mode Effect and Criticality Analysis), fault trees, safety audits, etc. in order to establish a complete safety case. The safety case was completed in 2007. The required authority approvals to carry passengers were obtained in the spring of 2008.

Wherever applicable, the safety analysis has not only been done for the test track, "as built", but also with considerations for a large, commercial system. The overall safety targets have been verified for a sample large system by performing a QRA (Quantitative Risk Assessment).

The QRA includes 78 different sensitivity calculations to verify the criticality of different input factors. For the larger system that was modeled, passenger risk is quantified to 0.165 fatalities per billion person kilometers, which is well below the acceptance criterion of 0.3 fatalities per billion person kilometers.

RAM and LCC

The RAM-process has resulted in the following:

- Generation of a qualitative and quantitative model of the PRT system, modeling the availability and reliability characteristics of all the components of the system, vehicle, track, station, etc.
- Comparison of system RAM characteristics to target values
- Contribution to a balanced design by identifying the main factors in providing adequate availability of the PRT system
- Early assessment of needs and possibilities for changes and redesigns based on a (preliminary) design phase RAM analysis
- Follow-up of RAM characteristics during test operation (see Figure 10)
- Input for maintenance planning, including definition of a necessary spare-part inventory

Figure 10. A full set of maintenance procedures has been developed that are verified during test track operation.

All relevant components in the system have been identified and each component has been assigned values for failure rate, MTTR (mean time to repair) and failure criticality. Wherever possible, sub-supplier data is used. However, most of the data is collected from generic data sources (IEEE STD-500, NPRD, etc.) and (in some cases) based on engineering judgment. For the control system, reliability is calculated using the reliability model RDF 2000.

The analysis for a typical commercial application has shown results in both station and line availability higher than 99%. System reliability is also very good, with corrective maintenance consisting of about 10% of overall maintenance. The vehicles account for about 75% of the total maintenance of the system. Items such as the on-board batteries (24V for on-board control), speed encoders and other sensors on the vehicle, doors, and some of the hydraulics are the main contributors.

A simplified life cycle cost model has been done as well. In this model, parameters can be chosen for system size, operation hours, mileage, cost of labor, energy, etc. and a good quality estimate of the operational cost can be obtained. It uses the RAM analysis combined with spare-part costs as input for all corrective maintenance. It also incorporates the actual maintenance plan with costs for consumables as a basis for planned maintenance.

Operational Experience

One of the key challenges of PRT is to run vehicles with a short headway, and even more so, to be able to merge vehicles in a junction while still maintaining a short headway. With three vehicles on the test track this has been successfully demonstrated at nominal speeds.

Extensive testing has been carried out on all systems. Operation in winter conditions has been one key test area (see Figure 11). In quite dramatic conditions with extensive amounts of snow, and also in conditions with very difficult ice crust situations on the steel track surfaces, the vehicles have cleared their way and run with minimal problems (to the contrary of most other public transport in similar situations). Most importantly of all, the brakes have also functioned under these extreme climatic conditions, and decelerations of more than 4 m/s^2 have been achieved even with the guide rail almost totally covered with solid ice.

Another important issue has been external noise. Measurements according to ISO 3095 (7.5 m to the side and 1.5m above top of rail) have been performed. Noise levels are just below 70 dBA at nominal speed, significantly lower than what is normally registered in LRTs and buses.

The propulsion system with the in-track linear motors has shown an actual performance very close to the initial calculations. Maintaining air gap has not been an issue at all and there has not been any need for re-adjustment or corrections of the LIM height settings in the track. No noticeable wear has occurred yet on the running wheels.

The main challenges that have been encountered at the test track have primarily been a result of geometrical constraints due to the limited size of the test track. The overall length of the track is relatively short, resulting in only short stretches of straight track, and it is not possible to have proper transition curves. The curve radii combined with the design speed require super-elevation for passenger comfort. This increases the transition-curve problem further. As a result of this, specific track alignment issues particular to the test track have been a complication in the testing and evaluation of some parameters.

The reliability and availability targets for the system have been set very high. So far, with approximately two years of trial operation, there have been very few failures actually requiring replacement of a component. As for the propulsion and electrical

Figure 11. Vehicle clearing its way in severe ice and snow at 5m/s.

systems, there has been only one faulty component, a communication board for a CAN bus (Controller Area Network), which was automatically diagnosed and repaired in a matter of minutes. The fault had no real impact on operation of the track, and would have caused neither a stop nor a delay in a commercial application. The few components that have been necessary to replace on the vehicle have been speed encoders, some sensors, and batteries on a few occasions. There have also been the usual door problems, but so far these have been more of an engineering kind rather than actual reliability issues.

Conclusions

Vectus set out to build a test track for its concept for Personal Rapid Transit with several objectives in mind. One of the most important goals was to prove a control concept using distributed asynchronous control based on a dynamic moving-block vehicle protection system. Proving the function itself was not the only target, but also to make a complete safety case and to obtain approval from the relevant authorities. This has been successfully completed, along with safety approval for the other systems required for PRT, i.e. track, vehicles, station, etc., and also other aspects such as operation and maintenance.

The track, the vehicle, the switching mechanism, etc., have been designed to optimize an elevated PRT-system application. Safety, reliability and performance, with headways of 3 seconds at speeds of 45 km/h, have been key design targets taking into consideration weather conditions with snow and ice. The vehicles are captive to the track and employ positive mechanical guidance through switches as important safety characteristics. The Vectus PRT concept is adaptable to a variety of propulsion system solutions. The test track has been built using in-track linear motors for propulsion. Testing has proved that the functional and performance requirements set forth for a large scale PRT system can be achieved, including successful operation in a winter climate.

The test site has provided an opportunity for an increasing number of visitors to appreciate PRT's potential to provide an attractive and modern method of transportation which includes good passenger comfort, low external noise, and high levels of security. Ongoing activities include further testing to verify reliability and also evaluation of various operational aspects of the system.

Defining the Right Roles for Automated Guideway Transit Systems

Hal Lindsey, A.M. ASCE[1]

[1]Senior Associate, Lea+Elliott, Inc., 44965 Aviation Drive, Suite 290, Dulles, VA, 20166; PH (703) 968-7883; FAX (703) 968-7888; email: hlindsey@leaelliott.com

ABSTRACT

Much attention has been, and continues to be, given to technology selection in the context of planning, procuring and implementing a new transit system. The recommended technology family sometimes evokes strong reaction, both positive and negative. This strong reaction can particularly be the case when the choices include Automated People Mover (APM) and Personal Rapid Transit (PRT) systems. Both APM and PRT systems are driverless transit systems within the larger Automated Guideway Transit (AGT) family. Neither is the right answer for every transit need. Both have a place in sound transit planning. Each one, properly applied, can fit some transit needs better than any other transit alternative.

This paper provides valuable information for a transit authority or private sector owner who is considering the implementation of a transit system. The planning process should consider numerous technology families, including fully automated transit systems (specifically driverless rapid transit, AGT, and PRT). The goal of this paper is to bring clarity to the issue of application by defining each of these three transit systems based on typical transit planning characteristics and providing contemporary references and examples that the reader can investigate further. It is hoped that the thoughts shared here will complement and reinforce a rigorous and objective planning process that seeks to satisfy transportation needs in the most beneficial manner to the owners and users of such systems.

THE CONTINUUM OF TRANSIT TECHNOLOGIES

As transit professionals we are regularly confronted with a variety of transit needs and asked to recommend a solution. While the transportation planning process and ultimate "solution" is composed of many different aspects, the recommended technology is often the most visible, therefore the most controversial – and sometimes-contentious issue. Fortunately, the transit technology toolbox contains many different technologies and offers a wide range of options from which to choose. No single technology satisfies every transit need, and some transit needs can be satisfied well by several technology families. It is short-sighted to approach developing a transit need with a pre-determined answer. This does not serve the client, the customer, or the transit industry well, as it does not fairly assess and fully evaluate the many facets of a transit need and technology response.

The key to arriving at the right answer is to very clearly define the need and to have a clear understanding of the various technology options. In some cases, the need might best be satisfied by focusing on a single technology family – like high speed rail, light rail transit, or AGT. In other cases, it makes more sense to define the critical performance requirements and issue a performance specification that allows the transit technology supply industry to respond with whatever technology family or families they deem best based on the evaluation criteria that the owner has prescribed. For example, if the evaluation criteria indicate that life cycle costs are the most important criteria to the owner, then the suppliers will likely put forward the least expensive combined capital and O&M cost solution that satisfies the performance requirements. In the aforementioned example, the most optimal solution may not have the lowest capital cost. In some cases, the owner is predisposed to a particular technology family and the transit planning professional simply advises the client of the range of available alternatives and their potential benefits to confirm the client's conclusions or to reopen the discussion.

A common list of urban transit technology families usually includes commuter rail (locomotive hauled and self-propelled electric or diesel multiple units), rapid transit (also called heavy rail), light rail (electric and diesel), AGT (includes all driverless transit systems regardless of size, speed or capacity and is further defined below) and emerging technologies (typically those at the engineering or test track stage of development). There are many overlaps among the technologies in this list and many opinions about what the list of technology families should include. This list serves to provide a framework and baseline of discussion in this paper about the AGT technology family.

AGT BY DEFINITION

AGT refers to a family of technology that is driverless. (In the context of this paper, "AGT" and "driverless" are used interchangeably.) There are many opinions about the definition of "driverless", but in this paper it is defined literally: the transit vehicle functions without a driver in the vehicle or leading car of a train.

For a transit system to be operated without a driver the guideway/track must be fully protected. In other words, it must be free of any potential access or crossing by pedestrians, automobile and other road vehicles, other non-automated transit vehicles, and animals, as the vehicle cannot "see" what is ahead of it. Instead of a driver, the train is controlled by an automatic train control system. AGT guideways can be situated above the ground level (elevated), at ground level (at-grade), or underground (including in a cut or a tunnel), as shown in Figures 1, 2 and 3 on the following pages. Guideway protection is easily accomplished in an elevated or underground environment. It is more challenging at-grade, but can be accomplished through the use of fencing and barriers.

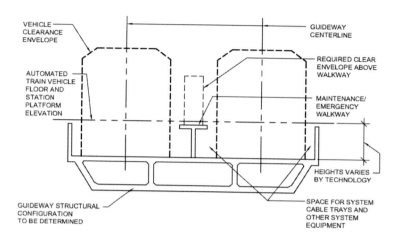

Figure 1. Typical AGT Cross Section.

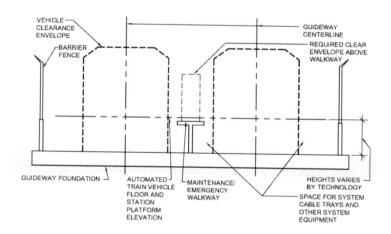

Figure 2. Typical AGT At-Grade Cross Section.

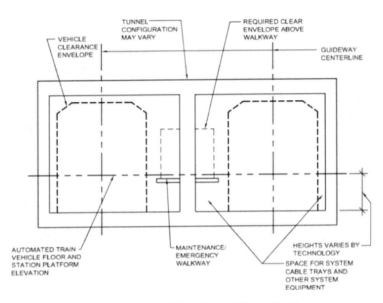

Figure 3. Typical AGT In Tunnel Cross Section.

Beyond the common requirement of full protection or separation, AGT systems come in many different shapes and sizes. They can ride on steel wheels, rubber-tires, air cushions, urethane wheels, or be magnetically levitated. They can be propelled by electric motors (rotary traction or linear) onboard the vehicle, motors or magnets embedded in the track, or pulled by a cable. Based on configuration (vehicle size and speed, headway and number of vehicles joined together) they can carry anywhere between a few hundred to over tens of thousand passengers per hour per direction.

Capital and O&M costs can also vary widely based on the location, alignment, and complexity of the guideway, degree of automation, size and number of stations, peak hour passenger-carrying requirements, and many other factors. It is a big and diverse family that was born in the 1960s with large and complex PRT and AGT systems such as Morgantown and DFW (the original Airtrans system) and this family continues to evolve and mature.

Today, there are about 100 AGT systems in service with many more in planning, procurement or under construction. Of this number, about 25 are driverless rapid transit lines or complete systems with numerous lines (designed to be driverless and those that have been retrofitted after starting out with a driver), 42 are located at airports worldwide, and the balance serve as downtown people mover systems or other urban and activity center circulators.

ADVANTAGES OF AGT SYSTEMS

There are many inherent advantages of AGT systems. A few of the most commonly acknowledged advantages are:

- Higher Average Speed – When a transit vehicle operates on a fully protected or grade-separated guideway, it can travel at a higher average speed since it does not share the guideway or track with automobiles, pedestrians or other impediments.
- Higher Capacity – Higher average speed means that the system can carry more passengers in a peak hour than a similarly configured system that has a lower average speed.
- Lower Life Cycle Costs – While the capital costs can be higher for the automatic train control system and separated guideway as opposed to a manually controlled system with conventional signaling, the operating costs of the system should be dramatically lower as drivers are not required. When these two factors are combined over the 20- to 30-year life of the system, it results in lower overall life cycle costs.
- Enhanced Operational Flexibility – When vehicles need to be added or removed from service to meet ridership demands, this is accomplished very simply by the Central Control operator using ready vehicles; manually driven systems need to schedule drivers in advance and pay them for their (often split) shifts.
- Frequent Service – By offering more frequent service, passengers have a shorter wait time and trains can be shorter in length. Shorter trains mean shorter station platforms. This translates into lower capital costs for the civil works.
- System Safety – AGT systems have a safer operating record than manually driven systems as human error is a major cause of transit accidents and fatalities.

MOST PREVALENT APPLICATIONS OF AGT

While virtually every technology family listed earlier could be designed to be driverless (by operating that technology on a separated/protected guideway and equipping it with an automatic train control system), driverless transit is most commonly found in rapid transit (a fairly recent and increasingly prevalent phenomenon) and in lower capacity APM and PRT systems. There are also many emerging technologies that propose to operate in a driverless mode and have yet to be launched. As the list of emerging technologies can be quite long, the focus of this paper will be technologies that have been launched into passenger-carrying operations.

The characteristics and examples of Driverless Rapid Transit, Automated People Mover and Personal Rapid Transit Systems are provided in the following sections.

Driverless Rapid Transit Systems. The following is a summary level description of these types of systems and their uses:

- Group Rapid Transit (as opposed to Personal Rapid Transit) – the vehicles/trains carry a large group of riders, travel along a pre-selected route and stop at each station on that route.
- Best suited for high capacity, line-haul applications (vs. feeder and circulator), yet there are some exceptions.
- Provides too much capacity, and often is inappropriate for alignment geometry and speed for most feeder and circulator applications.
- Large vehicles: 50 to 75 feet (15 to 23 m) long and 10 feet (3 m) wide for high capacity applications.
- Short-headway operation, typically in the range of 90 to 120 seconds.
- Long trains of five to ten cars.
- Peak-hour capacity can range from 10,000 to 50,000 passengers per hour per direction.
- Cruise speeds typically 50 to 60 mph (80 to 100 kph).
- Usually steel wheel – steel rail vehicles.
- Self-propelled by motorized vehicles.

Examples of Driverless Rapid Transit Systems are provided in the following photographs.

Left to right, top to bottom:

1. Vancouver Expo and Millennium Lines – Bombardier
2. Vancouver Canada Line – Rotem rapid transit vehicle, Thales train control
3. Singapore Northeast Line – Alstom rapid transit vehicle and train control
4. Paris Line 14 – Alstom vehicle with Siemens train control
5. Copenhagen – Ansaldo
6. Lille – Siemens
7. Lyon Line D - Siemens
8. Dubai Metro – Mitsubishi Heavy Industries
9. Nagoya Linimo Line - HSST

Automated People Mover Systems. The following is a summary level description of these types of systems and their uses:
- Group Rapid Transit (as opposed to Personal Rapid Transit) – the vehicles/trains carry a large group of riders, travel along a pre-selected route and stop at each station on that route.
- Best suited for feeder/distributor and circulator applications (not line haul).
- Vehicles are typically not large enough for line haul applications.
- Small to medium-sized vehicles: 20 to 45 feet (6 to 14 m) long and 8 to 9 feet (2.4 to 2.7 m) wide.
- Short headway operations, typically in the range of 60 to 90 seconds.
- Peak-hour capacity can exceed 10,000 passengers per hour per direction.
- Cruise speeds typically 25 to 40 mph (40 to 60 kph).
- Usually rubber tired vehicles, but also maglev and steel wheel-rail.
- Usually self-propelled, but increasingly cable-propelled systems are being used not only as shuttles, but also more complex configurations with switches and releasable grips.

Examples of Automated People Mover Systems are provided in the following photographs.

Left to right, top to bottom:
1. Miami Downtown People Mover - Bombardier
2. Detroit (ALRT I) Downtown People Mover - Bombardier
3. Las Vegas Monorail - Bombardier
4. Atlanta International Airport CONRAC APM – Sumitomo/MHI
5. Washington Dulles International Airport – Sumitomo/MHI
6. Chicago O'Hare International Airport – Siemens
7. Paris Charles de Gaulle International Airport - Siemens
8. Tokyo Yurikamome APM – IHI
9. Toronto International Airport – Doppelmayr
10. Mexico City International Airport – Doppelmayr
11. Perugia APM – Leitner-Poma
12. Indianapolis Clarian Hospital – Schwager-Davis

Personal Rapid Transit Systems. The following is a summary level description of these types of systems and their uses:
- Personal Rapid Transit (as opposed to Group Rapid Transit) - Passengers direct a vehicle to take them from their origin station to a specific destination station, bypassing all intermediate stations, thus offering point-to-point service.
- Best suited for many-to-many circulation or many-to-one feeder/distributer applications, but currently initial systems are few-to-one or few-to-few to test the concept.
- New generation vehicles will likely be too small for most line haul applications.

- Early PRT (Morgantown) had relatively large vehicles (20 passengers) and relatively few (4-6) stations; new generation PRT vehicles carry four to six passengers, all in the same group.
- New generation systems are mostly still in the conceptual or engineering stage, several have test track experience, and two are under construction.
- Smaller, lighter vehicles operate on a lighter, less intrusive aerial guideway and, for underground systems, in a smaller tunnel cross-section.
- New generation vehicles are 15 to 20 feet (4.6 to 6.1 m) long and typically 7 to 8 feet (2.1 to 2.4 m) wide.
- Short headway operations, typically in the range of 15 seconds. New generation systems are proposing headways lower than 15 seconds, which have yet to be tested beyond a few vehicles on test tracks.
- For new generation vehicles traveling along the guideway as single vehicles or in small platoons, peak-hour capacity is expected to be in the range of 2,000 to 4,000 passengers per hour per direction.
- New generation systems will have cruise speeds in the range of 15 to 20 mph (25 to 32 kph).
- These vehicles are usually rubber-tired. Newer generation vehicles are making broad use of battery power to promote Green initiatives.

Examples of PRT systems are provided in the following photographs.

Left to right, top to bottom:
1. Morgantown WVU – Boeing Vertol
2. DFW Airtrans – Vought, some PRT functionality, replaced in 2005
3. London Heathrow – ATS, under construction
4. Abu Dhabi Masdar City – 2gethere, first phase under construction

CONCLUSION

The transit industry has a wide range of technology available options for planning a new transit system. Beyond those that were listed in this paper are many others that are still in development; some of those will certainly come to fruition in the future. Selecting the "best" technology – family and specific technology – for an application can often include debate and spirited discussion. This is appropriate as long as the foundation of that debate is for the betterment of the system being considered (client and patrons) and the transit industry and not for parochial self-interest (whether it is for a specific AGT technology or non-automated systems). Working together, the AGT community can take advantage of new and not-so-new technology alike so that our industry can grow and we can serve our client's needs in a manner that promotes the greater transit industry. We need to: be candid about what works and what does not in different applications, share lessons learned, and help each other by cooperatively forging the best solution for the defined need. We have a tremendous opportunity to learn from one another and grow as an industry, and must capitalize on those opportunities.

ACKNOWLEDGMENTS

The author notes and appreciates the support of Harley Moore, Nicole Gray, Crystal Punzalan and Doug Draper in preparing this paper.

SOMEWHERE IN TIME – A HISTORY OF AUTOMATED PEOPLE MOVERS

William J. Sproule

Professor, Michigan Technological University, Department of Civil and Environmental Engineering, Houghton, MI 49931; phone: 906-487-2568; wsproule@mtu.edu

Abstract

The history of automated people movers is a fascinating story of innovation by governments, companies, entrepreneurs, transportation interest groups, researchers, and individuals. Some believe that the initial work began when the auto manufacturers were conducting in-house research on automated highways and other companies were developing systems using driverless vehicles on separate guideways. However the impetus for the development of these systems in the United States was provided by amendments to the Urban Mass Transportation Act of 1964. The amendments required that a project be undertaken to study and prepare a program of research, development, and demonstration of new systems of transportation. Extensive research studies were undertaken in the late 1960s and 1970s. Several manufacturers developed prototypes and early applications included installations at Tampa and Dallas-Fort Worth International Airports and in Morgantown, West Virginia. The Downtown People Mover studies generated considerable interest in the late 1970s. Research and development work was also underway in Canada, Europe, and Japan. Today there are over 130 installations of various types and configurations throughout the world and many more are under construction or are being considered. This paper travels somewhere in time to review some of the events in the development of this new transit technology.

Introduction

When one looks back in history, one is overwhelmed by the terminology, acronyms, and technologies that have evolved over the years. As a starting point, it is probably appropriate to define an Automated People Mover (APM). An automated people mover is a guided transit mode with fully automated operation in which driverless vehicles operate on fixed guideways in exclusive rights-of-way. The vehicles come in a variety of designs and they can rubber-tired, steel-wheeled, magnetically levitated, suspended or drawn by cables. The guideway structure can be constructed below grade in tunnels, at grade, or on elevated alignments. The specific design details will depend on the system but generally the guideway consists of steel or reinforced concrete sections.

The Early Years

Some believe that the initial work on automated transit technology began in the 1950s when General Motors was doing in-house research on automated highways and other companies were developing ideas on systems using driverless vehicles on separate guideways. In the late 1950s, the New York City Transit Authority experimented with automated operation for rapid transit in a project called the "Shuttle Automatic Motorman" (SAM). The system operated for about two years in the early 1960s on the 42nd Shuttle between Times Square and Grand Central Terminal.

In 1958, Alan Hewes of Cape May, New Jersey formed Universal Design Limited to develop a straddle beam monorail. His system was installed in ten amusement parks, fairgrounds and zoos before being acquired by the Westinghouse Air Brake Company (WABCO). In the late 1960s, WABCO engineers developed a fully automated version which was installed at the Houston Airport in 1972.

During the same period, Charles Paine formed the American Crane Hoist Company and one of the objectives of his company was to develop a suspended monorail system for the Los Angeles Fairgrounds in 1962 and the 1964-65 New York World's Fair. Out of his experience came the Braniff Airlines' Jetrail system. It was a fully automated suspended monorail system at Dallas Love Field Airport that connected a remote parking lot with the terminal building.

Meanwhile across the Atlantic, Habegger Limited, a small family owned Swiss firm was independently developing a "straddle beam monorail" for the 1964 Swiss National Exhibition in Lausanne. Numerous applications followed around the world and the system was first automated for Expo'67 – the world exposition in Montreal, Canada. The design proved durable and popular and was the genesis of monorails that are now offered by several companies.

These pioneering efforts initiated by small entrepreneurial firms were all low-speed systems marketed primarily for special purpose applications at expositions, fairgrounds, and zoos. Early attempts to use these simple system technologies for serious urban transit application were unsuccessful. The story might have ended had not the U.S. federal government got involved. The U.S. government began supporting automated transit systems by providing a grant to Westinghouse in the early 1960s to assist in the construction of a test facility in South Park, near Pittsburgh, for a system known as "Skybus" or "Transit Expressway". The system featured the first automated rubber-tired vehicles capable of operation at 60-second headways. The vehicles had a capacity of approximated 100 passengers and a top speed of 50 mph (80 km/hr.).

The Role of the U.S. Government

Significant impetus for the development of automated transit systems in the United States was provided in 1966 by the Reuss-Tydings Amendments to the Urban Mass Transportation Act of 1964. These amendments required that the Secretary of Housing and Urban Development to:

"undertake a project to study and prepare a program of research, development, and demonstration of new systems of urban transportation that will carry people and goods within the metropolitan area speedily, safety, without polluting the air, and in a manner that will contribute to sound city planning. The program shall concern itself with all aspects of new systems of urban transportation for metropolitan areas of various sizes, including technological, financial, economic, governmental, and social aspects; take into account the most advanced available technologies and materials; and provide national leadership to efforts of states, localities, private industry, universities, and foundations."

The resulting 1968 report to Congress, "Tomorrow's Transportation: New Systems for the Urban Future" set the tone for UMTA's (Urban Mass Transportation Administration, U.S. Department of Transportation) research and development program for the next ten years. The study, which became known as the *New Systems Study Project* also popularized such terms as "major activity centers", "dial-a-bus", "dual mode", and Automated Guideway Transit (AGT). The use of the term Automated People Mover (APM) came later.

AGT concepts were identified to move people and goods in major activity centers such as airports, shopping centers, industrial parks, central business districts, and universities. New terms were introduced to describe these systems:

Shuttle-Loop Transit (SLT) – This is the simplest type of AGT system in which vehicles would move along fixed paths with few or no switches. The vehicles simply shuttle back and forth on a single guideway, the horizontal equivalent of an automatic elevator. They may or may not make intermediate stops. Vehicles in a loop system move around a closed path stopping at any number of stations. In both shuttle and loop systems, the vehicles may vary considerably in size and may travel singularly or coupled in trains depending on the system manufacturer.

Group Rapid Transit (GRT) – This category would have more extensive use of switching. Stations may be located on sidings off the main guideway permitting trough traffic to bypass and service could be provided on several routes.

Personal Rapid Transit (PRT) – A system that could carry one person or a group of up to six persons in vehicles that operate with short headways. Operation would be fully automated to provide an optimum route over a network of guideways from origin to destination without intermediate stops. Small, unobtrusive guideways would form a gird throughout the service area, and stations would be off-line to allow through service.

Downtown People Movers (DPM) – A category related to the application of an automated system operating in the central business district.

In 1969, UMTA initiated the Morgantown project to develop an AGT system and demonstrate a system in revenue service. The system would operate on the West Virginia University campus in Morgantown. The objectives of the project were to demonstrate the feasibility of a fully automated urban transportation system, determine the potential application of such a system, and qualify the system as a candidate for use in other locations. The system incorporated features of the GRT category. Boeing Aerospace was selected as the system manufacturer. Project authorization was given in 1970, ground breaking was held in fall 1971, and the system went into revenue service in October 1975. The system was expanded in the late 1970s and it still operates on campus today.

Another federal initiative was the Transpo'72 Exposition at the Washington Dulles Airport in which four systems (the Bendix Dashaveyor, the Ford ACT, the Otis Hovair, and the Rohr Monocab) were demonstrated in limited configurations. Although the Morgantown system received some negative press during construction and in its early years of operation, these programs helped in changing the low-reliability park technologies to proven transit systems.

At about the same time, an innovative terminal design was being proposed for Tampa International Airport. To reduce walking distances, an airside-landside concept placed aircraft gates in satellite terminals that would be located on the apron and separated from a central terminal building. A key component of the concept was the need for a reliable transit system that would shuttle passengers between the satellite terminals and the central terminal building. The Westinghouse Transit Expressway was selected for the project and it went into service when the new terminal opened in 1971. A Westinghouse system was also used to link two satellite terminals to the main terminal in an expansion project at the Seattle-Tacoma International Airport in 1973. Both applications were very successful and airports quickly became, and continue to be, an important market for AGT.

Another airport project that incorporated an automated transit system in its development was the new Dallas-Fort Worth International Airport. The AirTrans system began in 1970 when an UMTA grant was made to the Dallas-Fort Worth Airport Board to finance studies and test tracks to evaluate two systems being considered for the new airport. A 1972 capital grant helped finance installation of the "AirTrans" system. An extensive network of overlapping linking four terminal buildings, a hotel, and remote parking went into service in 1974. The AirTrans system was manufactured by the Vought Corporation.

In 1974, the Transportation Subcommittee of the U.S. Senate Committee on Appropriations requested an assessment of PRT and other new systems. The work was undertaken by the Office of Technology Assessment and five areas were examined – current developments in the United States, international developments, operations and technology, social acceptability, and economic considerations. The final report, "Automated Guideway Transit: An Assessment of Personal Rapid Transit (PRT) and Other New Systems", was released in 1975. One of the report recommendations was to support an AGT demonstration project in a city to ascertain feasibility. Congress agreed and the Downtown People Mover (DPM) program was one of the results.

In 1976, UMTA solicited proposals nationwide for DPM projects. Although 68 cities responded with letters of interest, only 38 were able to submit proposals. Four cities were selected as DPM demonstration sites – Cleveland, Houston, Los Angeles, and St. Paul. Three other cities – Baltimore, Miami, and Detroit – were advised that they could divert funds from existing transit funding commitments for their proposed DPM systems. In 1977, Congress directed UMTA to consider funding for Jacksonville, St. Louis, Baltimore, and Indianapolis. Three DPM projects were eventually built in Miami, Detroit, and Jacksonville in the 1980s.

While technological development continued, the severe operational problems encountered by the deployed systems in early revenue service eroded the confidence of these systems in solving urban transportation problems. It was felt that government sponsorship of research on the critical problems of automated transit systems and an assessment of existing AGT designs was required. In response, UNTA initiated the Automated Guideway Transit Supporting technology program in 1975. The program included numerous projects aimed at specific problem areas including systems operation, safety and passenger security, vehicle longitudinal control and reliability, vehicle lateral control and switching, and guideway and station technology.

Concurrent with the DPM demonstration program, UMTA also funded the development of a new AGT technology known as Advanced Group Rapid Transit (AGRT). This program examined several advances in technology including magnetic levitation, high speed switching, and new command and control capabilities to permit short-headway operations in complex networks. As a result of substantial cost increases in the program, a review of the project's feasibility as well as its relationship to the overall goals of the Department's mass transportation program was undertaken in 1978. Among the findings in the 1980 final report was that the federal programs underestimated the complex institutional, economic, and technical barriers to innovation. Neither transit operations nor local public officials were anxious to volunteer their communities as laboratories for transit experiments unless the federal government was prepared to underwrite the financial risks. Potential transit system suppliers found it difficult to justify major investments in transit innovation given a history of uncertain federal support, tight development timetables, complex institutional barriers, and the lack of stable markets.

Activities Without U.S. Government Support

In 1981, the U.S. federal government decided to reduce its role in the research, development, and support of AGT systems. Committed DPM projects were completed in Miami (Metromover, 1986), Detroit (Detroit People Mover, 1987), and Jacksonville (Automated Skyway Express, 1989), but much of the new AGT activity in the United States shifted to applications in major activity centers. Although the specific origin is not known, it was about this time that the term Automated People Movers (APM) started to appear in the literature.

Airports continued to be a major application for AGT and today, there are over 40 airports with AGT systems and they have become a standard component of large airports. Systems have been used in amusement parks, zoos, expositions, museums, universities, hospitals, shopping centers, hotels, resorts, and casinos. Other papers presented at this conference will describe many of these systems and the planning for future installations.

One of the exciting concepts presented in the 1968 New Systems Study Project was Personal Rapid Transit (PRT). Although it was defined and research was undertaken, no systems have been built. In the early 1990s, the Raytheon Corporation and Northeastern Illinois Regional Transportation Authority (RTA) announced plans to develop a system called PRT 2000. A test track was completed and extensive research and development work was undertaken, and exciting plans were prepared for a system that would operate in Rosemount, Illinois, adjacent to the Chicago O'Hare International Airport. The program was abandoned in the late 1990s but there are several in the United States that continue to do research and examine potential PRT opportunities. One group that has been a strong promoter of PRT is the Advanced Transit Association (ATRA). The ATRA (www.advancedtransit.org) is a professional organization has been a leader in the investigation and development of advanced transit technologies and applications.

International Activities

Research and development work was also being done in Canada, Europe, and Japan. Although the initial impetus for the development of AGT systems in the United States was a desire to develop less labor intensive and innovative solutions for urban transit, most of the applications have been in major activity centers. However in other countries automated operation has been used for mass transit systems.

In Canada, the Ontario Ministry of Transportation initiated a program in the early 1970s to support the application of AGT systems. Following a review of systems, the German maglev system (Transurban) manufactured by Krauss Maffei was selected for a Toronto demonstration project. However the project was cancelled when the German government withdrew their support for the system. Another part of the Ontario transit initiative was the formation of the Ontario Transportation Development Corporation (UTDC) Limited with the task of developing new transit technologies. In an extensive research and development program, the UTDC identified the need for an intermediate capacity transit system (ICTS) to fill a void between high capacity rail and lower capacity buses and streetcars. A system was developed to serve urban transit requirements and to serve as a people mover system for airports and other major activity centers. Plans were developed for extensive systems in Toronto, Ottawa, and Hamilton, but none of the projects came to fruition. In 1981, Vancouver, British Columbia, took a bold step by selecting the ICTS technology over a conventional light rail system for an urban rapid transit line to serve the region and support the transportation theme for the 1986 World's Fair in Vancouver. The system consisted of automated trains operating in line haul service on an exclusive right-of-way. The original Vancouver Skytrain opened in 1986 with 13 miles (21.4 km) of guideway and 15 stations. Several extensions have been added to the system. In Toronto, a light rail extension was being planned between the main east-west subway line and the community of Scarborough, but in 1982 these plans were changed and the decision was made to use the ICTS technology as a manned AGT system on the Scarborough RT line. Service began in 1985. The ICTS technology was also selected for the Detroit DPM system which opened in 1987. The UTDC is now part of the Bombardier organization.

During the planning of a new town in France, it was decided that a rapid transit service was needed to link the new town of Villeneuve d'Ascq and Lille. As a result, Lille proposed that a fully automated transit line be built. Matra proposed a new system called VAL and it was selected. Revenue service began in 1983 with 13.2 km of guideway and 18 stations and the system has been expanded. The Lille system was the first automated system in line haul service. Similar systems are now operating in Toulousse and Rennes. Automated driverless trains have been introduced on rail transit lines in Europe and these will be discussed in the conference paper and presentation, *Urban Mass Transit Goes Driveless* by Mr. Gerard Yelloz.

In Japan an interest in APMs by government and industrial organizations began in the early 70s when Vought Aerospace licensed its technology to Niigata Engineering. Niigata made several improvements and the Japanese government adopted this technology as its standard and other suppliers entered the APM business. Japanese companies have become major competitors in the worldwide APM marketplace. The Japanese have also been leaders in the application of monorail systems in urban applications and they have used automated people movers as feeder systems to link new development areas with regional rail networks. Examples can be found in Kobe, Komaki, Omiya, Osaka, Sakara, Tokyo, and Yokohama.

There is also heighted interest in PRT in Europe and several papers that are being presented at this conference describe exciting projects. One example is the ULTra PRT prototype supplied by Advanced Transport Systems that is being implemented at London's Heathrow International Airport. A PRT with automated four-passenger vehicles will link automobile parking lots with the terminals. On-demand, non-stop service will be provided. A second example of PRT development activities is underway in Sweden where test tracks for SkyCab and Vectus technologies have been built in Hofors and Uppala respectively, and several planning studies are underway.

ASCE Conferences

As automated transit systems evolved, the American Society of Civil assumed a leadership role in organizing APM conferences. Murthy Bondada and Edward Neumann were the first chairs of the ASCE Automated People Movers Committee and they provided the initiative and organizational skills for the first ASCE conference on automated people movers in 1985 in Miami, Florida. This year marks the seventh conference that has been organized and sponsored by ASCE in the United States.

1985 – Miami, Florida
1989 – Miami, Florida
1993 – Irving (Las Colinas), Texas
1997 – Las Vegas, Nevada
2001 – San Francisco, California
2005 – Orlando, Florida
2009 – Atlanta, Georgia

Through the efforts of members of the ASCE Automated People Movers Committee, international conferences have been held in between the U.S. conferences. ASCE has been a cooperating agency, and members have assisted national engineering societies to organize these APM conferences.

1991 – Yokohama, Japan
1996 – Paris, France
1999 – Copenhagen, Denmark
2003 – Singapore
2007 – Vienna, Austria

APM Standards

The American Society of Civil Engineers has also taken an important role in the development of standards. Under the initial leadership of Tom McGean, the ASCE Automated People Mover Standards Committee has established safety and performance standards for an APM system. All standards are developed by a consensus process managed by ASCE that includes balloting of the standards committee and ASCE members, and balloting by the public. The first standards were released in 1997 and since then revisions and a re-balloting process has been completed, so today there are four parts.

Automated People Mover Standards – Part 1, ASCE Standard No. 21-05, American Society of Civil Engineers, Reston, VA, 2006 – minimum requirements for safety and performance of APM systems (a comprehensive revision of ASCE Standard 21-96)

Automated People Mover Standards – Part 2, ANSI/ASCE/TD&I Standard No. 21.2-08, American Society of Civil Engineers, Reston, VA, 2008 – vehicles, propulsion, and braking systems

Automated People Mover Standards – Part 3, ANSI/ASCE/TD&I Standard No. 21.3-08, American Society of Civil Engineers, Reston, VA, 2008 – electric equipment, stations, and guideways

Automated People Mover Standards – Part 4, ANSI/ASCE/TD&I Standard No. 21.4-08, American Society of Civil Engineers, Reston, VA, 2006 – security, emergency preparedness, system verification and demonstration, operations, maintenance and training, and operational monitoring

The web site for APM standards activities is www.apmstandards.org.

APM Guidebooks

Airports have been an important application for APMs. In 2003, the Airport Cooperative Research Program (ACRP) began to fund various airport research activities that have not been addressed in other programs. ACRP is sponsored by the FAA and managed through the Transportation Research Board. Two research projects have focused on airport automated people movers and final reports are expected to be published in summer 2009. One is a guidebook for planning and implementing automated people mover systems at airports and the second is a guidebook for measuring the performance of people mover systems at airports. These reports will be invaluable for future work. A third project is examining a variety of other conveyance systems, like moving walkways and escalators, in airport applications.

Guidebook for Planning and Implementing Automated People Mover Systems at Airports, ACRP Project 03-06, expected release – summer 2009

A Guidebook for Measuring Performance of People Mover Systems at Airports, ACRP Project 03-07, expected release – summer 2009

Airport Passenger Conveyance System Usage/Throughput, ACRP Project 03-14, expected release – fall 2009

Conclusion

Although APMs were initially envisioned as new urban transportation systems, they have found very special and unique applications in major activity centers in the United States while the original vision of automated systems for urban transit has occurred in other countries. Today, there are over 130 installations of various types and configurations operating throughout the world, and many more are under construction or are being planned. As one looks back who could have predicted the development of automated people movers as they are today.

Historic AGT Reports

Tomorrow's Transportation: New Systems for the Urban Future, a report from President Johnson to the United States Congress, Washington, DC. May 1968

Automated Guideway Transit – An Assessment of PRT and Other New Systems, United States Congress, Office of Technology Assessment, Washington, DC, June 1975

Impact of Advanced Group Rapid Transit Technology, United States Congress, Office of Technology Assessment, Washington, DC, January 1980

Automated Guideway Transit Technology Development, G. Daniel and others, prepared for the U.S. Department of Transportation, Report No. UMTA-VA-06-0056-80-1, Washington, DC, March 1980

ASCE APM Conference Proceedings

Automated People Movers – Engineering and Management in Major Activity Centers, Proceedings of an International Conference on Automated People Movers, Miami, FL, Edited by Edward S. Neumann and Murthy V.A. Bondada, American Society of Civil Engineers, Reston, VA, March 1985

Automated People Movers II – New Links for Land Use: Automated People Mover Opportunities for Major Activity Centers, Proceedings of Second International Conference on Automated People Movers, Miami, FL, Edited by Murthy V.A. Bondada, William J. Sproule, and Edward S. Neumann, American Society of Civil Engineers, Reston, VA, March 1989

Automated People Movers Movers III – Future Prospects for APMs, Proceedings of Third International Conference on Automated People Movers, Yokohama, Japan, Edited by Takashi Inouye, Takeshi Kurokawa, and William J. Sproule, Japan Society of Civil Engineers and Japan Transportation Planning Association, Tokyo, Japan, October 1991

Automated People Movers IV – Enhancing Values in Major Activity Centers, Proceedings of Fourth International Conference on Automated People Movers, Irving, TX, Edited by William J. Sproule, Edward S. Neumann, and Murthy V.A. Bondada, American Society of Civil Engineers, Reston, VA, March 1993

Automated People Movers V – APMs Toward the 21st Century, Proceedings of Fifth International Conference on Automated People Movers, Paris, France, Chairman Y. David, AFCET and ASCE, June 1996

Automated People Movers VI – Creative Access for Major Activity Centers, Proceedings of Sixth International Conference on Automated People Movers, Las Vegas, NV, Edited by William J. Sproule, Edward S. Neumann, and Stanford W. Lynch, American Society of Civil Engineers, Reston, VA, April 1997

Automated People Movers VII – APMs in Urban Development, Proceedings of the Seventh International Conference on Automated People Movers, Copenhagen, Denmark, Chairman L.K. Eriksen, The Society of Danish Engineers, May 1999

Automated People Movers 2001 – Moving Through the Millennium, Proceedings of the Eighth International Conference on Automated People Movers, San Francisco,

CA, Edited by Robert R. Griebenow and Ramakrishna R. Tadi, CD-ROM, American Society of Civil Engineers, July 2001

Automated People Movers – Connecting People and Places, Proceedings of the Ninth International Conference on Automated People Movers, Singapore, CD-ROM, Association of Consulting Engineers - Singapore and ASCE, September 2003

Automated People Movers 2005 – Moving to Mainstream, Proceedings of the Tenth International Conference on Automated People Movers, Orlando, FL, CD-ROM, American Society of Civil Engineers, Reston, VA, May 2005

Automated People Movers 2007 – The Sound of Moving People, Proceedings of the Eleventh International Conference on Automated People Movers, Vienna, Austria, CD-ROM, Arch+Ing Akademie and ASCE, April 2007

Eco-Industrial Design for Cityval, Siemens' new AGT

Marc Zuber, Siemens Transportation Systems

Marketing and sales support Director, Siemens Transportation Systems, 150 avenue de la République, BP101, F92320 Châtillon cedex, France
marc.zuber@siemens.com

Abstract

Automation for driverless systems is a clear market trend. Actual achievements of full automation transportation systems make it possible to reduce dwell time, adapt traffic to the demand and eliminate human errors and thus increase line capacity and network efficiency and safety.

Today these requirements of advanced passenger service and more efficient transportation means converge into a more global approach with the objective to highlight over-all benefits in term of sustainable development and reduction of carbon emission and pollution.

Cityval, Siemens newly developed rubber tired AGT, is following very strict roads in the area of Eco-Industrial Design and sustainable development. This paper is aimed at presenting these new features of our transportation solutions while focusing in three main area :

- Eco-Industrial Design
- Energy efficiency
- Modularity and operation flexibility

1. Cityval Eco-Industrial Design

The sustainable development is a reference criterion for public and decision-makers. Its principles are part of the international commitments and are about to become an essential part of the legal requirements in many countries.

Cityval is designed according to the eco-industrial development (EID) principles. Their baseline is to integrate the needs of suppliers, designers, customers and consumers since the design and to aim at a long system lifespan with reduced ecological impact.

Taking into account sustainability by the eco-design, Siemens Transportation Systems scope of work is widely speaking the scope of a system integrator. In this context, sustainability implies the introduction of clear criteria and quantifiable objectives concerning the choice of resources (materials, energy, components, facilities, subsystems, etc.) as well as a lasting performance.

These criteria and objectives are identified at design phase and are incorporated in the suppliers' specifications. Selection of each sub-system or equipment is then widely based on the compliance with Siemens EID requirements..

2. Energy efficiency

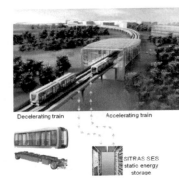

The reduction of consumption is the fastest means to reduce greenhouse effect gas emission and to minimize the ecological impact.

The energy efficiency is an important line in the design of Cityval. Efforts were made at three levels: new technologies, optimized energy management and energy recuperation.

The vehicle propulsion uses integrated equipment that limits the mechanical elements and increases the efficiency by reducing consumption and saving volume for the passengers.

The train control system, Siemens Trainguard MT CBTC contributes to the same principle: the radio communication decreases notably the number of fixed components, eliminates track wiring and reduces the maintenance effort. Furthermore, the moving block minimizes the headway between the trains and so increases the infrastructure efficiency.

Cityval benefits from the automatic traffic management optimization that reduces energy by 15 to 20%. GTT the Turin Val operator reports a reduction of about 1,500

MWhs per year equivalent to 620 tons of the CO^2 per year. Cityval recovers almost 40% of the kinetic energy when braking. The energy is reinjected into the system to be used by another train in line. The mechanical brakes are only applied for emergency braking saving the wear of brake and reducing to a minimum the particles diffusion.

3. Cityval modular concept

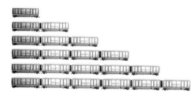

The modular concept was a major development focus during product definition. The Cityval system was designed for 1 to 6 vehicles trainset.

This innovative concept handles passenger flows from 1,000 to 30,000 pphpd (passengers per hour per direction with 4 pass./m²) without additional development costs.

The attractiveness of the service is one of the success keys in an eco-industrial approach. Two major aspects contribute to it: flexibility and quality.

The driverless CBTC Automatic Train Control provides a very wide range of operating modes between any stations on the line: loop operation, shuttle operation, pinched loop operation with turnaround before or after stations.

This increased flexibility in traffic regulation increases reactivity to exceptional events and unexpected transport demand: more frequent trains at peak times and maintaining operation off peak time or "on-demand service" at night. This adaptability makes Cityval one of the most flexible systems available for 24/7 operation.

Conclusion

The design of Cityval answers concretely to the objectives of sustainable development and has been acknowledged at European level for its innovating character and priority given to sustainable development. For these reasons, it has been awarded development funds by the French government.

Guideway Design and Construction

Brian K. Adams[1], John A. Heath[2] and Gary B. Lineback[3]

[1]Heath & Lineback Engineers, Inc., 209 Corporate Drive, Suite 300, Carrollton, Georgia 30117; PH (770)722-7631; FAX (770)424-2907; email: badams@heath-lineback.com
[2]Heath & Lineback Engineers, Inc., 2390 Canton Road, Bldg. 200, Marietta, Georgia 30066; PH (770)424-1668; FAX (770)424-2907; email: jheath@heath-lineback.com
[3]Heath & Lineback Engineers, Inc., 2390 Canton Road, Bldg. 200, Marietta, Georgia 30066; PH (770)424-1668; FAX (770)424-2907; email: glineback@heath-lineback.com

ABSTRACT

The City of Atlanta is in the midst of a massive expansion at Hartsfield-Jackson Atlanta International Airport, including an Automated People Mover (APM) operating on a 1.5 mile (2.4 kilometer) long overhead guideway and linking the main terminal with an off-site Consolidated Rental Car Facility. The system includes two terminal and one intermediate station and an elevated maintenance and storage facility. The City elected to procure the APM through the design-build-operate-maintain (DBOM) procurement method. The design of the system was constrained by the performance and aesthetic specifications set by the Owner, and clearance requirements for the major facilities crossed. The structural design was developed by Archer Western and Heath & Lineback (H&L) to produce the most efficient solution to the various constraints and to maximize the strengths and expertise of the construction company and led to a unique solution including a variety of structural arrangements. The basic superstructure type is a simple span, single cell precast prestressed concrete box beam. The deck is cast with the box and the deck edges follow the curved alignment of the guideway. The webs and bottom flange of the box are chorded to fit from pier to pier. Several spans had to reach in excess of 120 feet (36.6 meters) and in these locations the box beams were "stretched" increasing the amount of prestressing strands. The lifting and shipping weight had to be limited to 250,000 lbs (1,112 kN) however, so lightweight concrete was prescribed for spans in excess of 120 feet (36.6 meters) to a maximum of 140 feet (42.7 meters). Spans in excess of 140 feet (42.7 meters) up to a maximum of 165 feet (50.3 meters) utilized steel "tub" girders designed to the same shape as the precast beams. Several of the spans were in tight horizontal curvature (radius < 520 feet (158.5 meters)). At these radii it was not feasible to maintain the concept of chording the spans. Two or three span continuous units of cast-in-place post-tensioned box girders were detailed for these locations. The typical substructure element is cast-in-place with a single column pier and hammerhead cap aligning with the box beam soffit.

INTRODUCTION

The City of Atlanta is in the midst of a massive expansion at Hartsfield-Jackson Atlanta International Airport. The work completed to date includes the new 5th Runway and improvements to the arrivals and departures driveway. Work continues on the site preparation for the proposed International Terminal. One critical component of the expansion is that of moving all Rental Car facilities to a remote site that is accessed by an Automated People Mover (APM) vehicle operating on an overhead guideway.

The City will build two parking garages for approximately 8700 vehicles at the new site of the Consolidated Rental Car Facility (CONRAC) which is located approximately 1.5 miles (2.4 kilometers) from the main terminal. All rental car facilities including parking, maintenance and ticketing will be located at the facility. Passengers arriving at the airport will travel on the APM system to CONRAC, complete the rental paperwork and leave via new access roadways to I-85 to be built as part of the CONRAC site contract.

GENERAL LAYOUT

Beginning at the airport, an elevated station is proposed immediately adjacent to the existing MARTA (transit) station and the arrivals/departures main terminal. This Central Passenger Terminal Complex (CPTC) Station will receive passengers at ground level, with the platforms elevated above (Figure 1).

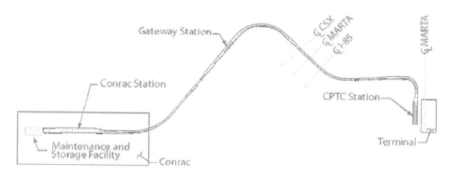

Figure 1 – General Layout

From CPTC, the system guideway will run as a pinched loop system along an all elevated curvilinear alignment that is generally north and west crossing the N. Terminal Parkway, parking lots, Airport Boulevard, I-85, MARTA, CSX Railroad, West Point Avenue, US 29/Roosevelt Highway and Convention Center Concourse on a 1.5 mile (2.4 kilometer) alignment to an elevated station at CONRAC. At approximately mid length of the alignment, an intermediate elevated platform at Gateway Station will provide access to the existing Georgia International Convention

Center and proposed hotel complex. Beyond the CONRAC Station, the alignment terminates at an elevated Maintenance & Storage facility. The entire guideway is comprised of a total of 73 spans for a total bridge length of 8,112 feet (2,472.5 meters). There are 58 precast girder spans for 5,970 feet (1,819.7 meters), 9 cast-in-place girder spans for 1,189 feet (362.4 meters) and 6 steel tub girder spans for 953 feet (290.5 meters).

The CONRAC site comprises two main parking decks wrapping to either side of the APM guideway, together with the various access roadways and the rental car operational facilities.

PROCUREMENT

The City elected to procure the APM through the design-build-operate-maintain (DBOM) procurement method. The procurement included all aspects of the structure, vehicle guidance and operational systems as well as the vehicle itself and included the Maintenance & Storage facility as well as structural work (platforms) at the Gateway and CPTC structures. A five year operations and maintenance period was included.

Upon receipt of all responses to the RFQ the City prequalified three teams and after receiving priced proposals in response to the RFP the Archer Western/Mitsubishi team was selected and negotiation proceeded to achieve agreement in the best and final offer. The complete design/build team was then:

Archer Western, Ltd/Capitol Contractors	Civil/Structural Construction
Mitsubishi/Sumitomo	Vehicle/Systems & Operations
Heath & Lineback Engineers, Inc.	Civil/Structural Design
PB Americas	Systems Design
The Architecture Group	Architectural Design
Accura/United Consulting Group	Geotechnical/QA Inspection
Street Smarts	Civil/Survey

CONSTRAINTS

The design of the system was constrained by the performance and aesthetic specification set by the Owner, and clearance requirements for the major facilities crossed. Mitsubishi offered their "Crystal Mover" vehicle for the project. The Crystal Mover is a rubber tired electric driven vehicle that can operate in a two-car or four-car configuration. Initial configuration and fleet size was based on operating two-car systems, but all components had to be designed for eventual expansion to a four-car arrangement. Mitsubishi provided the vehicle operational envelope, systems and loadings for the structural team. The allowable live load deflection was set at span length divided by 800 with L/1000 being desirable. The structure was designed in accordance with the AASHTO Standard Specifications for Highway Bridges, ASCE 21 – Automated People Mover Standards and project specific specifications.

DESIGN DEVELOPMENT

The structural design was developed by Archer Western and Heath & Lineback (H&L) to produce the most efficient solution to the various constraints and to maximize the strengths and expertise of the construction company and led to a unique solution including a variety of structural arrangements.

PRECAST BOX BEAMS

The basic superstructure type is a simple span, single cell precast prestressed concrete box beam. The section of the cell was sized so that the webs are centered under the running plinths for the vehicle. The design was optimized for tangent section of single track guideway at a span length of 120 feet (36.6 meters). In this arrangement the box depth of 5 feet (1.5 meters) gave optimum efficiency for casting, delivery and erection. Concrete strengths of 7000 psi (48.3 MPa) at 28 days and 5000 psi (34.5 MPa) at release were required. A minimum of 20 and maximum of 102 – 0.60 in. (15.2 mm) diameter straight strands were used with no more than 40% debonded.

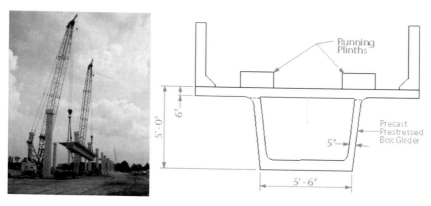

Figure 2 – Precast Box Beams

The basic box design was modified for length of track with gentle horizontal curvature (radius > 785 feet (239.3 meters)). At these radii the box was built with straight (parallel) webs but the deck slab was curved. The structure as erected is therefore a chorded structure but the deck and guidance system follow the true curved alignment.

For significant lengths of the alignment the two parallel guideways run close together (14 feet (4.3 meters) centers). In this configuration the two parallel box beams were made continuous across the width of deck slab, by means of a cast-in-place closure pour (Figure 3).

The live load deflection criteria set forth on the project specifications governed the design of many of the Precast Box Beam spans.

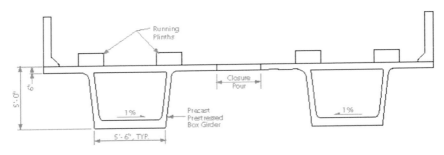

Figure 3 – Parallel Guideways

Several spans had to reach in excess of 120 feet (36.6 meters) to clear road and rail facilities. In these locations the box beams were "stretched" increasing the amount of prestressing strands. The lifting and shipping weight had to be limited to 250,000 lbs (1,112 kN) however and so lightweight concrete was prescribed for spans in excess of 120 feet (36.6 meters) to a maximum of 140 feet (42.7 meters). A unit weight of 120 pcf (1,922 kg/m^3) was used for the lightweight concrete girders. The box depth could not be varied for aesthetic reasons.

CAST-IN-PLACE BOX GIRDERS

Several of the spans were in tight horizontal curvature (radius < 520 feet (158.5 meters)). At these radii it was not feasible to maintain the concept of chording the spans. Two or three span continuous units of cast-in-place post-tensioned box girders were detailed for these locations. The box girder was designed as separate single cell structures with full width diaphragms at each pier (Figure 4).

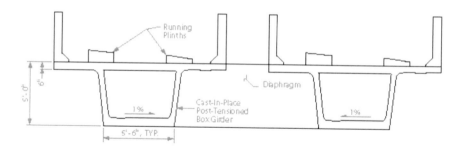

Figure 4 – Cast-in-place Girders

STEEL "TUB" Girders

The spans over the I-85/CSX/MARTA Transportation corridor and over Airport Boulevard are in excess of 140 feet (42.7 meters) (with a maximum of 165 feet (50.3 meters)) and out of the range of the precast solutions because of weight restrictions, were too high and too difficult to build on falsework for cast-in-place solutions, and there was not enough structure to justify a segmental solution. For these spans steel welded plate box girders "tubs" with cast-in-place deck were detailed (Figure 5). A four-span continuous unit in horizontal curve was used over the transportation corridor with a two-span continuous unit in tangent over Airport Boulevard.

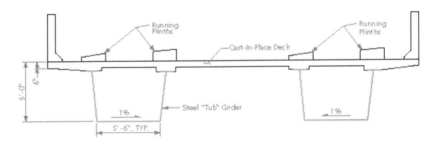

Figure 5 – Steel "Tub" Girders

BRIDGE ARTICULATION

The majority of the superstructure is supported on laminated neoprene pads with a single 1 in. diameter steel dowel engaged in either a slotted or round hole in the bottom flange of the box girder. The steel dowel is designed to keep the box girder aligned in the proper location while allowing for thermal movements when used in conjunction with a slotted hole in the bottom beam flange.

SUBSTRUCTURE

The typical substructure element is cast-in-place with a single column pier and hammerhead cap aligning with the box beam soffit. Double boxes use a single pier shaft with full hammerhead. All piers are bull nosed for aesthetic purposes and were founded on heavy steel H piling (typically HP driven to 280 tons (2,491 kN)).

The typical column dimensions used are 6'-6" (2 meters) by 3'-6" (1 meter) with a maximum column size of 9'-6" (2.9 meters) by 4'-6" (1.4 meters). The footings are typically 10'-0" (3 meters) by 15'-0" (4.6 meters). The largest hammerhead cap is 31 feet (9.4 meters) wide and 7'-7" (2.3 meters) deep at the haunch.

The lateral wind loading requirements set forth in AASHTO and the project specifications governed most of the foundation designs. The lateral deflection limits set forth on the project specifications dictated the column dimensions.

CONCLUSION

Design for APM Systems provides the structural engineer with some unique challenges. Most challenging is to meet the exacting geometrical and deflection criteria that are needed to provide a high quality smooth ride. The high visibility of the elevated system requires attention to aesthetics to ensure that the facility provides an exciting and vibrant portal. Precast concrete tub girders make this project economical while maintaining and enhancing the required aesthetics and contributing to the overall success of this rewarding project.

The completion of the Hartsfield APM, anticipated for 2009, will offer a major improvement to the Atlanta Airport facility.

HOW to DESIGN a PRT GUIDEWAY

J. Edward Anderson*

*Managing Director, PRT International, LLC, 5164 Rainier Pass NE, Minneapolis, MN 55421-1338; (763) 586-0877; jeanderson@prtinternational.com.

Abstract

The guideway is the most expensive item in a PRT system. Yet in all but a few cases the design of the guideway was more or less an afterthought – something that did not require a great deal of attention. This is a major reason many PRT systems have not survived. Primary attention had to be placed on the development and design of the control system because it was the single technological advance that made consideration of PRT possible. With limited resources, control downgraded the importance of everything else about a PRT system. During the long history of PRT development and design, guideways have been designed for Veyar, Monocab, TTI, StaRRcar, Uniflo, Dashaveyor, Morgantown, The Aerospace Corporation PRT System, Cabintaxi, CVS, Aramis, ELAN-SIG, VEC, Swede Track, Mitchell, SkyCab, Taxi 2000, PRT 2000, Microrail, Skytran, MonicPRT, ULTra, Vectus, and others. This plethora of designs likely has had much to do with the reluctance of city planners to recommend PRT. No two of these guideway designs are very close to each other. Now that the control problem is well understood, it is time to turn more attention to the guideway. The purpose of this paper is to stress the importance of adequate consideration of guideway design requirements and criteria as the basis for the design of guideways that have the potential of becoming standardized and widely deployed.

Introduction

As an engineering professor working on PRT for 13 years with no commitment to any particular system, I was privileged to visit the inventors and developers of Veyar, Monocab, TTI, StaRRcar, Uniflo, Dashaveyor, Morgantown, The Aerospace Corporation PRT System, Cabintaxi, CVS, Aramis, ELAN-SIG, VEC, Swede Track, Mitchell as well as other AGT systems then under development including Westinghouse Skybus, Jetrail, Airtrans, Ford-ACT, UTDC, Universal Mobility, H-Bahn, Krauss-Maffei, VAL, and AGRT. Later I developed Taxi 2000 and watched in dismay as it degraded into PRT 2000, mainly because guideway design was not taken seriously. Later I learned of Austran, Cybertran, SkyCab, Microrail, Skytran, MonicPRT, ULTra, and Vectus. Now there are many more offerings than I can name. Some of these systems were on paper only, some were built as test tracks, and some were built as applications, but they all provided opportunities to become aware of the variety of guideway designs.

At the University of Minnesota early in my work on PRT I coordinated a Task Force on New Concepts in Urban Transportation. We conducted planning studies of PRT for Minneapolis, St. Paul, and Duluth and soon saw that such studies were mandatory to real understanding of the problems of designing and installing a PRT system, including its guideway. We discussed our work with many public officials, planners, and interested citizens not only in Minnesota, but in many locations around the United States, Canada, Europe, and Asia. We reviewed the work of the many government-funded studies related to AGT design. The most helpful for guideway design were [Snyder, 1975], [Stevens, 1979], and [Murtoh, 1984]. Out of this experience, I was able to write down a hopefully comprehensive set of requirements and criteria for the design of a PRT guideway, and subsequently found a design configuration that met them all. The discussion in this paper applies to <u>elevated</u> guideway structures for the simple reason that after trading off underground, surface-level, and elevated systems planners almost always opt for elevated systems.

As overall guidance for guideway design I find it difficult to improve on the following statement [Pushkarev, 1982] by Louis J. Gambaccini, New Jersey Transportation Commissioner and creator of the nation's first statewide public transit agency.

> "Fixed guideway transit is not a universal solution nor should it be applied in all urban areas. Fixed guideway is a potential strategy, as is the bus, the ferry boat, the car pool or the van pool. In many possible applications, fixed guideway is a superior strategy. But whatever strategy is finally selected, each should be evaluated not in the narrow context of transportation alone, nor solely in the framework of accounting. It should be measured in the broader context of its contribution to the overall long-term aspirations of the urban society it is supposed to serve."

Our challenge today is to design and build PRT systems even more able to "contribute to the overall long-term aspirations of the urban society" than Mr. Gambaccini could imagine thirty years ago.

Definitions

From the Oxford American Dictionary:

A Need: A circumstance in which a thing or a course of action is required.
A Criterion: A standard of judgment.
An Attribute: A quality that is characteristic of a person or thing.

From Wikipedia:

A Requirement: A necessary attribute, capability, characteristic, or quality of a system in order for it to have value and utility to a user.

Design Process

After decades of experience in the practice and teaching of engineering design I realized that the first step in a design process is to study deeply and follow rigorously a comprehensive set of rules of engineering design. I make no claim that my set

[Anderson, 2007a] of such rules is complete, and I welcome collaboration with other experienced engineering designers to develop a more comprehensive set. But I have observed that the less successful PRT guideway designs have resulted primarily from violating one or more of these rules. What is now commonly called "risk management" consists mainly in following rigorously such a set of rules. My contribution was inspired by reading, as a young design engineer, the Rules of Engineering of W. J. King [King, 1944]. Beginning with these rules, the design processes I used to arrive at my conclusions about the design of a PRT system are summarized in a DVD [Anderson, 2008d].

The next step is to write down a simple statement answering this question: What does a PRT guideway really need to do if it is to win competitions? Here is my short answer:

> A PRT guideway must carry vehicles containing people safely, reliably, and comfortably in all reasonable environmental conditions for up to 50 years over curves, hills, and straight sections at an acceptable range of speeds, acceptable cost, and acceptable visual impact.

But, we need to be more specific. Only by long experience in the design of whole PRT systems can one unearth all of the requirements and criteria for guideway design. Designing a PRT guideway cannot be done successfully without a great detail of development work on the whole system because the guideway design depends on other system features and other system features depend on guideway specifics [Anderson, 2000, 2008a]. In the following section, in no particular order, I give my list of guideway design requirements. All are important. To be successful, none can be ignored. For clarity and ease of reading, I list the requirements for the design of an elevated guideway without comment and without quantification. I then discuss alternative system issues and tradeoffs that in some cases affect guideway design and in others are influenced by the guideway-design requirements. Next, I list three guideway-design tradeoffs. Then, I suggest design criteria. Finally I state how, by using this process, I arrived at my guideway design. My bottom line goal for decades has been to design a system of urban transportation that can recover all of its costs from revenue – to turn urban transportation into a profitable enterprise.

PRT Guideway Design Requirements

1. The guideway must assure an acceptably high level of safety for the passengers that ride in the vehicles mounted on it in all reasonable circumstances.

2. Consistent with other requirements, the guideway must have minimum size, weight and capital cost.

3. The appearance of the guideway must be acceptable and variable to suit the community.

4. The switching concept for merge and diverge sections of the guideway must be straightforward, easily explained, and one of the first items to clarify while developing the configuration.

5. Accommodation of hills, valleys, and horizontal curves must be straightforward.

6. The design must permit straightforward manufacturability and installation.

7. Ride comfort must be acceptable.

8. The design must be compatible with the Americans with Disabilities Act.

9. The guideway must be designed to minimize operating cost.
10. The minimum span length must be determined from careful city planning.

11. The guideway must be designed for long life under the variable vertical, lateral, and longitudinal loads that can reasonably be expected.

12. The guideway must be designed to withstand reasonable earthquake loads.

13. There can be no passenger injury due to collisions of street vehicles with support posts, falling trees, etc. if such events may be possible.

14. The system must be designed to operate in the presence of wind, rain, snow, ice, lightning, dust, salt and other airborne corrosive substances, nesting birds and insects, i.e. in a general outdoor environment.

15. The guideway must be designed so that under winter conditions, guideway heating will not be necessary, except for systems not intended to operate under winter conditions.

16. The guideway must be easy to erect, change, expand, or remove.

17. The guideway design must permit access for maintenance.

18. The guideway must be designed for relief of thermal stresses.

19. The guideway must be designed for competitive operating speeds.

20. The guideway design must permit the system to expand indefinitely.

21. If power rails are used, the guideway must be designed so that frost will not form on them.

22. It must be very difficult if not impossible for anyone to be electrocuted by the system.

23. The guideway must be designed with adequate torsional stiffness.

24. It must be very difficult if not impossible to walk on the guideway.

25. The guideway design must liberalize the required post-settling tolerance.

26. The guideway design must eliminate slope discontinuities.

27. There must be space in the guideway for the communication means.

28. The design must minimize electromagnetic interference.

29. The design must minimize acoustical noise.

30. The design must minimize the potential for vandalism or sabotage

31. Provision must be made in the guideway design to prevent corrosion.

32. There must be no place in the guideway for water accumulation.

33. The design must provide for vibration damping.

Issues and Tradeoffs in PRT System Design

Early in my career at the University of Minnesota, I was privileged to hear a lecture by California Institute of Technology Professor Fritz Zwicky, in which he stressed "the morphological approach which attempts to view all problems in their totality and without prejudice." During World War II, he was deeply engaged in the design of jet engines, in which process, before any detailed design was begun, he and his colleagues wrote down in chart form every way they could conceive that a jet engine could be designed. The process described in his book [Zwicky, 1962] is general. It is a useful guide to the design of anything, and it strongly influenced the way I taught engineering design and in the methodology I practiced in the design of my PRT system. Zwicky's influence is present in the preceding and following discussion. One makes progress by "standing on the shoulders of giants." Zwicky was one of the giants. Here are some of the results of morphological thinking:

1. Safety issues. These issues are mentioned because they need to be treated as part of the overall PRT system design process. Neglecting any one of them can result in rejection. Discussion of the details is, however, beyond the scope of this paper. [Irving, 1978; Anderson, 1978a; Anderson, 1994]
    a. How can the control system be designed for maximum practical safety?
    b. How can the vehicles be designed for maximum practical safety?
    c. What should be the minimum operational headway?
    d. Should seat belts, air bags, or neither be required?
    e. Should shock-absorbing bumpers be designed into the vehicles?

    f. How should potential collisions with street vehicles or other objects be handled?
    g. How can people be prevented from walking on the guideway?
    h. How can the possibility of electrocution be prevented?
    i. How should fire safety issues be handled? NFPA 130.
    j. How should evacuation and rescue be handled?

2. Is the system predominately elevated, at grade, or underground? The issues are
    a. Congestion relief
    b. Safety
    c. Land requirements
    d. Costs

3. Is a walkway along the guideway necessary?
This issue has been debated for a long time [NFPA 150, Anderson, 1978b]. If one or more vehicles are stranded on the guideway, how should passengers be rescued? The requirement of a walkway will make the guideway larger and more expensive, for which reason the guideway designer would like not to be required to include walkways. There are two essential subsidiary considerations that must be understood:

    a. Can all kinds of people including the elderly and the disabled in all reasonable kinds of weather use a walkway? Could a walkway be acceptable in rainy, snowy, or windy conditions? A little reflection shows that a walkway would be usable for the more able bodied people in a warm and dry climate, and thus, if PRT is to be acceptable for all people, it must be possible to design the system in such a way that the mean time between incidents in which a walkway would be desirable is long enough to be acceptable [Anderson, 2006], and in the remote situation in which someone might need to be rescued a means other than a walkway is acceptable.

    b. Can the system be designed in such a way that the mean time between circumstances in which a walkway would be useful is so rare that other rescue means become acceptable?

These questions were studied in sufficient detail in the Chicago PRT Design Study[1] that it was concluded that walkways would not be required except in circumstances such as river crossings. When there is ground underneath the guideway, the preferred alternative rescue means would be a fire truck or a cherry picker. Even when crossing rivers, detailed work on analysis of hazards and potential failures and their effects [Stone & Webster, 1991] resulted in the conclusion that rescue could best be accomplished by means other than a walkway. The study team concluded that PRT systems can be designed to be sufficiently simple and reliable that walkways will not be needed.

---

[1] Formally, the Northeastern Illinois Regional Transportation Authority PRT Design Study of 1990.

4. Should the system be dual mode or single mode, i.e., with vehicles captive to the guideway? This question has been studied [Irving, 1978; Anderson, 2007b] in sufficient detail to convince us that we should concentrate on single-mode PRT systems. We considered many issues including
   a. The effect on community development patterns.
   b. The effect on system cost and ridership.
   c. The effect on capacity.
   d. The effect on those who cannot, should not, or prefer not to drive.

5. Should the vehicles be supported above the guideway or should they hang below? This is a complex tradeoff that I have examined in increasing detail [Anderson, 2008b]. The issues are:
   a. Visual impact
   b. System cost
   c. Natural frequency
   d. Ease of switching
   e. Rider security
   f. All-weather operation
   g. Torsion in curves

6. How should the vehicles be suspended? [Anderson, 2008c]
   a. Wheels
   b. Air cushions
   c. Magnetic fields

7. How should the vehicles be propelled? [Anderson, 1994; 2008d]
   a. Rotary motors
   b. Linear motors
      i. Induction
      ii. Synchronous
      iii. Air
      iv. Rope

8. What should be the people-carrying capacity of the vehicles? [Anderson, 1986]
   a. Understand the size of groups in which people travel.
   b. Understand the ease of taking two or more vehicles.
   c. Understand the effect of vehicle size on system cost.
   d. Need to accommodate wheelchair + attendant, bicycle, baby stroller, or luggage.

9. Assuming electric motors, should they be rotary or linear? [Anderson, 1994]

10. Should the motors be on board the vehicles or at wayside? [Anderson, 2008d]

11. If the motors are on board, should they draw power from batteries or power rails? [Anderson, 2008d]

All of these tradeoffs and more will affect the cost and performance of the system and should be studied very carefully before detailed design is initiated.

Tradeoffs in PRT Guideway Design

1. Cross sectional dimensions: The minimum-weight cross section should be used. [Anderson, 1978, Chapter 10; 1997; 2007c]
2. Material: Steel, concrete, composite?
3. Truss or plate or pipe?

PRT Guideway Design Criteria

1. <u>Vertical and Lateral Design Loads</u>. This is the only set of criteria considered by Moutoh, 1984. One must consider dynamic loading due to vehicles moving at speed, wind loads, earthquake loads, longitudinal loads due to braking vehicles, and loads due to street vehicles crashing into the support posts, if that is to be permitted. The best study I have seen on dynamic loads is one done in the M. I. T. Mechanical Engineering Department by Snyder, Wormley, and Richardson [Snyder, 1975]. In their computer studies, they simulated vehicles of various weights operating at various speeds and various headways, and running over guideways of various span lengths. By placing their results in dimensionless form, the usefulness was extended considerably. I studied their results [Anderson, 1978a] and noted that the shorter the minimum headway the smaller was the difference between dynamic and static deflection, and in the theoretical limit of zero spacing between vehicles the dynamic and static deflection are the same, i.e., the guideway cannot tell the difference. Assuming PRT vehicles operating at a minimum headway of half a second, I found that the maximum dynamic guideway deflection and stress with vehicles operating at line speed was less than the maximum deflection and stress with vehicles nose-to-tail on the guideway. Therefore the maximum possible vertical load becomes a uniform load and it is easiest to calculate. The loading criteria used in the Chicago PRT design study were

    1) Fully loaded vehicles nose to tail on span + 30 m/s (70 mph) crosswind.
    2) No vehicles + 54 m/s (120 mph) crosswind. [I now assume 80 m/s (180 mph)]

    The maximum wind load on a guideway can be substantially reduced by reducing its drag coefficient based on known wind-tunnel data [Hoerner, 1965], [Scraton, 1971].

2. <u>Longitudinal loads</u>. The criterion is based on vehicles operating at minimum headway all stopping simultaneously at 0.5 g. I found this load to be less than the maximum wind load.

3. <u>Earthquake load</u>. There is debate on the maximum horizontal acceleration measured due to an earthquake. In a presentation at a Society of American Military Engineers conference in San Diego in the last week of March, 1994, shortly after the Los Angeles earthquake, an Army Major General who had been placed in charge of rebuilding the Los Angeles freeways told his audience that the maximum horizontal acceleration measured was 1.6g, which is higher than any figure I have seen in print. The bottom line, though, is that the lighter the elevated structure, the easier it is to design foundations to withstand such loads. I have found that for the guideway I designed a horizontal acceleration of the ground of 0.86 g is equivalent to a wind load of 80 m/s (180 mph). A PRT guideway must be designed to the local earthquake code, which varies considerably from one region to another.

4. <u>Design stress</u> – The designer must use standard values for the selected material.
    a. Specify corrosion protection for the life of the structure.
    b. Prevent water accumulation.
    c. Plan to clean out any bird droppings, which are corrosive.
    d. Design to account for material fatigue over the specified life.
    e. Design to relieve thermal stresses.

5. <u>Maximum allowable deflection</u>. The standard for steel transit guideways is span/1000 whereas the AASHTO bridge standard has been span/800.

6. <u>Minimum allowable span</u>. The Chicago PRT design study conclusion: 28 m (90 ft)

7. <u>Ride Comfort</u>
    a. Observe the ISO standards for acceleration vs. frequency
    b. Observe the ISO standard acceptable constant acceleration and jerk for normal and emergency operation, which are also given in the ASCE APM Standards.
    c. Crossing frequency of vehicles out of phase with natural frequency of guideway to prevent resonance.

8. <u>System Life</u>. The Chicago RTA specified 50 years.

9. <u>Compliance with the Americans with Disabilities Act (ADA)</u>.
    a. Must accommodate a wheelchair with an attendant.
    b. In the Chicago study, the disability community strongly demanded access to every vehicle, with the wheelchair facing forward.
    c. Must provide for visual and hearing disabilities.

10. The <u>minimum line headway</u> needs to be specified at the beginning of the design program based on detailed site-specific planning studies. When it is not, as has usually been the case, the system may be destined for a limited range of applica-

tions. Based on many independent studies we have designed for a minimum headway of half a second. [Anderson, 1994]

11. Design for the <u>expected environment</u>
    a. Rain, ice, snow of a given rate of accumulation.
    b. Ambient temperature range, typically -40°C to +50°C.
    c. Lightning protection.
    d. Sun.
    e. Dust, sand, salt.
    f. Nesting bees, birds, squirrels, etc.
    g. Earthquakes – Design to maximum expected horizontal acceleration at the site.
    h. Fire. [NFPA 130]
    i. Vehicles crashing into posts. [Anderson, 2006, Appendix A]
    j. Interference from other elements of the urban scene.
    k. Ice build up on power rails due to clear winter night sky.

12. <u>Speed range</u>. Select the cruising speed to minimize cost per passenger per unit of distance.
    Consider that turn radii, stopping distance, kinetic energy, and the energy needed to overcome air drag all increase as the square of speed; and that energy use depends on streamlining, low road resistance, and propulsion efficiency. Consider that the maximum operational speed for acceptable ride comfort is proportional to the guideway natural frequency, which depends on guideway stiffness and the type of support. [Anderson, 1997]

13. <u>Costs</u>. The design team should aim for costs sufficiently low to be recoverable in fares, i.e., the system should be designed to be a profitable private enterprise. Such a conclusion clearly cannot be reached without a great deal of development work, but by striving for this goal the design team will insure its future.

14. Require a small amount of <u>vibration damping</u> in the guideway.

15. <u>Acoustical noise</u> should be less than the noise of automobiles on streets.

16. <u>Electromagnetic noise</u> generated cannot interfere with existing devices.

17. <u>Communication</u> means must be accommodated.

18. <u>Expansion</u>. Design so that the system can be expanded indefinitely.

19. Design to minimize the effects of <u>vandalism and sabotage</u>.
    a. Assign young engineers to study ways to vandalize the system and how to prevent it.
    b. The spread-out nature of a PRT system provides no inviting target.

My Conclusions [Anderson, 2007c, 2008a, 2008d]

1. Resolving the basic tradeoffs related to the guideway, I reached the following conclusions:
    a. The guideway will be mostly elevated.
    b. Single mode.
    c. Supported vehicles.
    d. Wheeled suspension.
    e. Linear-induction-motor propulsion.
    f. Motors on board, powered via power rails.

2. Before designing the guideway, determine the vehicle maximum weight with careful weight-minimization design.

3. Use the optimum guideway cross section for minimum weight and cross sectional area.
    a. The optimum guideway is narrower than it is deep.
    b. A vertical chassis is required.
    c. Careful attention must be given to the attachment of the cabin to the chassis. Detailed finite-element analysis gives a practical solution.

4. The minimum-weight, minimum-size guideway is a steel truss.
    a. Robotic welding is required for acceptable cost.
    b. Corrosion protection is required.
    c. The guideway should be clamped to the posts for maximum stiffness.
    d. Expansion joints should be placed at the point of zero bending moment in uniformly loaded spans.

5. Cover the truss with composite covers, opened 10 cm (4 in) at top, 20 cm (8 in) at bottom, with curve radii at top and bottom $1/6^{th}$ guideway height, hinged at bottom and latched at top, with a thin aluminum layer and sound-deadening material on the inside [Anderson, 2008a]. The benefits are:
    a. The interior of guideway is protected from all but very minimum snow and ice.
    b. The interior is protected from effects of the sun on the tires and other equipment.
    c. Differential thermal expansion is eliminated.
    d. The exterior environment is shielded from electromagnetic and acoustic noise.
    e. The power rails (if used) are protected from the winter night sky, which prevents ice accumulation.
    f. Wind drag is 40 % of that on an opened truss. [Scraton, 1971]
    g. The interior of the guideway can be accessed for maintenance.
    h. The appearance of the guideway can be selected to suit the community.

General Conclusions

I studied the field of PRT pro and con for several years before becoming sufficiently convinced that within this technology, properly optimized, would someday be a means of realizing urban environments of a quality far superior to that possible using only conventional technology, and thus serious involvement would be a worthy use of my time. A very important finding of the UMTA studies of 1968 was that if only conventional transit would be deployed, congestion would continually increase, but if the new personal transit systems would be deployed, congestion could be contained. [Hamilton, 1969]

I envisioned more and more clearly as the years of my involvement progressed that when PRT in some form becomes accepted by the transportation-planning community it will be open technology, studied in regular engineering classes in universities. There will have to be sufficient commonality in these systems so that planners will be convinced that they can be expanded and supported with assurance that multiple suppliers will be available decades hence. Ideally, the preferred designs will be determined by the market place. The details of the designs will be found in the open literature. Vehicles will be supplied by various companies, will be selected by competent engineering companies in consultation with the client, and will operate on standard guideways. Control systems including their software will be studied in universities and trade schools, and a number of companies will supply them. This is exactly the way all civil works are designed, bid and built. Good engineers and engineering companies will do well in PRT based on competence, as is true of those who specialize in other public works. The all-too-common thought of inventing a unique system that will make one fabulously rich is not only a fallacy but may be a major deterrent to full commercialization of this technology. Those of us interested in realizing this technology need to work together for the betterment of society.

Quoting the Engineers' Creed [NSPE, 1954]: *"The engineer places service before profit, the honor and standing of the profession before personal advantage, and the public welfare above all other considerations."* Roads are funded by governments. The rights of way for PRT guideways will have to be provided by governments and their designs will have to be approved by governments if the potential of PRT is to be realized. Paraphrasing Gambaccini: *"PRT should be measured in the broader context of its contribution to the overall long-term aspirations of the urban society it is supposed to serve."*

References

Anderson, J. E. 1978a. *Transit Systems Theory*. Lexington Books, D. C. Heath and Company.
    Available on www.advancedtransit.org.
    Chapter 7, Requirements for Safe Operation

Chapter 8, Life Cycle Cost and Reliability Allocation.
Chapter 9. Redundancy, Failure Modes & Effects, and Reliability Allocation.
Chapter 10, Guideway Structures.
Chapter 11, Design for Maximum Cost Effectiveness.

Anderson, J. E. 1978b. Get Out on the Guideway and Walk. *Advanced Transit News*, 2:5.

Anderson, J. E. 1986. Automated Transit Vehicle Size Considerations, *Journal of Advanced Transportation*, 20:2:97-105

Anderson, J. E. 1994. Safe Design of Personal Rapid Transit Systems. *Journal of Advanced Transportation*, 28:1:1-15.

Anderson, J. E. 1997. The Design of Guideways for PRT Systems. www.archive.org, Enter www.taxi2000.com. Click on any date in 2001. Click on Publications. Paper #18.

Anderson, J. E. 2000. A Review of the State of the Art of Personal Rapid Transit. *Journal of Advanced Transportation*, 34:1:3-29.

Anderson, J. E. 2006. Failure Modes and Effects Analysis and Minimum Headway in PRT. www.prtnz.com.

Anderson, J. E. 2007a. Fifteen Rules of Engineering Design. www.prtnz.com.

Anderson, J. E. 2007b. How does Dual Mode Compare with Personal Rapid Transit? www.prtinternational.com

Anderson, J. E. 2007c. The Structural Properties of a PRT Guideway. Available on request.

Anderson, J. E. 2008a. An Intelligent Transportation Network System. www.prtinternational.com.

Anderson, J. E. 2008b. The Tradeoff between Supported vs. Hanging Vehicles. www.prtnz.com.

Anderson, J. E. 2008c. Maglev vs. Wheeled PRT. www.prtnz.com.

Anderson, J. E. 2008d. Solving Urban Transportation Problems through Innovation. A video. www.prtinternational.com.

Hamilton, W. R. and Nance, D. K. 1969. Systems Analysis of Urban Transportation. *Scientific American*, July 1969.

Hoerner, F. 1965. Fluid Dynamic Drag. Amazon Books.

Irving, J. H., Bernstein, H., Olson, C. L., and Buyan, J. 1978. *Fundamentals of Personal Rapid Transit*, Lexington Books, D. C. Heath and Company, Lexington, MA.

King, W. J. 1944. The Unwritten Laws of Engineering. The American Society of Mechanical Engineers. United Engineering Center. 345 East 47th Street, New York, NY 10017.

Moutoh, D. U. 1984. Investigation of Structural Design Criteria for Automated Transit Aerial Guideways. N. D. Lea & Associates, Inc. Report No. UMTA-IT-06-0311-84-1.

NSPE. 1954. Engineers' Creed. National Society of Professional Engineers.

Pushkarev, B. S., Zupan, J. M., and Cumella, R. S. 1982. Urban Rail in America; An Exploration of Criteria for Fixed-Guideway Transit. Indiana University Press. Bloomington, Indiana.

Scraton, C. and Rogers, E. W. E. 1971. Steady and Unsteady Wind Loading. *Phil. Trans. Roy. Soc.* London a. 269:353-379.

Snyder, J. E., III, Wormley, D. N., and Richardson, H. H. 1975. Automated Guideway Transit Systems Vehicle-Elevated Guideway Dynamics: Multiple-Vehicle Single Span System. Report No. UMTA MA-11-0023-75-1

Stevens, R. D., Silletto, J. G., Wormley, D. N., and Hedrick, J. K. 1979. AGT Guideway and Station Technology, Volume 6, Dynamic Model. Report No. UMTA-IT-06-0152-79-5.

Stone & Webster. 1991. System Design Report, Section 7, Policy Issues, pp. 113-4. Personal Rapid Transit Program, Regional Transportation Authority, Chicago, Illinois.

Zwicky, F. 1962. Morphology of Propulsive Power. Society for Morphological Research, Pasadena, California.

Heathrow PRT Guideway, Lessons Learned

A D Kerr[1] and R J Oates[2]

[1] BE, MSc, DIC, CEng, MICE, Director, Ove Arup & Partners Ltd,13 Fitzroy Street, London, W1T 4BQ, United Kingdom, tony.kerr@arup.com, Phone +44 (0) 20 7636 1531, Fax +44 (0) 20 7580 3924
[2] MEng, CEng, MICE, MIStructE, Senior Engineer, Ove Arup & Partners Ltd,13 Fitzroy Street, London, W1T 4BQ, United Kingdom, rachel.oates@arup.com, Phone +44 (0) 20 7636 1531, Fax +44 (0) 20 7580 3924

ABSTRACT

The construction of the guideway for the Heathrow Airport pilot PRT system is complete and the opportunity has been taken to re-assess some of the critical design decisions. The objective of the designers has been to develop a family of modular elements which can be manufactured in a factory, installed using simple and fast construction methods, and which link to provide a visually coherent linear feature. As this is the first application of PRT the design decisions will also inform subsequent design codes or standards and for this reason there is value in reassessing some design outcomes.

1   INTRODUCTION

The construction of the guideway infrastructure for the Heathrow Airport (LHR) pilot ULTra PRT system is substantially complete and the opportunity has been taken to re-assess some of the critical design decisions. At the outset the objective of the designers had been to develop a family of modular elements which could be manufactured efficiently and to reliable tolerances, installed using simple, quick construction methods, and such that the elements provided a visually coherent linear track or guideway feature.

The LHR guideway includes elevated and at-grade sections, a station on part of the floor of a multi storey carpark adjacent to the new Terminal 5, and two at-grade stations in a surface car park beyond the end of the north runway. The two-way at-grade section is some 440m in length; the elevated guideway is some 1360m in length, comprising both single and two-way elements, giving a total route length of 2.3km, excluding station and depot paths. A route plan is shown in Figure 1.

The LHR guideway construction has confirmed that the tolerance standards adopted by the requirements for vehicle control and passenger ride comfort are at the limit of that which the UK civil engineering construction industry can achieve, even through the application of factory manufacture techniques for standard steel and concrete elements. This finding is being further evaluated to establish the relationship between operations and the alignment and the ride quality aspirations.

The constraints at LHR required that the overall vertical and horizontal alignments be carefully threaded between fixed points, whilst at the same time respecting ride comfort standards. This has involved specifying the three-dimensional location of points on the final alignment, with a specified variation or tolerance.

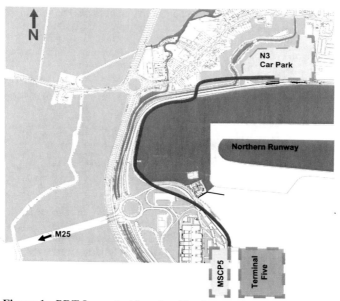

Figure 1: PRT Layout at London Heathrow Airport

As this is the first application of PRT, the design decisions have informed a Design Code developed in parallel with the design, and prepared as a joint effort between system and infrastructure designers. The objective of this Code has been to capture decisions made with their reasons and to provide an authoritative basis for the design. This was intended to limit the need for future checkers and reviewers to take questions back to first principles in order to be satisfied that the solution offered would deliver a safe, reliable and appropriate guideway. Where relevant, the Code refers to existing UK or USA codes or standards, in particular with relation to the expected properties of steel, concrete and corrosion protection. The important consideration however has been to capture relevant aspects of operations and maintenance for their particular requirements of the guideway. The outcome from reassessment of the design approach will be reflected in Design Code revisions.

The issue of a passenger comfort standard has been considered and parameters developed with the LHR guideway design based on ASCE APM standards. This has resulted in relations linking speed and alignment radius based on moving vehicle mechanics, a limiting lateral acceleration of 2.5m/s and an angular velocity limit of 0.5 rad/s. This has been translated into lengths of transition curves at entry and exit to circular curves. Standards have been specified for surface regularity and steps at adjoining running planks, which relate to a jerk standard and experience of the transmission of irregularities through the vehicle suspension system to the passengers. An early design decision was to avoid super-elevation in the alignment and this has influenced the speed of travel around route turning radii.

On the basis of experience from the design and construction of the LHR pilot guideway for PRT, three topics have been identified for further discussion in this paper, towards refining the approach to the next application, and confirming appropriate design parameters. These topics are discussed in detail in the following sections of this paper and are:

- *Modular design for elevated guideway*
  The objective of modular design has been to maximise the repetition of design solutions, to maximise the application of routine factory fabrication, and to simplify the effort of site erection. At the same time the design should be as structurally efficient as practical within the limits of serviceability.

Photo 1: elevated guideway steelwork

- *Guardrail design*
  The guardrail has been added to provide protection to construction and maintenance workers and to protect passengers who evacuate from a failed vehicle. A formal risk assessment justifies a rail but what standard should apply to its design?

Photo 2: construction guardrail

- *At-grade construction*
  The at-grade section is robust to ensure little or no differential movement between elements and continuity of the running surface provided by the concrete planks. The design, which resulted from this approach, warrants review.

Photo 3: full-speed at-grade guideway

2   MODULAR DESIGN FOR ELEVATED GUIDEWAY.

2.1   Background

The basic form for the elevated LHR Guideway had been developed in 2002 for the Cardiff Bay Test Track, as a one way route comprising a pair of side beams with cross members at regular 2m intervals. The cross members support the running surface, cable tray, and drainage channels.

The side beams provide the simply supported spanning elements, as well as an upstand to contain an errant vehicle and provide navigational direction. The side beams also support the guardrail, control and

Photo 4: elevated single track (Cardiff Test Track)

safety equipment. The running surface is a pair of 450mm wide pre-cast concrete planks. Evaluation from the Test Track had indicated that this form was simple and easy to construct and erect, and for the 18m span was efficient in the utilisation of materials, and provided the very shallow profile and cross section suitable for constrained urban areas.

The 18m length can be transported in one piece and will span over a typical UK urban road at right angles. For these reasons the basic concept for the elevated structures was considered appropriate for the LHR application. The decision to adopt a simply supported span resting on bearings at each end avoids on-site connections for a continuous structure, and is considered to be efficient in terms of utilisation of a foundation and column to support the ends of adjacent spans.

There is a study to be carried out relating to the cost of spans to cost of foundations and columns, to inform the optimum span length, but this is also a function of ground conditions and local constraints.

Different forms of construction were considered for the Test Track including fabricated trusses, pre-cast and in-situ concrete and composites. Information from steelwork fabricators indicated that the cost of cutting and welding for a fully fabricated element can amount to 50% of the total cost. Concrete lacks flexibility where a variety of radii are used and involves high mould set up costs. On this basis the assumption has been that the least cost option would be to construct from steel using as little fabrication as possible.

The following discussion assesses the design considerations applied to the LHR pilot guideway.

2.2 LHR Components

The alignment has been determined by consideration of local constraints, aesthetics regarding the setting and road approach to T5, as well as operational and ride comfort requirements. The approach to design has been to consider a range of issues in parallel. It is not practical for any one consideration to take precedence over others and hence design is a complex and iterative process.

The LHR pilot comprises the following components of elevated guideway;

- From T5 (MSCP5) to at-grade section
  571m of double track in 30 spans whose lengths vary from 8m to 36m
- From at-grade section to carpark N3
  326m of double track in 19 spans whose lengths vary from 14m to 30m
  490m of single track in 35 spans whose lengths vary from 4m to 19m

These results are summarised in table 1:

Table 1. Summary of Elevated Track Components

	Single		Double	
	Number	Length (m)	Number	Length (m)
Straight 18m standard	3	54	20	360
Straight same design	13	192	7	100
Other straights	-	-	9	247
Curved as standard	11	128	9	146
Fabricated curve	-	-	3	54
Merge/diverge element	5	68	1	12
Totals	**32**	**442**	**49**	**919**

From this summary it can be seen that alignment constraints from fitting a route into a constrained site dominate and repeated use of the "standard" module amounts to only 28% of total elements used. However the same "design" is used in 78% of the total number of elements. The remaining 22% of elements have been either fabricated from plate or take another form to meet the requirements of the span to be achieved.

2.2 *Structural Performance*

The simply supported beam performance is determined in part by the wall thickness of the rolled hollow section (RHS) which forms the side beam. The selection of a wall thickness is a consequence of bringing together load, design life, fatigue and welding considerations, in general structural analysis.

The sensitivity of this selection is demonstrated in figure 2, in which span length is plotted against wall thickness excluding and including fatigue effects. This relates to the 450mm by 250mm rolled hollow section adopted for the main girders in the LHR design.

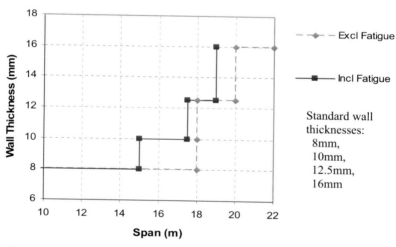

Figure 2: Comparison of Span / Wall thickness (450mmx250mm hollow section)

Structural analysis confirmed that the critical location in the model was at the welded junction of the cross member with the side beam. The wall thickness of the main girders was therefore dictated over their whole length to cater for welding on the cross-girders at 2m intervals. This would appear to increase the cost of the steelwork as more material is employed, however the use of a thinner wall section would require internal stiffeners and additional fabrication costs, outweighing any saving in material.

2.3 Procurement of Sections

Sections are generally manufactured in rolling mills to conform with dimensions established by the industry or through national standards. Since rolling is a mechanical process tolerances are allowed for overall linear dimensions, section shape and wall thickness. These variations must be allowed for in the assembly of spans in addition to assembly tolerances on site. Rolling mills have output limitations on length and straightness. In the UK the principal mills produce as standard 15m lengths, and hence the 18m design standard module has to be formed with at least one butt weld to lengthen the side beam.

2.4 Fabrication

Off site factory fabrication has brought benefits of work in a controlled environment with access to lifting, rolling and automated welding. For this application it was found that the supplied steel reliably complied with UK codes and standards. The design has sought to minimise fabrication rolling for alignment curves (horizontal and vertical) in more than one dimension, on any one element.

Tolerances also apply to fabrication and when added together with mill rolling tolerances establish the range of likely outcomes for the constructed guideway. For key components of the guideway tolerances were tightened as shown in Table 2.

Table 2. Table of standards from steelwork specification (extract)

Tolerance description		Recommended tolerance - ULTra
Cross Girder level alignment	**BS5400 pt6**	
1. Levels between cross girder under consideration and the two adjacent cross girders in either direction	$G = L1 + L2$; $\Delta C = G/500$ or 3mm whichever is the greater	More stringent tolerance required to achieve vertical ride quality
Typical cross girders at 2m centres	$G = 4m$; $\Delta C = 8mm$	$G = 4m$, $\Delta C = 3mm$
2. Difference in level at each end of same cross-girder	Not given	$\Delta = 3mm$

2.5 Transport and Installation

One of the determinants of the selection of the standard module length has been the UK road vehicle regulations. Loads up to 18m in length can be transported without escort or special timetabling provisions. The same regulations indicate that loads up to 4m in width are permitted without special provisions. The double track is 3.99m wide overall.

Once on site the responsibility of dealing with the size and weight is assigned to the contractor. Limitations will be a function of availability of cranes or other lifting equipment.

The 18m standard module weight characteristics are summarised in table 3, illustrating that the addition of the concrete running surface planks is a significant component of the total weight.

Photo 5: 18m spans before painting

Table 3. Weight Characteristics

Track Type		
18m Straight	Single (tonnes)	Double (tonnes)
Steel	5.3	8.2
Concrete planks	4.3	9.5
Total (tonnes)	**9.6**	**17.7**

These can be added after erection of steel work, giving contractors the option to assemble on the ground and lift as one assembly or to order a lighter lift and place the planks once the guideway structure is in place. At LHR it was not possible to use cranes to install certain spans below an already constructed road ramp. This particular lifting problem was solved by the use of a transporter.

Photos 6 & 7: installing spans beneath existing structures

2.6 Maintenance

Avoidance of ledges where dirt and birds can collect has been a consideration in design. Hollow elements are the most efficient structural form, but introduced the risk of internal corrosion and this has been dealt with by ensuring that each has been sealed with a welded closing plate. This also acts as a stiffener and therefore is a small additional cost. These considerations confirmed the use of the rolled hollow section for the side beams and the cross members.

Photo 8: elevated/at-grade abutment

A minimum clearance of steelwork above ground has been incorporated to ensure safe access to the underside of all elements for inspection and maintenance, and to avoid dead spaces where litter could gather. There is also a security issue in an airport environment and avoidance of semi-enclosed spaces is important. As a consequence, "abutments" are provided at the transition from elevated to at-grade guideway.

2.7 Summary and Conclusions

Whilst the amount of repetition of an 18m standard module has been lower than anticipated in the early stages of design development, reassessment of the design has not identified a more efficient basic assembly. Review of the selection of concrete as the running surface might change this conclusion as the concrete is a significant element of the total dead load. (It also provides an element of dynamic damping so the use of a lighter running surface would introduce wider considerations). Overall the average steel weight per metre of elevated route over the 2.3 km system is 0.28tonnes or 144kg/m^2 of route plan area.

3 GUARDRAIL DESIGN

3.1 Background

PRT is a form of on-demand one to four person, personal public transport system. It is a new form of public transport and, in UK terms, is neither a "railway" nor a road based transit system. As a Guided Transit System it falls within the applicability of the Railways and Other Guided Transport (Safety) Regulations 2006 (ROGS). In the UK these safety regulations are administered, by Her Majesties' Railway Inspectorate (HMRI), based in the Office of the Railway Regulator (ORR).

At speeds below 40kph the regulations indicate that "self regulation procedures" should be applied through a Safety Verification Team (SVT) comprising Competent Persons.

For the Heathrow scheme, one of the outcomes of the work of the SVT has been the inclusion of a guardrail along the whole length of the elevated route. The SVT requirements have been recorded in the Design Code, however, the justification for a guardrail it is not covered by existing standards. The guardrail has a major cost and visual impact and a review of the design is warranted in this discussion.

3.2 Risks

Much of the PRT guideway is located at height, typically around 6m above the adjacent ground level in order to provide clearance to vehicles travelling on roads beneath. This results in a risk of fall from height for personnel involved in cleaning or maintaining the guideway, or in the case of evacuation of a PRT vehicle.

Photo 9: safe access - edge protection

UK Health and Safety Regulations (specifically the CDM regulations) require that infrastructure must be designed to be safe to build, maintain, operate, use and demolish. All members of the design team must take safety into consideration in their designs, and protection from falling is a key area where serious and often fatal accidents can be avoided by use of sensible precautions such as edge protection.

Specific Health and Safety risks associated with the automated nature of PRT must also be considered. For the ULTra system at Heathrow airport the SVT have reviewed the design and provided advice to ensure that these particular risks have been adequately mitigated by the designers.

The ULTra vehicles are designed to be very safe, for example they are able to self-diagnose problems and programme themselves for maintenance before they break down, they run on electric batteries so do not have a conventional fuel tank. It is anticipated that there will be relatively few situations where a vehicle makes an unscheduled stop on the guideway. In the event of an incident a rescue machine would drive back up the guideway and hook up to the stranded ULTra vehicle towing it and its passengers to the nearest emergency disembarkation point. The emergency operation strategy is intended that in the majority of cases of a vehicle failure the passengers would remain in the vehicle – the safest place for them. In a small number of cases (such as a fire within the vehicle) it would be necessary for passengers to exit through the escape hatch at the front of the vehicle, the instructions would then be for the passengers to sit down on the guideway and wait for assistance. However it is recognised that people do not always behave as expected, especially in stressful circumstances and may decide to walk along the guideway to 'escape'. Children and people with mobility or visual impairments must also be able to be evacuated safely. Finally, in certain weather conditions and at night an evacuated passenger may not realise how high up they are.

A key concern to all parties is that PRT is as safe as reasonably can be achieved with the technology available. The negative publicity arising from any serious accident on this emerging form of transport could be enough to severely damage the system's credentials as a viable form of transport, not to mention the moral and ethical implications of designing infrastructure that is not safe to use.

When the elevated structure was developed for the ULTra test track in Cardiff fall prevention was not originally provided. At the time this was deemed acceptable as the system was in development, was not in regular use and was being operated by a small number of personnel who were very familiar with the system. The length of elevated guideway under consideration was also relatively short at around 50m. However, it became obvious during the early design stages of the Heathrow project when risk assessments were undertaken that for a commercial system which would be used by a greater number of people, some form of fall prevention would be required.

3.3 Options
Four options for edge protection were considered: No protection at all, provision of mansafe latchways, provision of a minimum edge restraint and provision of full footbridge style parapets.

Varying levels of edge protection: Photo 10 left: Cardiff Test Track; Photo 11 centre: mansafe and basic guardrail; Photo 12 right: typical UK footbridge parapet

To provide no edge protection at all was considered to be unacceptable, as there would be too great a risk of fatal injury from falling for operatives and evacuated passengers. Provision of a fall arrest system consisting of a harness worn by a person who is clipped to a cable running the length of the guideway would provide minimum safety protection to operatives undertaking maintenance tasks. However, an evacuated passenger could not be expected to understand how to use a latchway system and so would be afforded no protection from falling. In the UK use of fall arrest systems is considered to be a last resort if no other protection can be provided as it requires correct use and training to be effective, falls are not prevented, only the severity of the injury is reduced. Regular maintenance and inspection is required to ensure that the system remains safe to use.

At the other extreme full footbridge parapets designed to UK highways standards could be used to provide protection from falling for both operatives and evacuated passengers they would also provide additional protection from objects – such as tools falling from the guideway potentially onto traffic or pedestrians beneath. This option would also give a real impression of security to passengers and the public. However, it would also have a significant visual impact and could be argued to be over-designed as footbridge parapets must be able to withstand significant crowd loadings. As a PRT vehicle is typically designed to carry four passengers crowding is an unlikely scenario. A summary of the risk assessment is shown in Table 4.

Table 4. Risk Assessment Summary

	Risk	Severity	Outcome	Comments
Do nothing	Passengers: medium fall likelihood *	Passengers: high likelihood of severe or fatal injury	Not acceptable	Risk and severity too high for all users.
	Operatives: high fall likelihood	Operatives: high likelihood of severe or fatal injury	Not acceptable	
Provide training and information (i.e. tell people to stay away from the edge)	Passengers: medium fall likelihood *	Passengers: high likelihood of severe or fatal injury	Not acceptable	Risk and severity too high for all users. No additional protection for
	Operatives: high fall likelihood	Operatives: high likelihood of severe or fatal injury	Not acceptable	
Provide mansafe latchway edge protection for operatives	Passengers: medium fall likelihood *	Passengers: high likelihood of severe or fatal injury	Not acceptable	Risk and severity too high for all users. No additional protection for passengers, operatives not prevented from falling
	Operatives: high fall likelihood	Operatives: medium likelihood of severe or fatal injury	Not acceptable	
Provide minimum edge restraint	Passengers & Operatives: Low fall likelihood	Passengers and Operatives: low likelihood of severe or fatal injury	Acceptable	Risk and severity acceptable for all users.
Provide full pedestrian parapet edge restraint	Passengers &Operatives: Low fall likelihood	Passengers and Operatives: low likelihood of severe or fatal injury	Acceptable	Risk and severity acceptable for all users. But may be excessive due to infrequent access.

(* medium risk considered for passengers as the likelihood of a passenger being on the guideway is very low to begin with)

The option chosen was to provide a minimum edge restraint designed for loadings appropriate to the low frequency of people on the elevated structures but offering fall prevention so that both operatives and evacuated passengers would be protected from falling off the structure.

3.4 Standards

Existing UK design standards for edge protection were consulted and used to inform the approach chosen. Design loads for edge protection were found to be fairly consistent for a range of edge protection scenarios from balconies in buildings to bridge parapets. A summary of loadings are given in Table 4.

Table 5. Guardrail Loadings

Horizontal Imposed Load	Characteristic Load Qch [kN/m]	Safety Factor γm	Design Load Qo [kN/m]
Low frequency use (PRT)	0.74	1.5	1.11
High frequency use (footbridge)	1.5	1.5	

The design standards also give guidance on the maximum deflection that edge protection should exhibit under load. This is to ensure that a guardrail can assist a person in maintaining their balance at height – and also in order to provide the feeling of safety. A guardrail that feels rigid when grasped is much more reassuring than one which deflects.

Various options for the design were considered. Initially a simple tubular top and mid rail system was proposed as shown in figure 3, however this did not provide sufficient mitigation against the risk of children or small objects falling from the guideway. The visual appearance was also deemed to detract from the overall slender appearance of the rest of the guideway infrastructure, although this option had the advantage of being relatively low in cost.

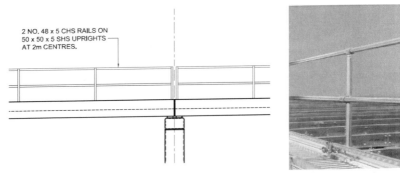

Figure 3: simple tubular edge protection system   **Photo 13: example**

In order to reduce the risk of objects falling through the edge protection options were considered that used tensioned wires at close spacing to infill the gap, however these could be used to provide footholds to assist a person attempting to climb over the guardrail, so were considered unsuitable.

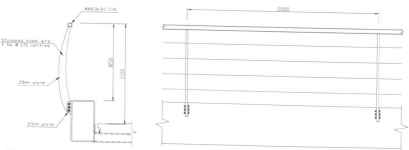

Figure 4: tensioned wire with tubular top rail

There was also significant debate whether a tubular top rail was required to the edge protection, or whether a tensioned wire was sufficient. There was a strong desire not to provide a tubular/rigid top rail to the edge protection to minimise the visual intrusiveness of the system, however UK standards imply that a rigid top member must be provided to edge protection for permanent situations, although this can be relaxed for edge protection on construction sites.

Figure 5: wire and mesh option

The use of the rigid top rail was eventually agreed when the design showed that significant forces would be required to keep the tensioned wires taut resulting in additional bracing requirements, as the top rail also provided a compression strut to resist tension forces in the wires.

3.5 Agreed Guardrail Solution

The agreed guardrail design proposed for the Heathrow PRT system and shown in figure 6 used a galvanised wire mesh fixed between two horizontal, tensioned wires, with a rigid tubular top rail. This provided suitable fall protection including from small objects falling off the guideway, but was still considered to provide a lightweight and relatively unobtrusive visual solution.

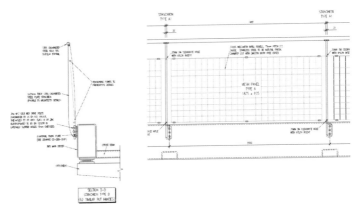

Figure 6: agreed guardrail solution

3.6 Discussion

It was intended in the original design that the guardrail would be installed onto the superstructure span before it was lifted into place. This guardrail would then provide fall protection during both the temporary construction phase and the operational situation. In practice this was not achieved for the LHR project, and a temporary edge protection system was installed. Tender prices for the guardrail were higher than expected due in part to the requirement for tensioning the cables at regular intervals along the guideway, and it was decided that in order to make cost savings on this project the temporary edge protection would be retained for the duration of the pilot scheme at Heathrow.

This gives the opportunity to revisit the risk assessment and edge protection design once data is available on the reliability of PRT. Consideration will be given to reducing the edge protection requirements depending on actual data recorded on the frequency of vehicle evacuation and the behaviour of passengers evacuated on to the guideway, and future development of the guideway infrastructure will take this into account. At this stage in the

Photo 14: final guardrail installed

development of a relatively untested and unfamiliar form of transport it is prudent to provide a higher level of protection than may be required in practice. Perceived safety is as important as actual safety when encouraging the public to trial PRT.

In the future statistical evidence may show that instances of people needing to be on the elevated sections of the guide way from evacuation of vehicles and maintenance operations are rare. Therefore it may be possible to consider decreasing the level of protection provided through a guardrail.

4   AT-GRADE ROUTE DESIGN APPROACH

4.1   Design Requirements

In order to maintain the surface regularity standard along the at-grade section of the route, a design has been developed which exceeds the original expectation that ground bearing routes need be "no more complex than a footpath". This outcome requires review and the approach to the design is discussed in the following section.

The ULTra infrastructure is principally intended to be kept elevated in order to separate it from the existing transport network. However, in some instances the guideway is required to run at or close to ground level. This 'at-grade' guideway is typically required when approaching ground level stations. For the Heathrow specific application the guideway has to run close to ground level around the west end of the north runway, in order to pass beneath the Obstacle Limitation Surface.

The main requirements for the at-grade guideway infrastructure are as follows:

1. Continuity of ride quality between elevated and at-grade guideway (i.e. similar running surface / navigation upstand.)
2. Containment of ULTra vehicle (although there is no risk of a vehicle falling off the guideway, it must still be prevented from straying out of the bounds of the track.
3. Protection of ULTra vehicles from collision from other forms of transport – typically motor vehicles.
4. Prevention of unauthorised access by people on to the track, and the inherent risk of injury.

The Cardiff Test Track consists mostly of at-grade guideway located on an access controlled site, which was constructed from a reinforced concrete slab on compacted granular fill; initially simple straight road kerbs were used to provide the upstand. These have now been upgraded to an ULTra specific precast concrete upstand which provides smooth curves and a greater level of containment.

Key issues with the test track were poor ride quality arising from inaccuracies in the running surface concrete, and in the horizontal tolerance on the kerbs/upstands. Because of the experience at the test track, there were concerns that in-situ concrete slab guideway would not give a good enough running surface for the Heathrow application, especially at locations where full speed running would be required. The in-situ concrete was also relatively slow to construct, requiring a significant amount of weather dependent work.

4.2 Applications

Alternative construction methods for the at-grade guideway at Heathrow were therefore investigated. The at-grade guideway at Heathrow can be divided into three types:

1. Full speed running areas on the 'main route' – characterised by the section of track around the Western Perimeter Road at the end of the north runway.
2. Slow speed running areas around stations in surface car parks (N3 Car park).
3. Slow speed running areas around stations in structures (MSCP5 Car park) running on the concrete slab of an existing post-tensioned multi-storey car park. (This is specific to Heathrow and is not discussed further in this paper.)

4.3 Full Speed Running

The following options were considered in early design stages;

Type A – near-grade Steel Structure type, set at a minimum practical distance above ground to allow for maintenance (i.e. running surface one to two metres above ground level.)

Advantages:
- uses modular superstructure elements, same running surface
- any unevenness in existing ground surface can be largely ignored
- maintains same running surface

Disadvantages:
- steel cannot be placed directly on the ground (corrosion problems) but must be lifted slightly, resulting in a dead area beneath guideway (security and aesthetics issues)
- still requires large scale construction (large cranes etc to lift in steel spans, difficult in this particular location at end of runway (headroom issues)
- looks odd visually (out of proportion) and could cause sight line issues for vehicles travelling along adjacent public road
- elevated guideway requires greater distance to airside security fence than at-grade; insufficient space to move fence at end of runway

Type B – at-grade infrastructure, running surface set within typically 0 to +300mm of ground surface.

Option B1: precast concrete running surface with integral upstand (L and inverted T beams).

Advantages:
- modularity
- can be made with same running surface as elevated infrastructure

Disadvantages:
- complex shape to pre-cast to the tolerances required
- simple on straight sections, but more difficult to achieve curves / gradients, or vertical curves
- difficult to line and level
- 6m concrete sections have similar weight to standard 18m span – hence still requires large cranes at end of runway

Option B2: precast concrete running surface and upstands on in-situ concrete base.
Advantages:
- planks and upstands can be made modular
- can continue running surface
- uses much smaller constituent parts, easier to construct in low headroom areas
- easy to line and level
- relatively simple to make curves, gradients, vertical curves
- separate precast parts simpler to construct to tolerances required

Disadvantages:
- achieving suitable on-site fixity between kerbs/slab for impact resistance
- site installation needs monitoring to achieve good tolerances

Option B3: insitu concrete running surface and insitu concrete kerbs
Advantages:
- simple construction
- inexpensive
- easier to form curves

Disadvantages:
- difficult to build to tolerances required (based on test track experience)
- heavily reliant on good workmanship and site supervision
- weather dependent for construction
- different running surface to adjacent elevated guideway

Other options considered for the full speed sections, but discarded at an early stage due to various impracticalities:

- asphalt running surface (difficult to achieve required tolerances, guideway width not suitable for standard laying machines due to cable tray requirements, greater risk of settlement affecting ride quality long term), and still requires containment upstand
- Concrete running surface with steel crash barrier, has tolerance problems above and barrier needs to be at low level to match vehicle impact zone

4.4 Design

The option selected for use around the Western Perimeter Road at Heathrow was B2 – precast concrete running surface and upstands on an in-situ concrete base. This option was preferred by the Contractor for buildability issues, and also gave good continuity of the running surface. This option also dealt with the variable ground conditions by spreading load through the in-situ slab. On-site close supervision and monitoring of the installation was found to be essential in order to achieve the stringent tolerance requirements.

Photo 15: at-grade guideway

The design also set out to maintain the 'light weight' aesthetic feel of the guideway and provide visual continuity with the elevated spans. This added cost to the at-grade infrastructure – for example, the alignment is on a long radius bend around the Western Perimeter road and the decision was made for aesthetic reasons to curve the outside edges of the outer guideway upstands, in addition to curving the inside faces where a curved upstand is required for navigation purposes. The outside kerbs faces could have been left as a series of straight sections forming a faceted curve – this would have reduced pre-casting costs slightly but at the expense of the appearance of the guideway.

It is apparent from this review that the arrangement of slab, precast kerbs and running planks meets operational requirements, local requirements and objectives, and visual standards. In this application variation in the design would require relaxation of some of the objectives and standards. A typical cross-section through the at-grade guideway is shown in figure 7.

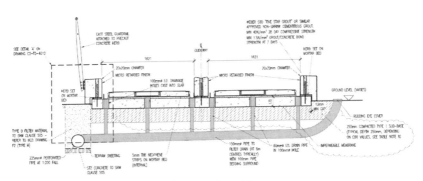

Figure 7: typical section through at-grade guideway

4.5   Low Speed Running – Surface Car Parks

In the N3 car park the guideway runs at-grade in the vicinity of the stations and a maintenance depot. In this at-grade area the guideway alignment is characterised by tight radii curves and some complex merges and diverges. The running speed is significantly reduced and means that slightly slacker tolerances can be used for the running surface. Another major difference with the full-speed guideway is that in the low speed at-grade areas close to the stations the cabling associated with the guideway can be run to the side of the infrastructure rather than in a cable tray between the running planks.

Thus it was impractical to simplify the design of the running surface and to use an asphalt running surface. The top of the existing car park surface was skimmed off or built up slightly to achieve the alignment levels required, and a new asphalt surface was laid.

It was possible to use in-situ mass concrete upstand kerbs as the lower speed of the vehicles resulted in reduced vehicle containment requirements. A separate vehicle barrier and pedestrian segregation fence was provided to protect the PRT system from errant road vehicles and trespassers.

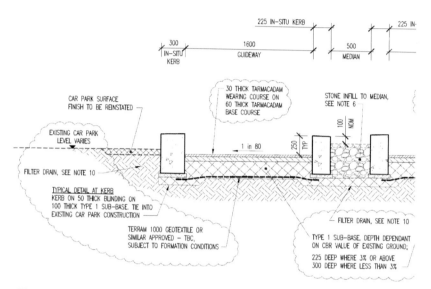

Figure 8: typical section through at-grade construction in N3 car park

This type of low speed at-grade guideway infrastructure has not been tested before and monitoring of the asphalt surface will be required to see how it responds to the passage of vehicles over time, and to establish that it can be laid to provide a smooth enough surface to satisfy ride quality requirements. It is noted that the PRT vehicles are consistent in the path they follow and rutting due to repeated passage of vehicles is a concern.

Photo 16: at-grade guideway - N3

5 CONCLUSION

This review has identified areas of reappraisal to be developed prior to commencing the design of the next project.

- **Tolerance Aspirations:** It has been established that it is difficult to use conventional civil engineering construction technology to provide a guideway which conforms with the tolerances in the final running surface as derived from ride comfort considerations. It is necessary to apply factory manufacture standards as much as possible, tighter (more stringent) tolerances than usual in civil engineering works and to have provision for fine lining the running surface after installation and all dead loads have been applied.

- **Modular design:** The application of modular design in a congested/constrained airport is a valid aspiration but its realisation will be limited by alignment constraints. This is particularly true when retrofitting a guideway to an existing airport. Modular design does not mean repetition of identical elements (although this brings some benefits) but use of a small family of solutions, either straights of common design or curves selected from a limited number of radii. The alignment with curves on inclines introduces particular requirements of fabrication, which is expensive in terms of labour in cutting and welding.

- **Guardrail Requirements:** Risk assessment has demonstrated that a Guardrail is needed, but operational experience should allow the design to be simplified, to reduce cost and visual impact.

- **At-grade Design:** At-grade design which provides for continuity of the running surface, anchorage for the upstand, visual continuity with the elevated structures, and with a facility to re-level the running surface after installation, sets up a minimum requirement for infrastructure. The design adopted is a reflection of local conditions, and no obvious basis for change has been identified.

- **Asphalt Surfacing use:** The performance of the asphalt surface is to be monitored during pilot system operation.

- **Actual Performance:** The LHR guideway has been constructed and the first vehicles are navigating the whole route. Experience from commissioning and initial operations will inform future design.

New Technology Integration for Older System Technologies

Steven M. Castaneda, P.E.[1]

[1]Associate, Jakes Associates, Inc., Jakes Plaza, 1940 The Alameda, Suite 200, San Jose, California, 95126, USA, Tel: (408) 249-7200; Fax: (408) 249-7296; E-mail: jakes@jakesassociates.com

Abstract

This paper examines the integration of newer subsystem technologies within older APM systems. These subsystems frequently include automated controls and propulsion. In addition, modern codes may pose particular challenges. Case study examples will include those for the CalExpo monorail, Miami Metromover, and Bellagio APM systems. This paper will further explore strategies to streamline and reduce the costs of rehabilitation while maximizing customer value.

Introduction

Out of the literally hundreds of Automated People Mover (APM) systems installed throughout the world over the last 100 years, there are dozens of systems in need of refurbishment. Given the variety of system types (small/large monorail, AGT, PRT, automated metro, other), technology suppliers, applications (airport inter/intra-terminal, circulators, mass transit feeders, activity centers, other), and operator types (airports, private entities, amusement parks, urban transit operators, other), the path toward integration of newer subsystem technologies into older APM systems through refurbishment is anything but clear. This paper examines the integration of newer subsystem technologies within older APM systems and identifies common challenges associated with typical integrations. This paper also discusses several approaches to APM system rehabilitation.

Typical Rehabilitation Goals and Challenges

There are many reasons why APM system technologies are upgraded, from guideway realignment, to safety/litigation concerns resulting from recent accidents, to full system failures due to obsolescence. These reasons generally have several root causes. Several of these causes are identified by the following:

- **Safety**
  - Does the system continue to provide safe transportation for riders?
  - Have there been system accidents due to flaws in the system design?
  - Have there been system accidents due to years of degradation of key subsystems?
  - Do current system conditions pose an increase potential for damage to existing subsystems?

- **Economic**
    - Does the system continue to provide an economic benefit either directly (through fare collection, advertising, other) or indirectly (movement of people to/from activity centers, enhanced customer experience, other)?
    - Are there opportunities to reduce operating/maintenance costs?

- **Maintainability**
    - Are maintenance costs becoming too high to sustain?
    - Are spare parts still available?
    - Is critical service from the original subsystem suppliers still available?

- **Reliability**
    - Is system availability suffering from decreased reliability?
    - Have system functions degraded over the last few years?

- **Performance**
    - Is the original system design capacity inadequate for current/future ridership levels?
    - Is system performance suffering?

- **Other**
    - Are there indirect reasons why the system needs refurbishment (system alignment change, travel patterns, political reasons, budgetary constraints, other)?

In nearly all cases, decisions about system rehabilitations and updates are based on several root causes and not one single issue.

Various Approaches to System Rehabilitation

Once a need (or several needs) for system rehabilitation have been determined, there are several paths a system owner can take. Typically, an owner will be faced with various options it could take. Below are several options JAI has identified for several systems rehabilitation projects.

- **Do Nothing** – This approach is only viable if the system estimated remaining service life is still measured in years and the system owner does not have a compelling reason to further invest in system rehabilitation. The system owner can expect continuously decreasing reliability, parts shortages and other challenges detrimental to system operation.

    Ultimately, this approach will directly lead to cessation of system operation, decommissioning, dismantling and scrapping. Nearly all APM systems have negligible salvage values since they are designed as discrete, unique systems and there is virtually no market for used constituent APM parts, components and equipment.

It is interesting to note that that the actual cost of system dismantling can amount to more than a modest investment in refurbishment and overhaul, assuming there is still a need for system operation.

- **Decommission a Portion of the System** – This approach involves the decommissioning of major system components of the existing system, such as vehicles, to further extend the life of the remaining system by utilizing the decommissioned components. An example would be if certain train control circuitry were no longer available – it may be more cost effective to use the circuit boards from an existing vehicle as spare parts for the remaining vehicles, rather than be forced to manufacture new custom circuit boards.

  This approach is only viable if the system's original technology supplier is no longer doing business and spare parts are no longer available; and if the system can function in a meaningful way after the loss of significant components. This approach, if possible, results in a short term solution for minimal cost. However, this approach adds little value or life extension to the system as the decommissioned portions are generally in poor condition to begin with.

- **Partial System Rehabilitation** - In this approach, rehabilitation/rebuild of several or all major subsystems are performed, including vehicles, train control systems, propulsion components, communication systems, and power distribution systems. See below for a more detailed discussion of subsystem upgrades.

  Advantages of this approach include reduced total system cost relative to procuring new subsystems. Disadvantages include an increased emphasis on system integration, potential warranty issues, increased construction management/oversight, and standard/code compliance issues (see below for a more detailed discussion on this topic).

  Owners may choose local suppliers, technicians, and other companies to perform the work on the original equipment. Further, typical APM vehicles are constructed of materials which could conceivably be restored by independent firms as well (subject to further standard and code compliance verification).

- **Partial System Replacement** - In this approach, replacement of several or all major subsystems are performed, including vehicles, train control systems, propulsion components, communication systems, and power distribution systems. Generally, major civil works are not replaced because of the enormous costs associated with modifying these major components (including guideway elements, piers, station platforms, other).

Advantages of this approach include increased reliability, warranty, and future continuity of spare parts as the latest technologies would be utilized. In terms of train control, the latest in automatic train operation, protection, and supervision (ATO, ATP, and ATS) technologies could be integrated along with central computer control (where applicable). Propulsion elements could include the latest in variable frequency drive (VFD) technology. Upgrades in communication elements may include wireless technology, integration of active information displays, live video capability, improved public address systems and others.

A primary advantage of this approach is the potential to integrate a fully reliable and fail safe train control system and the potential to upgrade the system to full compliance with current codes and standards.

- **Complete System Replacement** - In this approach, the owner would replace the entire APM system with a new and proven technology based on its specific performance requirements and unique application. Although this approach is generally the most ideal from a technical and code compliance standpoint, this is also the most costly approach.

The above listing is only a representative example of potential approaches a system owner may adopt in the rehab of an aging APM system. All system rehabilitations are distinct and should be based on key unique characteristics, specific needs, and challenges of the particular system and operation.

There will invariably be various challenges associated with bringing new technologies or components into an aging APM system. We have found that many of these challenges are project-specific and are manageable and resolvable as long as the owner and contractor work through the issues in the spirit of cooperation and understanding.

However, challenges in the key area of procurement have been encountered in many rehabilitation projects. For example, identifying qualified specialty contractors with APM knowledge has been especially challenging given that APM technology is a very small niche market. Another challenge includes defining specifications and terms and conditions that can be accepted by the contractor, including: liability (the existing technology, in many cases, is not the contractor's system technology); warranty (this is especially true where the contractor is not the O&M provider of the system); system demonstration; and liquidated damages (always difficult).

Typical Subsystem Upgrades

There are several key subsystems which are typically upgraded in the rehabilitation of an APM system. These systems include the following:

Automated Control Systems

When a new train control system is to be installed in an existing APM vehicle, a common cause is that the equipment is old and/or difficult for which to find parts or support. In this situation, to install a new system almost always includes a complete rewiring of the train. This is a good idea because the wiring/cabling of the system is usually old with cracked insulation and connectors may be in need of replacement. This presents a fire hazard and future availability risk. Another reason is that the interface from other new subsystems typically requires additional wiring installation anyway and instead of struggling with old wiring systems, the entity performing the upgrade will opt to start fresh with all new wiring.

Another critical reason why APM control systems are upgraded is to provide for increased capacity. Newer ATO/ATC system technologies, such as moving block technology, can provide for a marked increase in capacity over older, more conventional technologies, such as fixed block systems. Integration of these newer technologies will often alleviate the need for more vehicles, saving system owners millions.

Door Control Systems

APM doors, like their conventional transit counterparts, are typically a source of transit operator and maintenance personnel headaches. This is because door systems are one of the few onboard subsystems that interact directly with passengers and are subject to decades of passenger wear and abuse. Given that many APM suppliers in the past directly engineered door mechanisms for a variety of reasons (unique vehicle designs, perceived cost savings, other), door system support and/or replacement parts are often unavailable because the original system supplier itself is no longer in business. Because of these reasons, APM door control systems are typically included in system technology updates.

Propulsion and PDS Systems

Propulsion systems are also often updated and/or replaced due to years of mechanical wear and degradation, including corrosion. This equipment includes differentials, gearboxes, u-joints, drive shafts, braking systems, guidance equipment, suspension components, electric motors, seals and bearings, and others.

On-board electrical propulsion components, such as drive motors, inverters, and others are also typically replaced because, unlike their sheltered stationary PDS counterparts, they operate in a dynamic, mixed environment (oftentimes outdoors) and are subject to mechanical wear, thermal stresses and others resulting in shorts, degradation of key electrical connections and insulation, and overall loss of performance.

Stationary PDS System components are generally not replaced as often as other systems in typical refurbishment programs. This is because system elements, such as power rails, power collection shoes, hangers, and others, are designed to wear and are included in routine maintenance of APM systems. Larger stationary system components, such as transformers, inverters, and others, are typically not replaced as they are generally sheltered from the elements and other external forces that normally add additional wear to other subsystem components (such as doors).

Communications

Upgrade and/or replacement of communications equipment is typically performed to add new features, functionality, and reliability. Advancements in communications technologies over the last few decades have made this kind of APM system upgrade relatively inexpensive given the potential impact. Specific communications technologies include wireless train control technology, integration of active information displays, real-time security video capability, on-board video (to expose riders to relevant information and advertising), improved public address systems, wifi-internet, and others.

Latest Standards and Codes: To Adopt or Not to Adopt?

Most APM system operators would like to include all ASCE 21 requirements to limit their risk of exposure to liability. However, it often becomes the result of what is cost effective. For example, can emergency walkways be integrated at a reasonable cost?

The major challenges with standards upgrades include the fact that many systems were built before many standards (e.g. ASCE I-IV, NFPA 130 and other standards referenced within those standards) were adopted nationally. Many do not have emergency walkways (NFPA 130 requirement) and many vehicles are constructed with materials (e.g. panels, glass, floor structure, other) that are not compliant with current smoke and flammability requirements of NFPA. Further, many older train control systems and/or associated subsystems are not considered fail-safe according to today's standards.

Depending on the original intent of the rehabilitation scope of work, retrofitting a system to meet all of these standards, more often than not, is less cost-effective than buying new trains (see the Metromover example below). All of these issues need to be considered. However, in most states or jurisdictions, NFPA 130 and the ASCE standards are simply industry standards and guidelines and not part of local ordinances/codes. Therefore, many rehabilitation projects do not have to fully comply with the standards.

It is a good general practice to improve the reliability, availability, maintainability and safety of an aging system to current industry standard levels, but if the costs are prohibitive, it is better to improve on some rather than do nothing and allow the system to continue to deteriorate and become more of a safety risk.

Case Study Examples

The three cases below represent three extremes in APM rehabilitation projects operating in completely different settings and performing different functions. Although the decision to move forward with rehabilitation for each system was ultimately approved, the approaches to these rehab projects were as varied as the installations themselves.

CalExpo Monorail Rehabilitation

The California Exposition and State Fair (CalExpo) monorail is comprised of 7,758 feet of single lane guideway and functions purely as an amusement park attraction, or ride, for fair patrons during the annual Cal Exposition State Fair. During operation, the system runs in a continuous loop service comprising a single station. The system features four, eight-car trains. The system is unique as it typically only operates during the fair season, approximately 18 days during July/August every year. During this time, the system carries approximately 100,000 passengers. Figure: "CalExpo System Route Alignment" shows the alignment of the CalExpo Monorail.

The system was built in 1968 by Habegger, AG, had fulfilled its life cycle and was in urgent need of either major refurbishment or replacement. To implement a new, comparable monorail system, CalExpo would have had to spend at least $40 million. Based on this estimate and limited available funds, CalExpo officials opted to rehabilitate the system, potentially resulting in another 20-30 years of successful and convenient service.

Jakes Associates, Inc. conducted an on-site assessment and review of the existing conditions of the monorail system to determine the best options to extend the life of the system. Through the assessment, Jakes Associates, Inc. noted several key issues described below:

- The train control subsystem was not designed with a central command station;

- CalExpo had its on-board train control system rebuilt by Von Roll in 1990 (no longer in business);

- The propulsion and power conditioning equipment (wayside) appeared to be in good working order;

- Vehicle bogies were becoming more difficult to maintain resulting from their aged condition and use of obsolete components;

- The monorail's tires were obsolete and no longer made by Michelin. Jakes Associates, Inc. noted that other operators of the same monorail technology as that of CalExpo had experienced similar problems related to the unavailability of load tires;

- The guideway appeared to be in good structural condition with the exception of the conductor rails, associated hangars, and diode elements. Structural welds revealed good integrity and there was neither damage nor excessive corrosion to any guideway elements observed;

Figure: CalExpo System Route Alignment

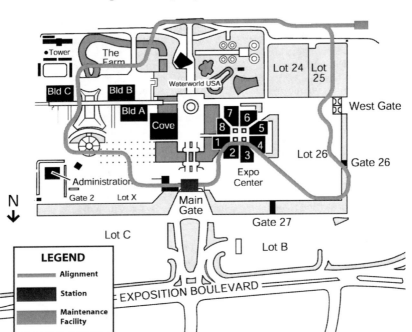

- Power rail contact shoes were lasting, on average, approximately 2-3 seasons, or roughly 1,500 miles, well short of the typical 3,000 - 5,000 mile range experienced by modern APM systems;

- The four existing monorail trains appeared to be in relatively good condition given their 35 year age. Areas which revealed deterioration included the under frame, propulsion and power conditioning equipment, and wiring/cabling. Further, body cracks and passenger compartment floor dry rot problems were noted during inspection; however, these were not considered critical. The under frame consisted of a welded tubular and rectangular beam steel structure;

- All electronic subsystems on the vehicle (automatic train control, tachogenerators, gen sets, other) were composed of obsolete components. Vehicle wiring and cabling appeared to be original;

- Other vehicle equipment, such as door operators and lighting, appeared to be serviceable and could be maintained on an on-going basis for several more years. However, it was noted that the existing door control system featured a design which was currently operating beyond its useful life.

Based upon this initial assessment, Jakes Associates, Inc. identified several options. However, CalExpo ultimately chose to perform a very limited rehabilitation project to rehabilitate the most critical system components, such as the train control system.

Miami Metromover Mid-Life Overhaul

Miami Dade Transit (Miami, Florida) operates one of the few urban downtown APM circulators in North America, known as the Metromover. Built in 1986, the free-to-ride system is currently served by a fleet of 29 Adtranz (now Bombardier) C-100 vehicles. There are 20 conveniently located wheelchair-accessible Metromover stations, one about every two blocks, serving the entire Miami downtown area by linking many of downtown Miami's major office buildings, hotels, and retail centers.

Miami-Dade Transit recently initiated a mid-life vehicle fleet overhaul and modernization project for the Metromover. The original intent of Phase 1 of the project entailed the refurbishment of the Metromover fleet, and was to include air conditioning repairs, floor restoration, and installation of new seats. The rehabilitation was to extend vehicle life for another 10 years.

However, opting for a more long-term, cost effective solution, MDT awarded a contract to Bombardier to supply 29 new CX-100 vehicles in lieu of rehabilitation services in January 2006. Miami-Dade Transit determined that the cost of rehabilitating the existing vehicles was almost as high as purchasing new vehicles outright.

Phase I of the project includes delivery of 12 vehicles (currently in progress) with Phase II seeing the delivery of 17 additional vehicles. MDT began receiving the new vehicles in the summer of 2008. These new vehicles include a more aerodynamic design, as well as an onboard CCTV system. Figure: "New Metromover Vehicles" shows the new vehicle futuristic look that includes a new design with slopped front ends.

The new vehicles will feature several new upgrades. For example, the new vehicles will have improved electronic-information displays that not only will show service schedules and stations, but will also display information about landmarks and tourist attractions close to Metromover stations. The same display units will relay emergency service-change information.

The new vehicles will feature a new wireless video-surveillance system that will allow the monitoring of activities in the vehicles by Central Control security personnel. The vehicles will also have a self-diagnostic system which will recognize problems before they start, which should reduce the risk of mechanical failures. A more robust air conditioning system will further be included to handle Miami's subtropical climate.

Figure: New Metromover Vehicles

Other system improvements include the installation of a new automatic carwash system, new station escalators (where needed) and canopies, new rubber platforms, and an upgrade to the existing PA system.

The recent rehabilitation of the Miami Metromover is a great example of proper integration of the latest in APM technology into an existing system and should extend the system's usability for decades to come.

Bellagio-Monte Carlo System Realignment

The Bellagio-Monte Carlo People Mover in Las Vegas, Nevada, performed extremely well during its brief operational life on the Las Vegas Strip. The system design featured a unique design known as Yantrak. For example, the vehicle represented a significant departure from conventional designs by hanging over the outer edge of the guideway by nearly half of its width. The vehicle suspension was unique, with two main support tires per vehicle rather than four, with a "rigidly attached lateral idler" (outrigger) to provide roll-stabilization. This design allowed for a narrower, less expensive, more attractive guideway compared to other systems with similar performance. Propulsion was provided by a belt with dozens of small AC induction motors distributed throughout the length of the guideway (providing extensive redundancy). The belt itself consisted of steel cables surrounded by a rubber jacket, yielding a smooth finished surface reducing noise and vibration, while allowing for operating speeds considerably higher than those used for most cable propelled systems.

The original and modified system connected Bellagio with the Monte Carlo. The trains consisted of three vehicles, each capable of holding approximately 40 passengers (up to 22 seated and 18 standees) for a total system capacity of approximately 3,000 passengers per hour per direction.

After only 6 years of operation, the alignment of this system was identified as an obstacle to the expansion of the Bellagio Resort property. The new expansion required the relocation of the Bellagio station and maintenance facility to the new Bellagio Spa Tower (a 32-story, 947 guestroom tower located west of the existing Bellagio garage). This work required new guideway section fabrication, major reconfiguration of the belt and pinched drive system, and train control and software changes.

Complexity of the system modification was further increased by the prototypical nature of the system installation. Many manufacturing processes had to be re-engineered, such as production of the custom beltway. Schwager Davis, Inc. (SDI) performed the system modification as the general contractor.

The initial scope included demolition of approximately one-third of the system length, including the Bellagio station. Table: "Bellagio-Monte Carlo People Mover System Characteristics" compares the system characteristics prior to and after the modification.

Table: Bellagio-Monte Carlo People Mover System Characteristics

System Characteristic	Original System	Modified System
Type	Dual-Lane Shuttle	
Location	Las Vegas, NV	
Train	3-Car Train	
Vehicle Capacity	40 People/Vehicle	
Total Length	2,400 feet	1,700 feet
Maximum Speed	45 ft/sec 30 mph	37 ft/sec 25 mph
Maximum Acceleration	2.3 ft/sq sec 0.07 g/sec	
Jerk Rate	0.065 ft/sq sec 0.003 g/sec	
Travel Time	95 seconds	68 seconds
Capacity	3,085 pphpd	4,000 pphpd

It is interesting to note that when the system was initially developed, the contract did not require a 25-30 year system life, as it was assumed from the very beginning that major system modifications were likely to occur at least every 10 years (unique to the Las Vegas environment). The initial approach was to develop a system flexible enough to undergo major modifications frequently. This was unheard of in the transit industry, but ultimately proved correct after only 6 years when realignment of the system was required.

The design phase for the system modification was completed by mid 2004 and the system opened for passenger service in early 2005. The new modified system operated for only 9 months before being dismantled to make way for the new, multi-billion dollar CityCenter project (currently under construction). The unique design performed well before and after modifications and upgrades, and proved that with the right approach, an APM system does not have to be dependent upon the original supplier.

OPTIMIZING APM FAILURE-MODE CAPACITIES

James W. Green, P.E.*

* Vice President, AECOM Transportation, 5757 Woodway, Houston, TX 77057, (713)-267-3115, Jim.Green@aecom.com

Abstract

APM systems have consistently enjoyed high levels of reliability and availability, but they are not completely free from failures. Even with typical 99.5 percent availability, the pesky 0.5 percent amounts to more than 3 hours per month with passenger service impacts. When reduced service capacities are offered during periods of heavy demands, the results can be significant, in terms of passenger queuing, congestion, and time delays, even if the disruptions are relatively short in duration.

This paper addresses the issues associated with conditions and opportunities that arise when failures occur in APM systems, but the focus of the paper is the development of system designs and operating strategies that enable the use of failure-mode routing of trains and other techniques to retain as much capacity over as much of the system as possible. The theories and techniques of failure-mode operation are given, along with case studies for several airport APM systems. A case is then made for a better industry response to the needs for higher failure-mode capacities, in terms of planning, procurement, design, implementation, and operations.

Introduction

While APMs have historically been highly reliable, there are many types of failures that can cause disruptions of service, resulting in reduced system capacities for durations that are long enough to cause serious queuing, congestions, and passenger delays. Typical failures include the following:

- Platform edge doors failing to open or close properly,
- Automatic train control (ATC) failures, such as false occupancies,
- Electrical power failures, such as short circuits and circuit breaker trips,
- Vehicle failures causing emergency braking that must be reset or totally stranded vehicles, such as over-speed detection, power controller failures, on-board ATC failures, broken power collectors, etc.,
- Trackwork failures, such as broken power rails, damaged running surfaces, etc.,
- And many other types, including owner caused shut downs not even covered in the contractual requirements (e.g., 99.5 percent).

Some of these failures may be rectified in relatively short periods of time, while others may be very extensive, with most somewhere in between.
Due to the redundancies built into APM systems, single point failures seldom shut down the entire system. Instead, some level of passenger service can typically be continued while failures are being repaired. The key issue addressed in this paper is the levels of system capacity that can be maintained during failures.

Failure-mode Operating Strategies

APM equipment failures are generally fairly localized, affecting only a small portion of the overall system. Therefore, when properly designed and implemented, normal operation throughout the remainder of the system can be continued, while bypassing the portion with the failure. For example, power system shorts, ATC false occupancies, and stranded vehicles can normally be isolated to individual sections of guideway, while other portions of the guideway function properly.

Failure-mode operating strategies are then used to reconfigure the APM system to operate around the failure. Such strategies may include the following, depending upon the type of system:

- Single shuttle mode in systems that normally operate as dual shuttles,
- Single loop operation in systems that normally operate as bi-directional loops,
- "Run-around" routes in pinched loop systems,
- Combinations of partial pinched loops and/or shuttles, requiring transfers of passengers at intermediate stations.

Another strategy that can be very helpful in conjunction with the other strategies listed above is "platooning" or running multiple trains in one direction through a restricted section before running multiple trains in the other direction. Since the typical headways between trains in a platoon are shorter than the times to traverse a failed section, the resulting platooning capacities can be greater than running single trains at a time through the failed section.

Consider a simple example of a dual shuttle system with round trip times on each guideway of 4 minutes, resulting in a 2 minute average headway. With 3-car trains and 60 passengers per car, the resulting normal capacity would be 5400 passengers per hour per direction (pphpd) (3 cars x 60 people per car x 60 minutes per hour / 2 minute headway). If a failure were to occur along one of the two guideways that prevented the use of that section of guideway, one approach might be to stop service on the affected guideway and continue service on the other guideway, resulting in half the normal capacity during the failure. The types of passenger loadings on each platform that would result from a failure duration of 30 minutes are shown in Figures 1 through 3, with varying levels of assumed demand during the failure. Figure 1 represents a condition where the failure occurred during a period in which the demand was 4000 pphpd (assumed constant throughout the hour). Figures 2 and 3 are similar conditions, except for the assumed levels of demand of 4500 and 5000 pphpd, respectively. In addition to the platform loadings during normal conditions and the conditions with a single shuttle service (half capacity), the figures also include the platform loadings associated with a 2-train platoon approach, in which both trains are operated, using both sides of both stations, but with both trains going one direction on the non-failed guideway before they return the opposite direction on the non-failed guideway. This operation, which could result in a capacity of over 4000 pphpd (75 percent of normal), is shown in Figure 4.

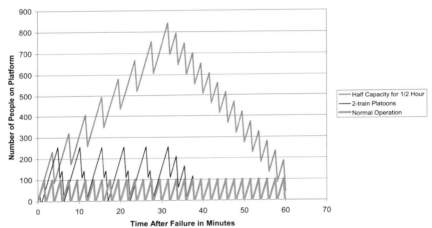

Figure 1 – Sample Dual Shuttle Failure Example – Demand of 4000 pphpd

At this level of demand and assumed capacities, normal operation results in peak platform loads of about 100 passengers before they board the next train. With one train out of service in single shuttle mode (half capacity), the platform loads would continue to build throughout the period of the failure, up to about 840 passengers within a half hour. Assuming full service was restored after 30 minutes, the platform load would gradually be reduced to normal levels in the next 30 minutes.

With the 2-train platooning approach, peak platform loads reach about 250 passengers, but are not continuing to grow during the failure and are reduced to normal levels shortly after full service is restored.

With 4500 pphpd demand, as assumed in Figure 2, the single shuttle peak loads reach about 1100 passengers and would require over an hour after full service was restored to once again return to normal levels. The 2-train platoon approach would reach a peak loading of about 500 passengers and return to normal loads in about 25 minutes after full service was restored. Quicker recovery could would occur if the demand in the following hour were less.

Similar improvements with platooning are shown in Figure 3 for 5000 pphpd demand.

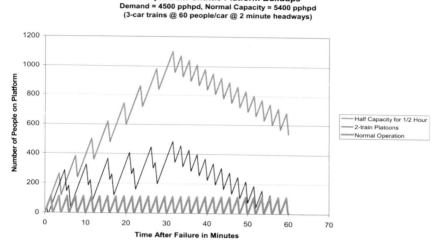

Figure 2 – Sample Dual Shuttle Failure Example – Demand of 4500 pphpd

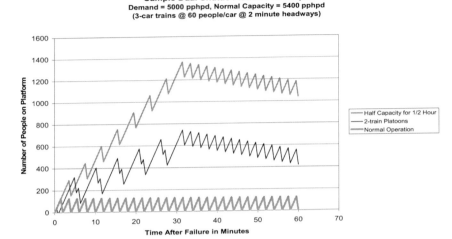

Figure 3 – Sample Dual Shuttle Failure Example – Demand of 5000 pphpd

Typically, the APM platforms would not safely hold these levels of passengers, so the flows would have to be managed, limiting access to the platforms, and queues would back up into the areas upstream of the station platforms.

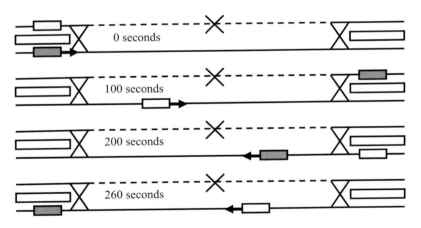

Figure 4 – Example of 2-train Platoon Sequence in a Normally Dual Shuttle System

An example of a pinched loop system with a failure in the center of the system and with a run-around route, is shown in Figure 5. The achievable capacity in this mode is dependent upon the time required to traverse the section along the failed section and the number of trains that can be platooned in each direction before trains in the opposite direction. If in this example the normal operation were again 3-car trains with 60 passengers per car at 120 second headways, the normal capacity would again be 5400 pphpd. If traversing the failed section took an average of 180 seconds in each direction (including clearance times at each end), the failure-mode capacity with single trains alternating through the common section would be 1800 pphpd. With 3-train platoons with 90 second headways between trains in a platoon, the failure-mode capacity would increase to 2700 pphpd.

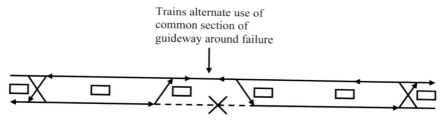

Figure 5 – Example of Run-around Route on a Pinched Loop System

In the example shown, the dwell times at the station near the failure and the associated acceleration and deceleration times would be included in the time on the common section, reducing the capacity compared with what could be achieved if the crossover that is shown on the left of the station would have been on the right of the station. If an assumed station related time of 60 seconds were removed, so that the average time to traverse the failed section were 120 seconds, the single-train capacity would increase to 2700 pphpd and the 3-train platoon capacity would increase to 3420 pphpd.

Dual loop systems with multiple stations can also benefit from enhanced routing strategies, such as run-around routes and platooning, compared with simply resorting to a single loop operation with only half the normal capacity.

Another strategy that can be used to help enhance failure-mode capacities is the inclusion of longer platforms than needed for normal operation (including extra platform edge doors). If longer than normal headways must be used during failure modes, some of the lost capacity can be regained during extended failure modes by using a longer than normal train consist. In the pinched loop example above, using 4-car trains instead of 3-car trains could increase the failure-mode capacity to 4560, or about 84 percent of the normal capacity, compared with about 33 percent for the simple run-around approach. This would provide even more dramatic improvement in platform loads, congestion, and delays than shown above for the dual shuttle example.

In relatively short systems, it is highly desirable to provide pedestrian corridors in parallel with the APM links, equipped with moving walkways. This allows a walk / ride choice during normal conditions and provides backup capacity during APM failures with reduced capacities.

Case Studies of Existing Systems

Most APM systems have some failure-mode capabilities, but few have the enhanced strategies described above.

Sample dual shuttle systems with single shuttle failure mode:

- Tampa airport
- Orlando airport
- Miami airport
- Cincinnati airport
- Sacramento airport (in development)

Sample dual loop systems with single loop failure modes:

- DFW airport
- Miami urban system
- Taipei airport (planned)

Sample pinched loop systems with run-around routes:

- Chicago airport
- San Francisco airport
- Phoenix airport (planned)
- Hong Kong airport (planned extensions)
- Many urban systems, such as Lille, Vancouver, and Copenhagen.

Sample APM systems with adjacent pedestrian corridors:

- Atlanta airport
- Cincinnati airport
- Los Angeles airport (planned)

Industry Support Needed for Enhanced Strategies

Too often in the past, enhanced failure-mode capacities have not received appropriate attention. Some members of the APM community have intentionally avoided requiring the use of enhanced techniques, with arguments such as the following:

- They add too much to the cost,
- They are too complicated (often translated: we don't understand them),
- The O&M staff won't use them anyway.

So opportunities to significantly reduce the amount of queuing, congestion, and delay during failures have been missed, even though the added costs would have been a minor percentage of the overall system costs.

Recommendations for improving this facet of the APM industry include the following:

- Owners should understand the potential impacts of reduced capacities during failure modes and expect their systems to be optimized to minimize the impacts,
- All suppliers should have double crossover options (X shaped, not sequential) to reduce the times in common sections during failure modes,
- Much greater use of double crossovers, crossovers beyond end stations, and run-around routes should be applied,
- The ability to analyze enhanced failure modes of operation should be standard tools for consultants and system designers, including the optimization of platooning techniques by system and by link,
- Procurement documents should require the development of enhanced failure-mode techniques, in response to detailed failure scenarios throughout the system, including the ability to automatically call up and transition to specific alternate routings with associated alternate nominal station dwells, routes, and associated train control features,
- Such automated responses should be integrated with techniques for recovery of stranded trains, to rescue related passengers and to minimize the time to restore full service,
- Specific capacity requirements should be defined, not just for normal mode of operation, but for all reasonable failure modes and locations, with both analysis and system acceptance test demonstrations used to verify compliance,
- O&M staff should be trained and tested regularly on the use of the full range of failure-mode techniques that are built into the system, and
- O&M payment incentives and penalties should include consideration for the actual responses to failure conditions and the extent to which capacities were optimized in the responses.

Orlando APM Running Surface Rehabilitation
Airsides 1 & 3

Sambit Bhattacharjee [1], P.E., M.ASCE (Primary Contact)
Dan McFadden [2], P.E., M.ASCE
Tuan Nguyen, P.E [3]

(1) Senior Associate, Lea+Elliott, Inc., 5200 Blue Lagoon Drive, Suite 250, Miami, FL 33126; PH 305 500 9390; Fax 305 500 9391; sambitb@leaelliott.com
(2) Associate Principal, Lea+Elliott, Inc., 5200 Blue Lagoon Drive, Suite 250, Miami, FL 33126; PH 305 500 9390; Fax 305 500 9391; dmcfadden@leaelliott.com
(3) Senior Project Manager, Planning & Engineering, GOAA, 5850 Cargo Road, Orlando International Airport, Orlando, FL 32827 PH 407-825-4662; Fax. 407-855-3531 ; TNguyen@goaa.org

Abstract

Rehabilitation of facility related elements, such as the guideway running surface of an Airport People Mover (APM) System should be undertaken in a proactive fashion. This would include development of several different repair alternatives for the owner's consideration, rather than a forced running surface replacement with associated emergency shutdown of the APM System. This paper will present how an extensive material testing/analysis allowed the team to evaluate the condition of existing concrete and develop a larger range of alternatives for the rehabilitation. The alternatives ranged from a durable rehabilitation with minimal impact to the service to a full replacement of the running surface. The paper examined features of an alternative evaluation matrix, developed as a decision-making tool for the Owner.

Introduction

Orlando International Airport (OIA) is operated by the Greater Orlando Aviation Authority (GOAA). The Airport utilizes four independent Automated People Mover (APM) systems connecting the main Terminal building to each of the remote Airside buildings. APM vehicles transport passengers between Terminals on elevated guideways. The APM running surfaces serving Airsides #1 and #3 are more than 28 years old. Over this time, the concrete running surface has experienced deterioration at several spans along the Airsides #1 and #3 APM guideways. These deficiencies include surface cracking, spalling of the running surface, and are a maintenance concern for the Greater Orlando Aviation Authority (GOAA). In this assignment, the Lea+Elliott team performed an evaluation of the guideway and potential rehabilitation alternatives. A general layout of Airside 1 and 3 is shown in Figure 1.1.

Each APM system operates approximately 16-20 hours per day with an average round-trip period of about three minutes or an operating headway of one and a half minutes. The maintenance hours for each train on a leg are staggered such that there is always one train available for the entire 24 hours of the day. Due to the must-ride nature of these systems, the rehabilitation solutions must have minimal impact on the APM service and be scheduled based on the operations of the airport utilizing the low-demand periods. The options were further coordinated with the airport operations to implement them with due regard to failure management considerations.

Figure 1.1: OIA - General Layout (Airsides and Guideway Designations)

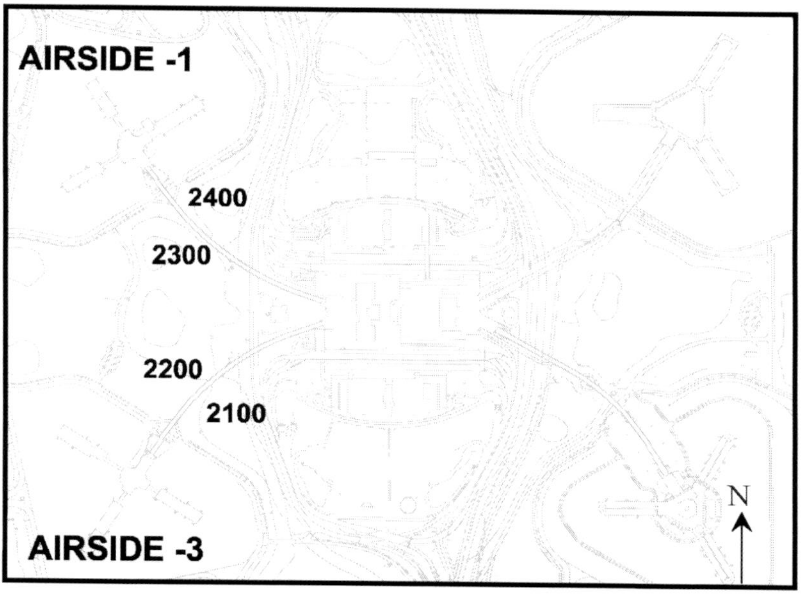

AIRSIDE -1

2400
2300
2200
2100

N

AIRSIDE -3

Background

The guideway and the running surface have exhibited a certain amount of deterioration over the last few years and several areas have been repaired with localized patching of concrete to respond to spalls and cracks on the running surface. Some repairs have performed well, whereas others have deteriorated quickly. The current procedure for monitoring and repairing the running surface on as needed basis is adequate, however this project was initiated to evaluate, analyze and present a comprehensive solution for long term rehabilitation for owners' consideration.

Understanding both the "must-ride" nature of the system and the ridership demands on the Airside APM legs is critical in determining the criteria for a comprehensive "best-value" evaluation of alternatives. The constraints, limitations and challenges that are critical to the project implementation are listed below:

- Improvements and rehabilitation should have minimal impact on passengers and ridership.

- Guideway repair/rehabilitation work duration should match the APM downtime (a 4-hour window in the night) is preferable.
- Improvements and rehabilitation activities that require more than 4 hours but less than 8 hours may be accommodated with adequate coordination with the owner. This is not desirable but may be scheduled during off-peak seasons.
- Span by span construction sequencing and use of quick setting material are considered essential for this project due to the above needs, even if these approaches are not the most cost effective.
- The tug roads under the APM system are essential and are required to remain open, specifically during heavy usage hours and peak seasons. This will be considered during the development of alternatives.
- Safety of people and equipment in the APM as well as in the area around the APM is critical.

Evaluation

The evaluation of rehabilitation alternatives was undertaken as a part of this study. These evaluations were categorized into Structural and Material Quality Assessments.

The structural assessment was conducted to review and record the surface condition of the APM running plinths. Additionally, the material quality assessment was used to identify the extent of material deterioration for the concrete plinths and thus devise a solution for the running surface. The analysis and assessment included:

- Transverse Hairline Cracking
- Open Transverse Cracking
- Map Cracking
- Longitudinal Cracking
- Corrosion Potential
- Field Carbonation
- Compressive Strength
- Chloride Ion Content
- Petrographic Analyses

The general condition of the concrete was found to be good, however the following issues were observed:

- The concrete running surface has developed widespread transverse cracking and some localized longitudinal cracking.
- Map cracking has developed in selected areas, but this is localized, and these cracks are generally tight (less than 0.010 in.).
- The concrete strength exceeds the design strength, and there is no sign of any durability-limiting chemical reaction occurring within the concrete itself.
- Some abrasion of the concrete surface has occurred. The abrasions are not yet to the point of affecting ride quality, but they are more pronounced at the station areas and at points where repairs have been done.

Alternatives

Three alternative concepts for the improving and rehabilitating the running surface were evaluated based on their ability to respond to the criteria established for the project. These concepts are:

Alternative-1   Basic Repair Option (Repair-as-you-go)

Alternative-2   Running Surface Improvements; Durable Rehabilitation Option (Crack Mitigation and Preventive Hydrophobic Coat)

Alternative-3   Replacement: This includes three potential alternatives- i) Full Concrete Replacement, ii) Concrete in Steel Channel Form and iii) Full Steel Replacement.

Evaluation Criteria

The following criteria were developed for the evaluation of alternatives.
- Impacts on APM System Safety
  - Operational safety pertains to the ability to safely stop the vehicle within its designed safe stopping distance.
  - Structural safety considerations pertain to ensuring that vehicles will be safely supported on the guideway and that there will be no hazards to personnel and equipment around or beneath the APM.
- Service Life of the Guideway Running Surface and Structure (Material Durability)
- Disruption to Traveling Passengers during Replacement/ Improvement
- Costs
  - Replacement and Improvement Costs
  - Service Interruption Costs
  - Life Cycle (25 years normalized) Costs
- Impact on Operational Conditions
  - Ride Quality
  - Capacity
  - Level of Service

Alternative Evaluation

The following tables provide a summary of ratings that were assigned to the various alternatives. Table 1.1 provides a qualitative rating, of the alternatives. These options were discussed with GOAA. Based on the review of the alternatives the following ratings were concurred and accepted by GOAA and the recommended alternative was developed.

Table 1.1: Alternative Evaluation Rating Matrix (qualitative rating)

Alternative Evaluation Rating Matrix — Draft Evaluation Rating — APM Replacement / Improvement Alternatives Ratings

Evaluation Rating	Alternative 1 Basic Repair Option	Alternative 1 Basic repairs with Miscellaneous Repairs	Alternative 2 Rehabilitation / Durability Option	Alternative 2 Rehabilitation/ Durability with Miscellaneous Repairs	Alternative 3 - (Replacements) 3.1 Concrete Full Replacement	Alternative 3 - (Replacements) 3.2 Concrete in Steel Channel Form Replacement	Alternative 3 - (Replacements) 3.3 Steel Full Replacement
Impact on APM System safety							
Operating System Safety	Neutral	Neutral	Neutral	Neutral	Neutral	Neutral	Poor
Structural System Safety	Poor	Neutral	Neutral	Neutral	Good	Good	Good
Service Life of Guideway Running Surface	Poor	Poor	Good	Good	Good	Good	Good
Disruption to Traveling Passengers during replacement / improvement	Neutral	Neutral	Neutral	Neutral	Poor	Poor	Poor
Cost							
Replacement/ Improvement Costs	Good	Good	Neutral	Neutral	Poor	Poor	Poor
Service Interruption Costs	Poor	Poor	Neutral	Neutral	Poor	Poor	Poor
Life Cycle (25 year normalized cost)	Poor	Poor	Good	Good	Neutral	Neutral	Neutral
Impacts on Operational Condition							
Ride Quality	Neutral	Neutral	Good	Good	Good	Good	Good
Capacity	Neutral	Neutral	Neutral	Neutral	Poor	Poor	Poor
Level of Service	Neutral	Neutral	Neutral	Neutral	Poor	Poor	Poor
Totals							

Preferred Alternative and Implementation Plan

The following implementation plan was considered the best value due to its capability to meet the project need and address the Owner's requirements. These measures are recommended along with increased inspections for the guideway.

> Based on the material condition and evaluation of alternatives; "Running Surface Improvement; Durable Rehabilitation Option" is considered the best value alternative using the following procedures:
> - Span by Span application.
> - Overnight application.

Backup and Standby Bus Options

Given the necessity to maintain continuous service on these "Must-Ride" APM systems, an operational back-up plan was developed. There is a possibility of using "backup buses" or using "forced restart of system," as a back up option during the resurfacing. This would be required only if the second APM of the leg became inoperable during the work. The appropriate costs have been identified.

- Provide Increased Maintenance on In-Service

- Busing Operation Between Airside 1 & Airside 3

- Forced Restart of System

- Provide a Temporary Covered Walkway to the Airside

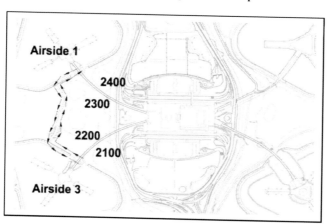

Figure 1.2: Backup Bus Route Map

Conclusion

A comprehensive approach has facilitated the success of the project through investigation, evaluation and alternative selection. The project is under design, with procurement scheduled for later part of the year. It is understood that the guideway rehabilitation task will be undertaken with minimal impact of the existing airside service. Most of the activities will be completed in tandem with the night maintenance of the system. However, some of the longer duration repairs will be completed during low demand period when one of the two shuttles will be out of service. The exact time and duration of the work will be coordinated with GOAA.

SYSTEM DEMONSTRATION: PREPARING FOR SUCCESS

Matthew Sturgell, P.E.*

* Senior Transportation Engineer, Lea+Elliott, Inc., 44965 Aviation Drive, Suite 290, Dulles, VA 20166; PH 703-537-7424; msturgell@leaelliott.com

Abstract

The System Demonstration period is a critical component of testing a new automated transit system. The primary purpose of System Demonstration is to assure the Owner that the System is truly ready for passenger service. This can be a stressful time for the Owner and Contractor alike. If System Demonstration takes longer than expected, or if the project is already behind schedule for other reasons, pressure to open the System intensifies. Adequate preparation is key to successfully completing the System Demonstration requirements in a timely manner. Therefore, it is important to understand what System Demonstration involves, what should be done to prepare for it, and how that preparation will lead to success.

System Demonstration

Automated People Movers and other automated train systems require a substantial amount of testing prior to opening to the public. This includes testing of the vehicles, automatic train control, power distribution system, communication system, and integration of all these components. One of the services Lea+Elliott provides system Owners is oversight of the testing phase. Lea+Elliott performs test witnessing to verify and recommend acceptance of the testing activities and results. As a culmination of the verification and acceptance activities and a prerequisite for carrying passengers, Lea+Elliott requires the Contractor to complete a System Demonstration. The ASCE APM Standards Committee recently published Part 4 of its APM Standards, which includes System Verification and Demonstration requirements. This standard recommends conducting a System Demonstration test for System acceptance.

System Demonstration, as the name implies, requires that the Contractor demonstrate the System by operating it as though it were in full passenger service. During System Demonstration the Contractor must adhere to the following operations and maintenance policies and procedures:

- The System must be operated in accordance with the accepted operations plan. It must be operated continuously on the specified daily and weekly schedule. The level of service specified, including headway and line capacity, for any peak and off-peak operating hours must be maintained.

- The System must be maintained in accordance with the accepted maintenance plan. To confirm that the maintenance plan can be accomplished and sustained during normal passenger service, neither additional nor fewer maintenance activities are allowed.

- The System must be operated and maintained in accordance with the accepted staffing plan. This requires the Contractor to use the same number and skill level of personnel that will be used during normal operations and maintenance after the System is open for passenger service.

- The System must be operated and maintained in accordance with the accepted operating procedures and maintenance manuals. These documents include the detailed operating and maintenance procedures for all components of the System.

The Contractor must conduct System Demonstration in accordance with the above policies and procedures until the specified System Service Availability has been achieved over a consecutive 30-day period. If the amount of downtime that occurs within the first 30 days causes the System Service Availability to fall below the specified value, System Demonstration must continue until the Contractor has achieved 30 consecutive days where the total downtime is less than the allowed limits. A high level of availability, and therefore a small amount of downtime, is necessary to successfully complete System Demonstration. The purpose of this requirement is to ensure that the System will perform reliability during passenger service.

System Service Availability

System Service Availability is a measure of the total quantity and quality of transportation service actually provided compared with that scheduled to be provided over a given time period. In other words, it is the percentage of time that all components of the System are operating as they should. For calculation purposes, Lea+Elliott defines System Service Availability as the product of Service Mode Availability, Fleet Availability, and Station Platform Availability.

Service Mode Availability accounts for any interruptions to the operating mode during a specific time period. Any deviations from the accepted operations plan, including changes to the route, headways, or line capacity, are considered downtime events.

Fleet Availability accounts for any vehicles that are not fully functional. Examples of issues that are considered downtime events include inoperable vehicle doors, issues with audio and visual passenger service announcements, and nonfunctioning lighting or HVAC.

Station Platform Availability considers problems associated with the station platform doors. Any platform door that does not operate when it should is considered a downtime event.

In addition to limiting the cumulative duration of downtime, the System Service Availability places limits on the number and duration of individual downtime events. Placing limits on individual events addresses the possibility that a large number of short downtime events could meet the availability requirement while causing frequent service disruptions.

To illustrate how System Service Availability affects System Demonstration, consider the following example. An APM system is scheduled to operate 18 hours per day, seven days per week, or a total of 540 hours in 30 days. The availability requirement to complete System Demonstration is 98.5%. This allows just over eight hours of total downtime during the 30-day test. Over the course of the first 24 days, minor disruptions total over seven hours of downtime. However, on day 25, a major issue disrupts service for 45 minutes. The next day, another major incident creates a 20-minute downtime event and drops the availability below the requirement. By day 30, the cumulative downtime is nine and one half hours. It takes 16 more relatively incident-free days before the cumulative downtime for 30 consecutive days drops below the allowed amount. Figure 1 illustrates this example. The cumulative downtime for days 17 through 46 is eight hours, achieving the availability requirement.

Figure 1: System Service Availability Example

Day	Daily Downtime (hrs:min)	30-day Cumulative Downtime (hrs:min)
1	0:20	
2	0:15	
3	0:25	
4	0:20	
5	0:10	
6	0:25	
7	0:15	
8	0:20	
9	0:15	
10	0:20	
11	0:15	
12	0:10	
13	0:20	
14	0:25	
15	0:20	
16	0:15	
17	0:20	
18	0:15	
19	0:10	
20	0:20	
21	0:20	
22	0:25	
23	0:15	
24	0:20	
25	0:45	
26	0:20	
27	0:20	
28	0:15	
29	0:20	
30	0:15	9:30
31	0:20	9:30
32	0:10	9:25
33	0:15	9:15
34	0:20	9:15
35	0:10	9:15
36	0:15	9:05
37	0:10	9:00
38	0:15	8:55
39	0:05	8:45
40	0:15	8:40
41	0:10	8:35
42	0:15	8:40
43	0:10	8:30
44	0:10	8:15
45	0:15	8:10
46	**0:05**	**8:00**

As you can see, the System Service Availability is affected by an array of factors. Operational, maintenance, and staffing issues can all directly or indirectly impact System performance. Being prepared on all these fronts is key to maintaining a high level of availability.

Operational Preparation

Most operational issues encountered during System Demonstration will create a downtime event and have a direct impact on System Service Availability. Even if these events are minor disruptions to service, they add up quickly and begin to consume the allowable downtime. There will always be unexpected issues that arise during System Demonstration and the Contractor should plan that these issues will use up some of the allowable downtime. It is prudent to have all other verification and acceptance activities complete and all known issues resolved before beginning System Demonstration. The intent of System Demonstration is that it be the final test, integrating all the System components and demonstrating the reliability of the System as a whole. Therefore, Lea+Elliott does not allow the Contractor to conduct other tests during System Demonstration. In addition, entering into System Demonstration with known issues, hoping to resolve them along the way, will likely result in System Demonstration taking longer than 30 days to complete. For all practical purposes, the System should be ready to carry passengers before System Demonstration begins. Being ready operationally will help ensure a successful System Demonstration.

One way to help ensure that the System is ready operationally is to complete the Safety Certification prior to System Demonstration. Safety Certification is an important contract requirement during which an independent party verifies all safety-related aspects of the System and certifies the System's safety. Not only is it desirable to have this certification before beginning System Demonstration, but the Safety Certification by its nature also helps expose issues that may otherwise be encountered during the test. Safety Certification takes time and requires that a certain level of testing be complete, so the timing and perquisites must be planned well in advance to obtain the certification before beginning System Demonstration.

The System operating procedures are one set of documents that are reviewed as part of the Safety Certification. These procedures should be complete sufficiently in advance of System Demonstration so that they can be reviewed during the Safety Certification process. It is also important that they be complete as early as possible so the staff can familiarize themselves with the operating procedures for all System components and be ready to implement those procedures during System Demonstration. Therefore, Lea+Elliott requires that a preliminary version of the operating procedures be submitted to the Owner for review six months prior to System Demonstration. This submittal schedule also provides sufficient time to address any questions or concerns the Owner has with the procedures.

Maintenance Preparation

Properly maintaining the system will help prevent operational issues from occurring. The first step in preparing to maintain the System is to have all the necessary maintenance equipment, spare parts, and expendables on site. While this may seem obvious, contractors often delay the purchase of equipment and parts until later in the testing phase or even just before the start of System Demonstration. However, obtaining the equipment and parts well in advance is beneficial. It allows ample time to inventory and organize everything, making maintenance more efficient while the System is in operation. It also allows necessary maintenance to be performed during the testing phase so that all the System components are in top condition heading into System Demonstration.

To help maintain an adequate inventory of spare parts and track maintenance schedules, Lea+Elliott's specifications require the use of Maintenance Management Information System (MMIS) software during testing and System Demonstration. This versatile software provides an organized and efficient way to manage the maintenance program. It is important that this software is up to date with the latest inventory and maintenance schedule information before System Demonstration begins.

While the MMIS helps manage maintenance of the System, the maintenance manuals contain all the procedures and part information necessary to actually perform the maintenance. Preparing these manuals and familiarizing the staff with them early is essential to safe and efficient operation and maintenance of the System. Similar to the operating procedures, Lea+Elliott's specifications require that a preliminary version of the maintenance manuals be submitted six months prior to System Demonstration.

Obtaining the equipment and parts, preparing the MMIS, and completing the maintenance manuals are all important, but they are of little benefit if the maintenance plan is not implemented prior to System Demonstration. In the rush to start System Demonstration, implementing the maintenance plan is sometimes overlooked. However, a System that is inadequately maintained during the testing phase can create issues that result in downtime during System Demonstration.

Staff Preparation

Being able to prevent problems through proper maintenance and respond to operational issues when they do occur is an important aspect of preparing for System Demonstration and requires sufficient, well-trained staff. Often, Contractors utilize implementation staff to test the System, then employ different staff to operate and maintain the System. While this philosophy can work, it is important to employ the O&M staff early. Therefore, Lea+Elliott requires the Contractor to implement the staffing plan and use the same number and skill level of personnel for System Demonstration that will be used during normal operations and maintenance after the

System is open for passenger service. In practice, the best way to do this is to employ all the permanent O&M staff prior to System Demonstration.

However, just having the O&M staff on board before System Demonstration is not enough. The permanent staff need to be brought in early enough to receive comprehensive, thorough training on all the components of the System that they will be responsible to operate and maintain. The Contractor must develop a structured training program that covers all aspects of the System and ensures all staff receive sufficient training for their job classification. The training must familiarize the staff with all the System components and cover the operating and maintenance procedures in detail. This is one of the most important aspects of being prepared for System Demonstration. Well-trained staff that are familiar with all the components and procedures will be able to address operating and maintenance issues that come up during the test. As such, Lea+Elliott requires that the permanent O&M personnel be tested for proficiency within their job classification.

While knowledge of the components and procedures is important, it is also advisable that staff have sufficient practice carrying out the operating and maintenance procedures. Employing and utilizing the staff early in the testing phase is one way to accomplish this. During the course of testing the staff will gain experience using many of these procedures. However, some procedures may be used for the first time during System Demonstration, so giving staff practice with them is also important. This probably means time in addition to the required testing, but it is a good idea to prevent staff from having to conduct procedures for the first time during System Demonstration to the extent possible. Practicing things like inserting and removing trains from service efficiently, troubleshooting issues out on the System, and recovering a train manually may be very beneficial. Time is of the essence when issues arise during System Demonstration. Sufficient, well-trained, and experienced staff will go a long way toward smooth operation of the System.

Preparing For Success

System Demonstration integrates all the different aspects of the System into one final test and proves how well the System as a whole will function in passenger service. System Demonstration can be a stressful time as the Contractor works to finish the project before the completion date. If System Demonstration carries on longer than the amount of time allocated in the schedule, or if the project was already behind schedule before entering into System Demonstration, the pressure to complete the test can be intense. With proper preparation, the stress of this phase can be greatly reduced. There are several things the Contractor should do to effectively prepare for System Demonstration:

- Complete all other acceptance activities
- Resolve all known issues
- Complete the Safety Certification process
- Complete the operating procedures

- Purchase and inventory the maintenance equipment, spare parts and expendables
- Utilize the MMIS software
- Complete the maintenance manuals
- Implement the maintenance plan
- Employ the permanent O&M staff early
- Train the O&M staff thoroughly
- Ensure the O&M staff have sufficient practice and experience

Adequate preparation on all these fronts significantly increases the likelihood of successfully completing System Demonstration on time.

Energy-efficient APM using High Performance Batteries

Masaya Mitake*, Hiroshi Ogawa** and Katsuaki Morita*

* Hiroshima Research & Development Center, Mitsubishi Heavy Industries, 1-1, Itozaki Minami 1-chome, Mihara Hiroshima, 729-0393, Japan; masaya_mitake@mhi.co.jp

** Plant and Transportation Systems Engineering & Construction Center, Mitsubishi Heavy Industries, 1-1 Itozaki Minami 1-chome, Mihara Hiroshima, 729-0393, Japan

Abstract

Mitsubishi Heavy Industries (MHI) is taking the lead and pioneering the transit industry in green technology implementation by developing more environmentally friendly transit solutions and advantages to customers. MHI is currently developing a battery-powered transportation system. The use of high-performance, high-capacity batteries will not only improve electrical power efficiency but also contribute to simplified guideway equipment, improved safety and reduced operation and maintenance activities. This paper describes MHI's next generation energy-efficient APM using High Performance Batteries.

Introduction

The use of on-board lithium-ion batteries ensures the highest power and capacity of the battery technologies currently available in the market. Lithium-ion batteries provide both high energy density and high energy output, and their use in the market has been rapidly expanding since the 1990's.

According to the background of the continually improving automobile industry's environmental strategies, the technological development and use of the lithium-ion battery in HEVs (hybrid electric vehicles) and EVs (electric vehicles) has also been increasing exponentially.

Major transportation companies have been announcing that lithium-ion type batteries will be applied for the commercial use of HEVs and EVs to take advantage of their environmentally-friendly characteristics. The batteries used for the Crystal Mover are standard, off-the-shelf components designed for the transit industry. Not only have lithium-ion batteries already been put into commercial use for hybrid electric buses and trucks, but this battery technology is also used in other industries.

Application of high-capacity batteries

Among the various types of high-performance, high-capacity batteries being developed for automotive applications, lithium-ion stands out as being capable of storing and instantaneously supplying high power, and allowing stable continuous charging and discharging at high currents. Lithium-ion batteries are the most suitable for APM vehicles with their operating patterns of repeated and frequent high rates of acceleration and deceleration.

Because of their high-energy density, lithium-ion batteries are small and compact to fit into the limited space available in proven Crystal Mover APM design.

On board battery system

Figure 1 shows the configuration of the battery system. Lithium-ion batteries are mounted under the vehicle floor and are connected to the main vehicle circuit

MHI proposes to employ an on board battery power system for propulsion utilizing lithium-ion type batteries, which are mounted on the vehicle. With lithium-ion batteries and the charging system, the battery powered system eliminates the conventional power supply from power rails. Consequently, the time for construction and maintenance can be reduced.

During braking, the kinetic energy is efficiently recharged back to the battery. This eliminates the need of either on-board or wayside brake resister banks that are typically required by standard power distribution systems with power rails to dissipate the braking energy as heat. As the proposed system is able to recover this energy by recharging the batteries instead of wasting it, the power consumption is typically reduced by 15% compared to the standard power rail systems. With the introduction of the lithium-ion battery system, MHI has achieved a compact, energy efficient, and more environmental friendly vehicle.

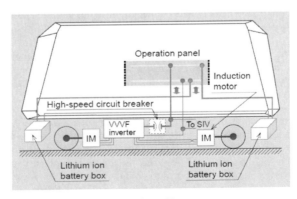

Fig1. Configuration of battery system

The batteries provide sufficient capacity to operate the train and all auxiliary power including lightings and HVAC for all operating modes in both normal and abnormal conditions, with redundancy. The batteries are configured as modules which are easy to handle and maintain.

Charging system

MHI has 2 types of battery charge system. One is a contact type charge system. With this system, charging energy is transferred from wayside power supply shoes to the power collector onboard the vehicle while it is stopped in the station. The advantages are simple configuration and high efficiency.

The other is a non-contact inductive power transfer system. This system is superior to the former in maintenance and safety, featuring robustness for outside environment.

Power management simulator

The design and the management of the battery unit operation are important for this system. MHI developed the analytical model applicable to lithium-ion batteries.

Using this model, MHI built the power management simulator to be used to design the best battery configuration for project-by-project basis. Expected life of batteries and power consumption are calculated. Flexible and easy system design can be achieved.

For example, number or distribution of charging stations will be optimized. The best pattern of battery operation for each customer can be proposed. Temperature influence to power consumption and lifetime of battery can also be simulated.

Production Vehicle

Summary

The newest, most environmentally friendly evolution of the proven Crystal Mover APM addresses the rapid growing need for energy efficient, safe, reliable and affordable transportation alternatives. MHI has worked on the development by incorporates high-performance, high-capacity battery technologies. The battery powered APM is an unsurpassed urban transportation system which meets the need of the customer. The new production vehicle incorporating the environmentally friendly concept has been manufactured and started demonstration operation on the test track.

Standards for Successful APM Implementation

Frank Culver – Manager, Mechanical Engineering
Mario Nuevo – South East Area Manager
URS – Washington Division
Miami, Florida, USA

1 Abstract

Today, an APM customer faces monumental investment in the design, manufacture, construction, and commissioning of an APM system, or APM system enhancement. Such procurements typically begin with the development of some manner of technical specification, general provisions, RFP document, and other associated processes.

Several alternatives exist for the approach to the technical specification. Such a specification can be highly detailed, or it can be very general in nature.

The approach to such a procurement can have significant cost, schedule, and quality impacts, to the degree to make a project not feasible.

The ASCE APM Standards Committee has developed four parts to a standard that can be very useful to APM owners when procurement of a system or system elements is being considered.

This paper will explore the background and alternatives associated with these considerations, and the tools available for developing a set of technical specifications for such programs.

2 Introduction

Since the early 1970s, when the Tampa Airport opened with the world's first automated people mover system, growth in APM transit has included airports, urban centers, downtown circulators, and business hubs. A variety of technologies have been developed utilizing different guideway systems, propulsion configurations, and vehicle sizes.

Over the past nearly 40 years, attempts have been made to include as many technology options as possible, while addressing the needs of the client.

The ASCE APM Standards Committee was formed in 1990 with the purpose of developing a consensus standard for Automated People Movers. With the support of manufacturers, owners, operators, and consultants, four parts to the ASCE APM Standard were published. The are:

    Part 1: Environment, Safety, and ATC
    Part 2: Vehicles, Propulsion, and Braking

Part 3: Electrical Equipment, Stations, and Guideways
Part 4: O&M, Training, and Manuals

The standard is subject to a regular review and reaffirmation process in order to assure that the standards cited and technologies are maintained as current as possible.

The standard is available from ASCE Publications, and has been utilized as the basis for specification for several APM procurements, including the fleet replacement for the Miami Downtown People Mover.

3 The need for Standards

Standards can fall into a variety of categories. Technical standards can be either performance based, or include detailed hardware specifications.

The wide range of technologies and system types in the Automated People Mover industry would render detailed hardware specifications either prohibited to some or many manufacturers, or be unmanageable, or both.

There is a need, however, to standardize elements of safety, performance, and comfort for APMs. For this reason, the ASCE APM Standards Committee utilized a philosophy of performance based standards for each of the four sections of the Standard.

These performance standards include such items as:

- Safety
- Reliability
- Failsafe Criteria
- Maintainability Criteria
- Ambient Conditions
- EMI
- Presence Detection
- Verification and Validation
- Structural Design Criteria

To name a few. These performance criteria were developed with all participating manufacturers, owners, operators, and consultants in a consensus environment. Any element of the standard that would prohibit a manufacturer's product was reviewed and revised until satisfactory to all involved.

The benefits of such standards are many. Particularly from a regulatory standpoint, standards of performance reduce the cost of specification and procurement document development, design review, inspection, certification, and test and validation. Further, such standards streamline the process for system manufacture and installation, as the details and process for the elements of system construction do not change.

Thus, manufacturers are able to improve their product cost, the Owner's cost of system procurement and installation follow are reduced, and reliability is increased due to the manufacturer's ability to "standardize" their product.

4 A Process Oriented Approach

The approach taken to the development of specifications for an APM involved two primary phases. The first phase would include those tasks leading up to the development of a technical specification and commercial requirements for a request for proposal, and award to the successful contractor. The second phase would involve the follow of all of the activities of the successful contractor, including design, manufacture, test, and commissioning.

The steps involved in Phase 1 include a technical assessment, the development of a concept report, preparation of technical specifications and an RFP, and finally, contract award. These steps are described in greater detail below.

4.1 Needs Assessment

To provide for a customer, one must first identify what the needs of that customer are. The assessment process is a process where the needs of the owner are identified and documented. This begins with high level performance and configuration features such as system length and orientation, major service areas or stations, power available, and anticipated passenger usage.

Once high level needs are identified, the assessment team can arrange for detailed discussions with the owner or the appropriate owner's representatives to develop details associated with each of the needs areas.

4.2 Concept Development

With the system needs assessment complete, the team began the concept process. This process begins with the data gathered in the needs assessment, and together with state of the art system information, lists and develops alternatives for the design alternatives for the system. This process includes rigorous visits with system contractors to describe the project, and garner various proposed high-level approaches to system design.

Once these alternatives are developed, cost analysis is performed to identify the most cost effective approach to each of the system approaches, including layout, service area, equipment layout, etc.. This is then reviewed in great detail with the Owner, such that consensus can be reached between the Owner, and all key constituents.

Key work scope areas can be identified in the Concept Report process are listed below.

4.2.1 Guideway

The guideway is the not only the primary cost element in any system's construction, it also defines the service route to stations, around geographical constraints, and fixed facilities. The guideway system will include footings, columns, spans, switches, and other elements.

Each system manufacturer will employ it's own guideway design. As there are often notable differences between the guideway of various technologies, the standards associated with guideway design must be performance based, and include such factors as wind design, factors of safety, emergency egress, and possibly aesthetics.

Where guideways are incorporated into a building structure, bridge, or other fixed facility, a detailed dialogue must take place between the guideway designer and the fixed facility designer.

In addition to acting as the path upon which the vehicle travels, the guideway is also the structural conduit for cabling, raceways, antennas, power rail, walkways, and lighting. Each of these must be noted in terms of application of standards for installation, maintainability, corrosion resistance, and accessibility.

Including performance standards in the guideway specification will help to insure that a guideway meeting the needs of the owner is constructed without undue cost impact.

4.2.2 Stations

Stations represent the interface between the adjacent urban or commercial area and the APM. Stations must be readily accessible, cleanable, vandal resistant, and structurally sound. Stations must be properly lit, and include signage that aids in passenger use and movement.

Stations can be at grade, or require some sort of elevation access for passengers, such as stairways, escalators, elevators, or ramps.

Each of these elements can be provided in a variety of ways, and cost bases. Standards associated with station elements are often tailored to the needs of the owner. Regardless, those details must be included in the specification. Baseline standards can be modified for increased aesthetic, communications, advertising, durability, or access.

Stations typically also serve as housing locations for equipment such as ATC hardware, power distribution equipment, and communications equipment.

These must all be considered in the initial design concept of the station, such that corresponding cost estimates provide a full and complete picture to the owner.

4.2.3 Power Distribution

Power distribution equipment can vary greatly between technologies. Some systems utilize a D.C. power distribution system, where others will utilize an A.C. power distribution system.

Some systems, particularly cable drawn systems, will have power provided directly to wayside motor rooms.

Systems with on-board propulsion will have power distribution equipment proving power to guideway mounted power rails.

In either case, the power distribution equipment must be provided and installed with environmental considerations, incoming power variation, duty cycle, and fault and safety considerations.

Standards for each of these factors are included in the ASCE APM Standard for reference in specification development.

4.2.4 Vehicles

The vehicle is probably the most visible and recognizable feature of an APM system. Vehicles are available in different configurations, sizes, styles, colors, and technologies. With all of these variations, again, performance standards must be utilized. Performance standards such as those associated with structural integrity, propulsion and braking acceleration limits, ride quality, interior heating/cooling performance, and materials and workmanship can be drafted that are not prohibitive to specific technologies.

This approach allows various contractors to propose their specific technology, while focusing on the real needs of the owner… moving people safely and efficiently.

4.2.5 ATC – Automatic Train Control

The automatic train control system is the operational and safety heart of any APM. Driverless operation requires a rigorously designed ATC system, one which has both hardware and software designs verified and validated.

Noting that virtually all ATC designs differ widely, it remains a critical requirement that the safety, reliability, and documentation standards for all designs be adhered do in a performance fashion.

4.2.6 Communications

Today's communications systems can include a whole host of subsystem elements, including radio, public address, advertising, fault annunciation, CCTV, and pre-recorded announcement functions. While a portion of these may not be desired by an owner, those functions that are provided must adhere to standards for proper systems integration, performance, reliability, maintainability, and even audibility.

Often, elements of the communications system are critical to passenger use of the system, and can be an important part of emergency operations.

Improvements in technology continue to push the limit of these electronic subsystems. Of note, CCTV systems are now available with included object recognition software which is capable of providing an annunciation to the Central Control operator if a suspect object has been left on-board a vehicle or at a station platform.

4.2.7 Lighting

The interior and exterior lighting for an APM vehicle can be integrated with the vehicle in a variety of ways. Further, various lighting systems are available from manufacturers.

Incandescent lighting fixtures are slowly being replaced by highly reliable LED units.

Requirements for lighting cite surface lighting standards in terms of lumens, allowing the contractor to utilize proven, yet advanced lighting hardware.

4.2.8 Maintenance Facility

Maintenance facilities, in every case, must be tailored to the vehicle technology. Some vehicles will include on-board propulsion, some will not. Some will have roof-mounted equipment, some will have equipment located in interior compartments and undercar lockers.

The maintenance facility must be designed for ready entry and release of vehicles before and after maintenance, along with any storage provisions required by the Operations Plan.

Standards for maintenance facility layout and equipment are general in nature, and revolve around those provisions that must be reviewed with the owner.

4.2.9 Operations

System Operations is the umbrella under which the hardware resides, utilizing the hardware to provide a service to riding passengers. The operations design begins with system design, and continues through the overall design process through to initiation of the Operations and Maintenance period. Operations plans and procedures are subject to change even after system operation has begun.

Operations plans and procedures detail how equipment and vehicles are utilized, in scheduled fashion, to provide the most safe, reliable service to passengers.

Standards associated with operational plans describe that documentation and review that is needed to secure full concurrence with the owner, and the customers served by the system.

Further, operations plans must include the input from emergency services such that any incidents requiring their service result in prompt and safe response.

5.3 Specification Development

Once a technical approach to the system design has been identified, a technical specification is prepared describing the work scope and details associated with the system and all subsystems. While the approach is technology driven, the technology must be qualified with service proven operation in order to minimize operational risk to the owner.

The specification should be performance based, with detailed requirements presented only as specifically required for the application for the owner.

This effort can also include the development of commercial terms for the project. Part of this was would be a detailed project schedule, intended to make the best match of contractor production capacity and the availability of owner controlled facilities for the work.

One key area associated with the schedule is consideration of state of the art equipment and software that may be specified. Following the development of software for unique systems associated with transit is key to the successful start up and operation of a new system.

Overall, the process of specification development with the owner requires a highly cooperative "partnering" environment with multiple in depth reviews with all levels of the owner's organization and other key constituents.

5.4 RFP

Special attention is given to the development of the Request for Proposals, or RFP. Key considerations of the RFP included schedule, the presence of world wide system contractors, and varying approaches to technology for such a performance based procurement.

It is not uncommon for a system procurement of this sort to include an industry review. The accelerated nature of these projects can combine the industry review phase into a comments review phase during specification release. The intent of this effort is to solicit comment from system contractors to improve the outcome of the procurement or to reduce the cost of elements of the procurement.

5.5 Contract Award

Contract award for the project can come in a variety of forms. Often, a technical evaluation is performed to assure that the systems being proposed are accurately and fairly evaluated against the needs of the owner. This is then followed by a commercial evaluation of the contactor's capabilities, financial position, and then price. This is intended to bring a balance between cost and quality, while reducing risk to the Owner.

6 Project Management

The successful implementation of a project requires that there be a rigorous program for managing all of the elements of the project, including design, procurement, manufacturing, quality assurance, installation, test, and commissioning. Standards for these elements are available which define the requirements for the management program that shall plan, schedule, review, control, and report on the Work.

Standards typically will cite tools for project management, including the following:

- Management plan
- Project schedule
- Progress reviews and reports
- Payment milestones
- Design and configuration control
- Design reviews
- Production baseline
- Test
- Acceptance and warranty

6.1 Management Plan

Prior to the initiation of a project, and often during the proposal phase, an owner will require that the bidding contractors provide a management plan. This plan will detail to the owner the manner in which the contractor plans to control the work.

6.2 Project Schedule

The project schedule is arguably the most critical tool available to the Project Manager. Standards for project scheduling exist in a variety of forms. It is important that the project schedule define all interfaces between the owner, the owner's representatives, the contractor, and any other elements of "input" or "output" to the project. Further, work completion associated with payment must also be included in the project schedule.

Beginning during the proposal stage, the project schedule is a living, working tool that is used, altered, and adjusted throughout the project. The project schedule, if not maintained, illustrating real occurrences and changes, is of no use to the project team.

6.3 Progress Reviews and Reports

Also key to any successful project is communication. While the project schedule is one document that illustrates the elements and relationship of those elements of the project, regular reviews between the Owner, the Contractor and other key constituents are required for smooth project implementation. These reviews can be complimented by regular project reports. In any event, a program of regular communication of all project elements and status is critical to project implementation.

6.4 Payment Milestones

As work progresses, resources and funds are expended by the Contractor. Prior to implementation of a final contract, a schedule of Payment Milestones should be developed, listing the activities for the project for which the Contractor must be reimbursed. Often, these milestones are a reflection of a positive cash-flow position for the Contractor for which he must have owner concurrence.

These payment milestones should be clearly reflected in the project schedule, with any changes or adjustments made promptly to that schedule.

6.5 Design Reviews

As the design of the system progresses, certain elements of the design will mature and serve as the foundation for further design elements and decisions.

Design status must be reviewed regularly with the owner and the owner's responsible parties such that a clear understanding of the design, its performance features, and interfaces is obtained.

In order to provide the owner with adequate notice of design features and progress before subsequent related design decisions are made, it is best to have a programmatic series of design reviews, allowing the owner not only input to the design, but to communicate elements of the design to related parties, such as utility and regulatory organizations.

6.6 Production Baseline

At the conclusion of the design review phase, the design of the system and its elements is documented. Often referred to as a "final release", the design is released for production/fabrication/construction.

Once the design is released for procurement to begin purchasing material, and production to begin preparing the production facility to build equipment and vehicles, a "production baseline" is set.

Any changes to the design production baseline must be communicated to affected parties as promptly as possible in order to minimize the cost and schedule impact of rework.

6.7 Design and Configuration Control

It is critical that, following design release, the design configuration be controlled such that any changes to a design element be communicated to all parties involved with that design element.

It is beneficial to all members of the project team to have a "design change notice" process which advises parties of a pending change prior to the change being implemented. This allows affected parties to provide input to the change should some adverse impact be pending as a result of said change.

Ultimately, Configuration Control should be a program in of itself to allow construction personnel, associated utilities, testing personnel, and other team members to know what the proper configuration of the design is at all times.

6.8 Testing

Any APM installation will involve unique features, performance requirements, and even possibly new subsystems.

Throughout the production phase, it is most effective to test subsystem elements as soon as possible to assure that performance requirements are met, and that interfaces, including software protocols, are as designed.

Full system testing should be performed to the greatest degree feasible prior to final installation at the Owner's location.

Then, full system testing, as installed, should be performed to confirm that the system operates in conformance with all requirements.

6.9 Acceptance and Warranty

With the conclusion of a successful installation and test phase, the project schedule will indicate a payment milestone for System Acceptance. This schedule item indicates that all elements of the system have been manufactured, installed, constructed, and tested in a satisfactory manner, and that that payment should be made for the Owner's acceptance of the system.

At this point, concurrent with initial operations, the warranty phase typically begins, and progresses as indicated in the contract documents.

Standards are available which highlight the characteristics of the Warranty period, including provisions for parts replacement, labor, and treatment of defects.

7 Contract Support

Finally, after system acceptance, the Operations phase begins. Depending upon the structure of the contract, this phase may include the contractor being responsible for Operations and Maintenance, or for some portion or all of that work to be taken on by the Owner.

In any event, the operations phase should be properly planned and documented such that resources and information are allocated for successful and cost effective system operation.

In addition, proper documentation for system operations and maintenance manuals and training must be considered.

7.1 Management Plan

A top level plan to describe and document the approach to systems management, operations, organizational hierarchy, and interfaces between the owner and the contractor is prepared as a baseline for agreement on system management.

7.2 Systems Operations Plan

The operation of the system can be very flexible, with variable headways, dwell times, and even routes for more complex systems. Marrying these features with the owner's planned schedule of service will yield the systems operations plan.

This plan then can serve as the foundation for staffing, maintenance schedules, cleaning schedules, and other elements of planning the work force.

7.3 Staffing Plan

A staffing plan is typically prepared to document the tasks required for proper operations and maintenance of the system, including parts procurement, cleaning, management, and administrative support. From such a task list, or work list, manpower to support these functions can be developed.

This plan must also take into account anticipated shift work, and shift overlap for proper communication of system issues from one shift to another.

7.4 Maintenance Plan

Maintaining an APM system requires a wide variety of disciplines and activities. The Maintenance Plan documents the maintenance activities required in accordance with the maintenance manuals, and presents a schedule, specific to the installation, of maintenance activities, intervals, and personnel.

7.5 Maintenance Manuals

Maintenance manuals are prepared in conjunction with all elements of the system, from electronic parts to fabricated parts to subcontracted assemblies and systems. These manuals typically will describe the system or component at a high level, and detail all elements of normal maintenance, cleaning, repair, and overhaul in order that the system or component can achieve its full design life in reliable fashion.

7.6 Failure Management

As an automated system, the APM is designed to serve passengers without human intervention.

In the unlikely event of a failure, however, prompt, safe, and effective response is required to return the system to normal automatic operation as quickly as possible.

Such response can take place at a variety of levels, beginning with basic Central Control Operator intervention, down to the level or dispatching recovery technicians to the vehicle or equipment room for problem resolution.

Failure management planning details all conceievable failures and lists appropriate responses to each.

Standards are available for the outline and content of such failure management plans to effectively manage the unplanned events associated with an automated system.

7.7 Training Plan

Well trained central control personnel, technicians, engineers, and recovery specialists are important to the smooth operation of an APM.

Planning for this training well in advance will help assure that the resources identified for these roles receive the proper training prior to full on the job activation.

While experienced system contractors will have service proven training plans for use, Standard outlines for training plans are available for both training plan development, and for owners to benchmark the data provided to them.

7.8 Training

At the first installation of the APM system, the contractor will typically have installation specialists who are very well versed in the details of the system, it's operation, maintenance, and failure recovery.

As the system begins operation with a portion of new staff, however, training is required to provide the new technicians with the knowledge and skill required to properly operate and maintain the system.

Standards outlining classroom, hands on, On-The-Job, and extra-curricular training are available to serve as a minimum baseline for APM system training.

It is important to acknowledge that an APM system is a complex integration of components from heavy steel structural members to electronic printed circuit boards, from refrigeration circuits and air handlers to high voltage propulsion control equipment. Even a capable technician will require training for the specific hardware on which he or she will work.

7.9 Manuals

Detailed and accurate operations and maintenance manuals are key for proper operations and maintenance of a system. Included therein should be the following:

- Systems descriptions
- Subsystems descriptions
- Operational procedures

- Maintenance philosophies
- Maintenance procedures
- Parts ordering details
- Exploded views of assemblies

Standards for the requirements for operations and maintenance manuals in the APM standard can serve as an effective baseline for manuals.

8 Conclusion

An Automated People Mover system spans a wide range of technology, and includes phases of design, manufacturing, construction, test, commissioning, and finally O&M.

In support of an owner's beginning the process of implementing an APM, and in the development of a specification for an APM, the ASCE APM Standards Committee has produced a four part standard addressing all elements of an APM.

With this consensus Standard, the industry is able to take advantage of cost and schedule benefits from procurement, to design, to project management, and finally systems operation.

ADVANCED COMPOSITE CARBODY SYSTEMS

Takaomi Inada[1], Genichirou Nagahara[2],
Seung-Cheol Lee[3], Dae-Hwan Kim[3],
Masaaki Kuwabara[2], Tsutomu Hoshii[2],

[1]Structural Analysis Group, Structural Strength Department Research Laboratory
[2]Transportation System Department, Logistics and Structures
IHI Corporation, Toyosu IHI Building,
Toyosu 3-Chome, Koto-Ku, Tokyo, 135-8710, Japan
Tel: +81-3-6204-7289, Fax: +81-3-6204-8683, e-mail: takaomi_inada@ihi.co.jp,
[3]Transit Division
Hankuk Fiber Glass Co., Ltd.
181-1 Yongji-Ri, Bubuk-Myun, Miryang-Si Kyungnam, 627-850, Republic of Korea

Abstract

IHI has successfully designed, constructed, and delivered several state-of-the-art Automated People Mover (APM) transit systems for demanding urban and airport transit applications throughout Asia and various overseas markets. Building upon this success, IHI is actively developing a new generation of APM vehicle system designed to effectively address the requirements of a rapidly changing industry and market. Composite materials are being applied to primary structures of the new IHI APM for weight reduction and for achievement of stylish design. This paper describes the development of the "Advanced composite carbody system" for the new IHI APM.

Introduction

Demand for APM systems are increasing for airport and urban transit applications. Highly desirable features of these APM applications include provisions for the highest level of system safety, reliability and passenger comfort and convenience.

IHI is developing a new APM vehicle to address the rapidly changing transportation market direction of recent years. For weight reduction and energy savings, composite materials are being utilized increasingly as primary structures in industries such as aerospace and automobile. As shown in Table: "Specification of the New IHI APM", size and passenger capacity of the new IHI APM has increased compared to the previous IHI vehicle for Tokyo Waterfront New Transit "Yurikamome". IHI and Hankuk Fiber Glass (HFG) began a joint program in 2006 to develop a novel carbody composed of laminate and sandwich panel with laminate skin for the new IHI APM[1]. This paper presents development details of the "Advanced composite carbody system" applied to the new IHI APM.

Table: Specification of the New IHI APM

	IHI APM	Yurikamome*
Car dimension (mm)	L12000 x W2850 x H3680	L9900 x W2490 x H3340
Passenger capacity	≥100	76
Weight per car	< 27 ton	19 ton

* Tokyo Waterfront New Transit

Overview of Advanced Composite Carbody System

The new IHI APM is developed using a hybrid design concept combined with sandwich and laminated composite structures. The body shell, which is a major part of the composite carbody, is a semi-monocoque structure which is cured in a large-sized autoclave. Our bodyshell fabrication process enables the carbody to have high quality surface, high specific strength, high specific stiffness and high heat resistance compared to carbodies fabricated using other types of fabrication methods, such as resin transfer molding and vacuum assisted resin transfer molding.

Figure: Advanced Composite Carbody of New IHI APM

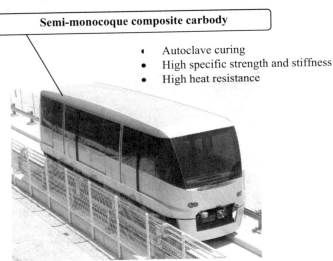

Development and Evaluation of Advanced Composite Carbody

We have designed and developed the composite carbody for the new IHI APM utilizing finite element analysis in conjunction with tests as shown in Figure: "Development and Evaluation Process of the Composite Carbody System". Details of the development and evaluation process of the composite carbody system are explained in the following sections.

Figure: Development and Evaluation Process of the Composite Carbody

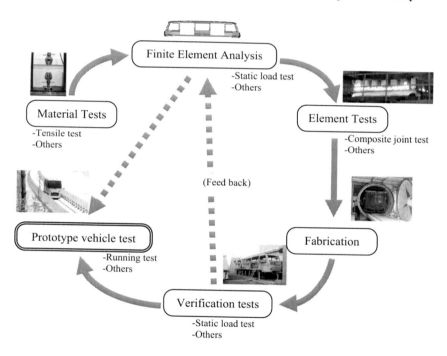

Material Tests of Composite Laminate Specimen and Sandwich Specimen

Composite laminate specimens for application to bodyshell and other structures were tested as shown in Figure: "Example of Material Test of Composite Laminate". Many kinds of material tests other than tensile tests and compression tests were conducted to obtain orthotropic material properties of the laminate. Material properties of the laminate skinned aluminum honeycomb core sandwich specimens were also obtained in the material tests. This material data was utilized in the finite element analysis of the new APM.

Figure: Example of Material Test of Composite Laminate

(a) Tensile   (b) Compression

Finite Element Analysis of the New APM

Finite Element Analysis (FEA) was applied to the structural design of the new APM and structural responses, such as deformation, stress and strain, were evaluated. Figure: "Example of a FEA Model for Static Load Test Analysis" shows an example of a FEA model for static load test analysis. The carbody was evaluated in numerous types of loading conditions that were decided based on design standards such as ASCE21[2] and JIS E7105[3], and final design was decided.

Figure: Example of a FEA Model for Static Load Test Analysis

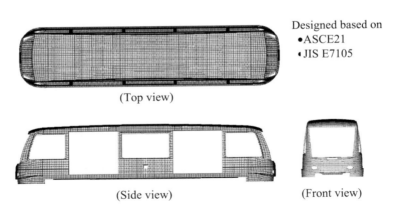

Designed based on
- ASCE21
- JIS E7105

(Top view)

(Side view)   (Front view)

Figure: "FEA Model for Collision Analysis" shows a FEA model for collision analysis. Major facilities under the floor including tires were modeled, and collision analysis was conducted. It was confirmed that the new APM vehicle with composite carbody has sufficient crashworthiness.

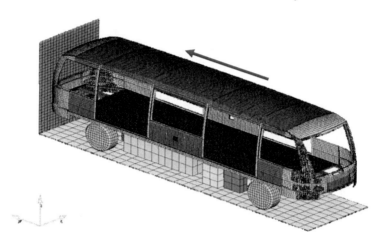

Figure: FEA Model for Collision Analysis

Element Tests of the Composite Carbody System

The composite carbody of the new APM is joined to the steel underframe with rock bolts. It is very important to evaluate its strength both in static and fatigue loading to assure adequate vehicle life. We conducted static load tests of the joining part specimens as shown in Figure: "Static Load Test of the Joint Part" and obtained bending strength in different loading conditions. Fatigue strength in bending tests was also obtained in the same manner.

It is also very important that the APM's carbody provides adequate fire resistance for passenger safety. We tested fire resistance of the floor assembly module according to NFPA 130 as shown in Figure: "Fire Resistant Testing". A floor assembly module specimen, on which concrete weights were loaded, was set on the furnace and subjected to high temperatures. Flames were observed from the edge of the specimen after 22 minutes, but the flames did not spread to the floor surface level. Test results meet NFPA 130 requirements and fire resistance of the floor assembly was validated.

Figure: Static Load Test of the Joint Part

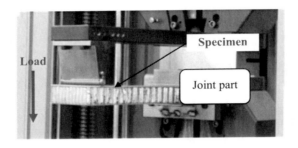

Figure: Fire Resistant Testing

(a) Specimen (Floor)

(b) Test result

Fabrication of Composite Carbody

Figure: "Composite Carbody Fabrication Process" shows the composite carbody fabrication process of the new APM. Glass fabric/epoxy prepregs are stacked on the inner surface of the composite molding combined with aluminum honeycomb core and adhesive films, and cured in the large-sized autoclave. The composite carbody is finished through de-molding from the composite mold and adhesion of reinforced frames and others.

Figure: Composite Carbody Fabrication Process

(a) Composite mold

(b) Autoclave cure

(c) Finished composite carbody

Verification Test - Full Size Carbody Load Test

In order to verify design accuracy and fabrication quality of the new APM, a composite carbody assembled with a steel underframe and a keystone plate floor was subjected to static load tests, such as bending, compressive and torsion tests, based on JIS E7105 as shown in Figure: "Full Size Carbody Load Test". The carbody was supported at the underframe and the load was applied to the carbody using hydraulic cylinders. Structural response data such as strain, displacement and acceleration data were acquired during tests and compared to analysis results for verification.

Figure: Full Size Carbody Load Test

- Bending test
- Compressive test
- Torsion test
- Others

(Strain, Displacement, Acceleration)

Prototype vehicle test

Through the development and evaluation processes described in the previous sections, we launched the prototype vehicle in April, 2008. We have been conducting prototype vehicle tests as shown in Figure: "Running Test of Prototype Vehicle" to evaluate total performance of the vehicle. Measured dynamic strain during running will also be compared to analysis results of the prototype vehicle.

Figure: Running Test of Prototype Vehicle

Conclusion

IHI and HFG have developed an advanced composite carbody system for the new IHI APM. The presented carbody is fabricated using autoclave and has superior structural strength and heat resistance compared to carbodies fabricated by other fabrication methods. Detail structural analyses, material tests and verification tests were conducted during development of the carbody, and a superior composite carbody for the New IHI APM has been achieved.

Acknowledgments

The authors express their deep appreciation of the computer analysis work done by Kwang-Bok Shin, Professor of HANBAT National University, Korea.

References

(1) Norihiro Takai, Genihirou Nagahara, Satoshi Suzuki, Masafumi Kawai and Shintarou Kitade, NEXT GENERATION APM: DESIGN FOR ADVANCED VEHICLE, Proceeding of "The ASCE APM2007 Technical Committees", (2007), ppt
(2) ANSI/ASCE/T&DI 21.2-08 Automated People Mover Standards-Part 2
(3) JIS E 7105 Rolling stock – Test methods of static load for body structures.

Advances in Passenger Convenience and Comfort

Kunihiro Tatecho[1], Masafumi Kawai[2],
Yuji Koike[3], Motoaki Tanaka[1],
Masaaki Kuwabara[1], Tsutomu Hoshii[1]

[1]Transportation Systems Department, Logistics and Structures
[2]Numerical Engineering Department, Research Laboratory
[3]Structural Strength Department, Research Laboratory
IHI Corporation, Toyosu IHI Building,
Toyosu 3-Chome, Koto-Ku, Tokyo, 135-8710, Japan
Tel: +81-3-6204-7289, Fax: +81-3-6204-8683, email: motoaki_tanaka@ihi.co.jp

Abstract

The new IHI APM system integrates numerous advancements in passenger convenience and comfort. Through detailed numerical analysis and data acquisition procedures performed on a dedicated IHI test track, passenger ride quality advancements now set a new market benchmark for rubber tired technology. Similarly, the HVAC subsystem integrates new 'smart' system features designed to optimize passenger comfort and reduce energy consumption. Information systems and a spacious interior cabin embody similar advancements. This paper explores these features in detail including the implications for future applications.

Introduction

There are various factors that contribute to vehicle passengers' comfort. Passengers experience the atmosphere of the vehicle according to vehicle appearance (both exterior and interior) particularly while the vehicle is parked at a station. During operation the ride comfort is more important. IHI has analyzed aspects of these experiences and summarized their impact in the following Table: 'Analysis of Factors Contributing to Passenger Comfort'.

Table: Analysis of Factors Contributing to Passenger Comfort

No	Importance/Feeling	Related System	Customer Exposure Time	Impact*
	Boarding vehicle			
1	Confirm destination	Information system	10sec	B
2	Interior images	Interior design	20sec	C
	Departure to next station			
3	Sign at departure	Information system	5sec	B
4	Door closing	DCU/Information system	4sec	C
	During operation			
5	Acceleration	Propulsion system	20sec	A
6	Vehicle vibration	Track/Carbody	60sec	A
7	Interior	Interior design	60sec	C
8	Outside view	Interior design	60sec	C
9	Site information	Information system	60sec	B
10	Deceleration	Propulsion/Brake system	30sec	A
11	Stopping point	Brake system	5sec	A
	Arriving at station			
12	Station information	Information system	10sec	B
	Throughout the entire experience			
13	Air conditioning	HVAC	90sec	A
14	Noise	All equipment	90sec	A

*A = Large, B = Medium, C = Small

IHI has identified the following items as having the most crucial influence to passengers' comfort through evaluation of the above criteria:

- Vehicle ride quality;
- HVAC system;
- Information systems.

Vehicle Ride Quality Improvement

The IHI vehicle is required to run at operational speeds of up to 44 mph (70 km/h), while typically conveying more than 100 passengers at a time. With the view of maintaining better ride quality than conventional APM vehicles, various design iterations have been applied through repeated operational testing. The initial target performance level (parameter analysis) was set to that of the successfully operating Yurikamome system in Japan[1].

To achieve such a performance level, it is important to reduce the disturbances from both the running surfaces and guide rails and to increase the rigidity of the car body. The running surfaces contain various irregularities and these can be the cause of vibrations especially in the vertical direction. The irregularities were minimized as much as possible through small machine grinding techniques.

Figure: Preliminary Power Spectrum Densities and Measured Time History Responses

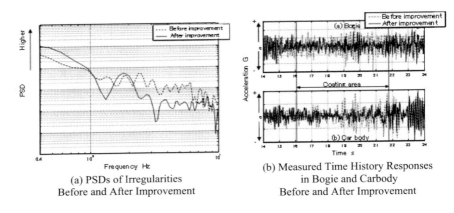

(a) PSDs of Irregularities Before and After Improvement

(b) Measured Time History Responses in Bogie and Carbody Before and After Improvement

The irregularities were measured before and after the various improvements and those properties were analyzed. The power spectrum densities (PSDs) were compared before and after surface coating in Part (a) of Figure: 'Preliminary Power Spectrum Densities and Measured Time History Responses'. The frequency has been transformed from a space frequency to time frequency (Hz) at a constant running speed of 34 mph (55 km/h). It is seen from Part (a) that the PSD after grinding has been reduced remarkably especially in the 3 to 9 Hz range and this fact is a major contributer to the reductions of disturbances from the running surfaces. The time history responses of bogie and carbody vertical vibrations are shown in Part (b) of Figure: 'Preliminary Power Spectrum Densities and Measured Time History Responses'. Bogie accelerations have been reduced and the influence on the carbody is mitigated effectively.

Adjustments to the guide rails were made in order to reduce lateral vibrations resulting from vehicle yaw motion. The distance between guide rollers on both sides of the test track were adjusted within a range of ±1 mm. The restoration springs used for steering tires were also adjusted and the neutral direction of the steering angle was determined. The preliminary time history responses of the yaw angle before and after adjustment are compared in Figure:
'Time History Responses in Yaw Angle Before and After Guide Rail Adjustment'. The maximum value is suppressed to less than half the values seen before the guide rail adjustments.

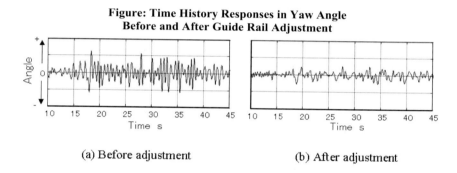

Figure: Time History Responses in Yaw Angle Before and After Guide Rail Adjustment

(a) Before adjustment (b) After adjustment

Lastly, the carbody vibration characteristic was studied. Elastic vibration modes tend to be excited by the flexible properties of a large body. Vertical resonant behavior is concentrated in a bending vibrational mode around the bogie system. This behavior can be explained by using the dynamic model shown in Part (a) of Figure: 'Dynamic Model and Body Transmissibility'. The system is modeled using three degrees of freedom. The fundamental bending mode is transmitted through a rigid car body/ bogie suspension system. The carbody transmissibility is shown in Part (b) of Figure: 'Dynamic Model and Body Transmissibility' against increasing elastic natural frequency. The abscissa shows the frequency ratio normalized by the secondary sprung mass natural frequency. The resonant behavior is induced highest at the bending natural frequency of f0. The transmissibility can be reduced by increasing the natural frequency.

Figure: Dynamic Model and Body Transmissibility

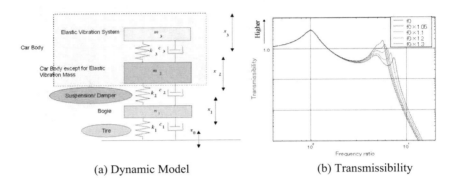

(a) Dynamic Model (b) Transmissibility

Figure: Increase of Bending Natural Frequency

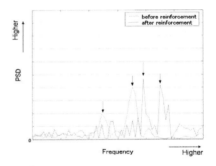

Thus, the carbody has been reinforced in order to avoid adverse resonant behavior. Specially designed plates have been manufactured and installed at the underframe. This effect is demonstrated by the natural frequency change before and after reinforcement in the spectrum of Figure: 'Increase of Bending Natural Frequency'. These frequencies have been obtained through free vibrations induced by obstacles arranged under the bogie tires. The Figure illustrates the resulting increase in the natural frequencies beyond the normal operating range.

The effects of the improvements described above have been confirmed by evaluating the ride quality performances. In analyzing the ride quality, the curve based on the "Japanese National Railways" standard[2] has been utilized, resulting in the acceleration spectrum curves as shown in Figure: 'Preliminary Measured Acceleration Spectrum Curves'. The data was collected at a constant running speed of 44 mph (70 km/h).

Figure: Preliminary Measured Acceleration Spectrum Curves

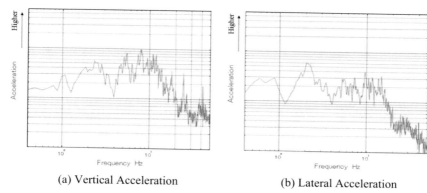

(a) Vertical Acceleration   (b) Lateral Acceleration

The preliminary ride quality levels for the above performances are further tabulated in Table: 'Preliminary Ride Quality Level Improvements'. They are calculated and compared with the Yurikamome system utilizing both the Japanese National Railways[2] and ISO 2631[3] standards. The values in parentheses denote the dB used reduction values compared with the ride quality levels before improvement. It can be seen that the ride quality level has reached the target level of Yurikamome and even surpassed it for lateral acceleration measurements.

Table: Preliminary Ride Quality Level Improvements

Target car	Ride quality level (Ratio to original condition before improvement)			
	Japanese National Railways		ISO 2631	
	Vertical	Lateral	Vertical	Lateral
Proto car	Good (7.8)	Excellent (7.4)	Fair (7.7)	Excellent (7.4)
YURIKAMOME	Good	Good	Fair	Excellent

Note: In parentheses are the dB used reduction values compared with the ride quality levels before improvement.

Heating, Ventilation and Air-Conditioning (HVAC) System Improvement

Concept

As described previously in reference (4), the new IHI APM adopts a unique heating, ventilation and air-conditioning (HVAC) system, which distributes fresh cabin air upwards from the window sills. The flow concept is shown in Figure: 'New HVAC Flow Concept'. The advantages of this configuration are as follows:

Figure: New HVAC Flow Concept

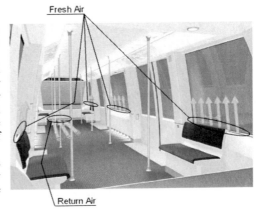

- Improved head clearance, which is essential for a more spacious cabin;
- HVAC units are mounted under-floor, which allows for a light-weight composite car body structure and clean exterior appearance while improving ease of maintenance;
- The new flow concept allows a more uniform flow environment in the cabin (see reference for further details).

Implementation

The HVAC unit exclusively designed and manufactured for the new IHI APM prototype is shown in Figure: 'HVAC Unit'. It is compact enough to be mounted between the bogie and the coupler. As a result, IHI can keep plenty of extra space between the bogies for miscellaneous components while maximizing passenger comfort.

Figure: HVAC Unit

Front View   Side View

Validation

In order to verify the adequacy of the new HVAC concept, measurements of not only temperature distribution but also flow rate, flow pattern, and interior noise levels have been carried out on the prototype. Part (a) of Figure: 'Prototype Measurements and

Temperature Distribution' shows some of these measurements being taken on the vehicle prototype at the IHI test track.

Part (b) of Figure: 'Prototype Measurements and Temperature Distribution' is an example of a preliminary temperature distribution taken from various measurements at the vehicle center section. Measurements such as these have helped confirm the validity of the IHI up-flow concept. In addition, the interior noise level was very low since the HVAC unit, a primary noise source, is mounted beneath the floor panels.

Figure: Prototype Measurements and Temperature Distribution

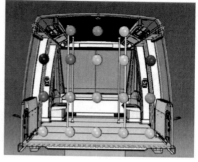

(a) Measurements on the Prototype Vehicle   (b) Temperature distribution

Evaluation

Through validation tests and several months of prototype test runs, IHI has evaluated the new HVAC concept with the following conclusions:

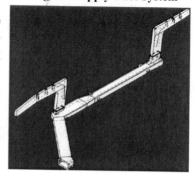

Figure: Supply Duct System

- The new up-flow concept is successful;
- The under-floor mounted location of the HVAC unit provides several advantages, including a carbody structural advantage, lower cabin noise and ease of daily maintenance;
- The HVAC supply duct network shown in Figure: 'Supply Duct System' exhibited somewhat higher pressure loss than originally estimated. IHI will continue to improve the duct design to lower overall pressure loss in the system.

Information Systems Improvement

The primary information systems installed on IHI's prototype APM vehicle include:

- ➤ LED panels;
- ➤ LCD panels;
- ➤ Speakers for automatic guidance.

The LED panels at each end of the vehicle provide basic updates in text format such as vehicle and station information. Supplementing this system is an LCD system consisting of four panels over each vehicle doorway. The panels can work together to display wide images and specialty impact messages such as shopping information. Further, speakers are installed on the sides of the LCD panels for a complete audio-visual information system. See Figure: 'LCD & LED Panel Arrangement' for the information systems arrangement inside the vehicle.

Figure: LCD & LED Panel Arrangement

IHI has achieved the effect of a brighter, more spacious and comfortable vehicle cabin in part by adopting large windows and doors. In certain situations, this can reduce the impact of the LCD wide screen panels by interfering with visibility. Table: 'Factors Affecting LCD Visibility' lists some of the factors that can result in reduced LCD visibility in the vehicle cabin.

Table: Factors Affecting LCD Visibility

No.	Factor	Equipment	Impact*
1	Shortage LCD brightness	LCD Panel	A
2	Reflection to protect plate	Plate material	B
3	Reduce transmission factor of protect plate	Plate material	C
4	Shortage LCD power voltage	Electrical	B
5	Coloring on contents	Contents	C

*A = Large, B = Medium, C = Small

The following improvements have been determined to have the greatest positive influence on LCD visibility including:

- Changing to high brightness LCD;
- Fitting low reflection sheet on protection plate;
- Increasing power voltage.

Figure: 'LCD Brightness Improvement' illustrates the improvement to LCD visibility by contrast and brightness from the original design. Table: 'LCD Brightness Comparison' quantifies the LCD brightness improvement.

Figure: LCD Brightness Improvement

(a) Before   (b) After

Table: LCD Brightness Comparison

	Brightness Improvement
Original Design	100
Increase Voltage	171
Adoption of High Brightness LCD	831

Note: Brightness improvement shows the ratio of LCD brightness after improvement from original design.

Conclusion

IHI implemented several improvements to enhance ride quality. IHI achieved improved ride quality in the vertical direction at a higher maximum speed than the Yurikamome baseline system (standard for APM vehicles in Japan) by reducing the irregularity of the running surface and adding carbody reinforcement. Also IHI achieved improved ride quality in the horizontal direction by adjusting the guide rail flatness. The new IHI APM now exhibits excellent ride quality at a higher speed than most conventional rubber tired APM vehicles. The ultimate effect is to allow improvements in system transportation capacity and headway.

The IHI HVAC system features a new up-flow concept installed under the APM vehicle frame. IHI has measured and evaluated several important criteria for this concept, resulting in improved flow patterns, reduced noise levels, and uniform temperature distribution in the cabin.

The merits of installing HVAC units under the frame include the following:

- Noise reduction vs. a ceiling-type unit;
- Improved maintenance vs. a ceiling-type unit;
- Reduced weight of carbody from eliminating structural needs of ceiling-type units.

To achieve a wide screen LCD panel design for passenger information systems, featuring a larger graphical image for advanced advertising impact, IHI improved the LCD screen visibility during normal daytime operations.

Acknowledgments

The authors express their deep appreciation of the unfailing earnest guidance accorded to the present study by Dr. Takehiko Fujioka, Associate Professor of University of Tokyo, and of the cooperation to improve the vehicle and use of the test track by Hankuk Fiber Glass Co., Ltd.

References

(1) Yuuji Koike, Motoaki Tanaka and Takehiko Fujioka, NEXT GENERATION APM: ANALYSIS ON RUNNING BEHAVIOR, Proceeding of "The ASCE APM2007 Technical Committees", (2007), ppt.

(2) Edit. by JSME, Dynamics of railway vehicles -Newest bogie technologies-, Denkishakenkyukai Co., Ltd., (1996), 70-71.

(3) INTERNATIONAL STANDARD ISO 2631-1 (Second edition 1997-05-1) and 2631-2 (Second edition 2003-04-1).

(4) Norihiro Takai, Genichiro Nagahara, Satoshi Suzuki, Masafumi Kawai, and Shintaro Kitade, NEXT GENERATION APM: DESIGN FOR ADVANCED VEHICLE, Proceeding of "The ASCE APM2007 Technical Committees", (2007).

Objectively Assessing Automatic vs. Manual Control for Transit Systems

John E. Joy, AIA[1]

[1]Associate Principal, Lea+Elliott, Inc. 1009 W. Randol Mill Road, Arlington, Texas 76021: PH (817) 261-1446: FAX (817) 861-3296; email: jejoy@leaelliott.com

ABSTRACT

Within the Transit Industry reside proponents and detractors of automatic control, as well as proponents and detractors of manual control. These factions are well versed in the aspects of both automatic and manual control, and produce and promote a variety of data, facts, and opinions via venues ranging from conferences to websites to internet blogs. Occasionally, ardent discussion or promotion of one control type occurs at the expense of the other control type. In these cases, contrasting claims may be made that increase the difficulty for prospective public owners and planners of transit systems to objectively assess automatic or manual control as it applies to their particular system.

This paper's purpose is to provide a high level, objective overview that can assist prospective owners and planners of transit systems with decisions regarding automatic and manual control by outlining salient points regarding both manual and automatic control. The paper summarizes the purpose of, and applications of, automation and presents automation rules of thumb. The paper defines both automatic train control and manual operation, then discusses other factors, including measurement of the cost effectiveness of automatic vs. manual control for transit systems. Lastly, conclusions and recommendations are made.

PURPOSE OF AUTOMATION

Transit professionals often debate the economic merits of automation. However, cost additions or cost savings regarding either capital, operational, or maintenance costs are not necessarily primary factors in the purpose of automation. Rather, the primary purpose of automation is to enable 1) the *precise* operation of the transit system that results in 2) the *efficient* operation of the transit system that ultimately results in 3) the ability to reliably provide *accurately managed headways*. It is the achievement of dependable *accurately managed headways* that constitutes the primary purpose of transit automation. The schedule-intense and mobility-intense environment of airports, particularly large International Airports, necessitate these dependable and accurately managed headways more so than do urban environments or leisure/entertainment environments. It is this fact that has made airports a major venue for APMs since their inception in the late 1960's. Regardless of venue, the accurately managed headways provide the potential for an automated transit system to have greater capacity than a comparable manual transit system.

In addition to accurately managed headways, automation may result in other favorable "byproducts" - the most significant of which is safety. Safety as applied to transit systems is a global concept encompassing countless specifics. It is widely accepted that the addition of automation to a transit system enhances safety in general terms. This is simply because automation removes the factor of human error from many physical and operational scenarios. Statistically, it has been historically documented that human error is the most prevalent cause of conditions resulting in harm to the transit system and/or its users, maintainers, and operators.

Cost differences may be a byproduct of automation. Although cost differences can vary widely and are project-specific, certain observations may be generally applied to people mover-scaled transit systems. Such observations are discussed later in this paper.

APPLICATIONS OF AUTOMATION

Automation is not specific to a particular transit type. Automation can be applied to Light Rail Transit just as it can be applied to People Movers, Commuter Rail, and other types of transit systems (Figures 1, 2, and 3). Automation can be retrofitted to existing transit systems or added at a later date to new transit systems. Such retrofitting of automation can involve total implementation, phased implementation, or partial implementation. Any retrofitting of automation will cause operational disruption of the system and will usually incur a cost penalty compared to the cost of having implemented the automation with the initial system installation.

Figure 1
Automated Monorail

Figure 2
Automated People Mover

Figure 3
Automated Light Rail

AUTOMATIC TRAIN CONTROL DEFINED

Automatic Train Control (ATC) is comprised of three major subsystems; 1) Automatic Train Supervision (ATS), 2) Automatic Train Operation (ATO), and 3) Automatic Train Protection (ATP). The function of these three ATC subsystems can be succinctly defined as follows:

- ATS provides *system management*.
- ATO provides *efficient operation*.
- ATP provides *safe operation*.

ATS and ATO overlap functionally depending upon the specifics of the installation. ATP may stand alone without ATS and ATO and the relevance of this fact is explained further in subsequent discussion. Explanation of the three subsystems is as follows.

ATS. Automatic Train Supervision provides system management. Examples of such management involve the ATS subsystem providing a link between the Central Control Operator (CCO) and the system, giving all pertinent information about the system and providing the means for the CCO to control various system functions. Other examples of ATS management include surveillance, communication, and graphics. Surveillance is via the system's closed circuit television cameras and monitoring screens. Communication devices include both wayside and onboard installations and are for the use of both the public and system operators depending upon the purpose of the device and the situation in which it is used. Graphics typically include dynamic informational and wayfinding signage as well as matching audio announcements in both onboard and wayside locations that are automatically tied to the operation of the trains. ATS can also "debunch" trains to ensure regular minimal headways are maintained.

ATO. Automatic Train Operation provides for the efficient operation of the system. This includes the automatic interfaces between the vehicles and stations that enable smooth, accurate berthing and the coordinated operation of the numerous automatic door sets. ATO also regulates train speed below certain limits.

ATP. Automatic Train Protection is a vital function that may stand alone without ATO and/or ATS. The function of ATP is to ensure the safe operation of the system by preventing trains from colliding with other trains and/or any other wayside appurtenances. ATP also prevents trains from overspeeding and ensures safe train/station door operations.

MANUAL OPERATION DEFINED

Comparisons regarding manual operation versus automatic operation of transit systems sometimes commence with the assumption that manual operation relies solely on drivers with no automation at all. Such pure manual operation can be defined as "Line of Sight" but is rarely used in transit systems due to safety considerations. Rather, "Manual with ATP" more accurately describes a typical manual operation. Both Line of Sight, and Manual with ATP operation are more fully defined as follows.

Line of Sight Operation. In a Line of Sight manual operation, the vehicle is free to move to the visually confirmed next obstruction so long as there is clear sight ahead.

An example of the next obstruction could be the next stop sign, next traffic signal, block signal, or an unexpected pedestrian or vehicular intrusion into the travel path. Through visual confirmation, the driver continually revises and regulates speed and progress. The most common example of this purely manual mode is the operation of over-the-road vehicles and in the case of transit, buses. However, even with a dedicated right-of-way free of wayside intrusions, trains guided by rail or guidebeam lack the maneuverability options of over-the-road vehicles in that they cannot "get out of the way of each other". Thus, there is virtually zero tolerance to human error on the train drivers' part which creates an inherently hazardous condition in all but the most simply configured systems.

Manual with ATP Operation. The most common way for manual operation in the context of transit systems to overcome the inherently hazardous conditions of line of sight is by employing Automatic Train Protection (ATP). Thus, "manual" operation of transit systems is typically not purely manual, but incorporates varying degrees of Automatic Train Protection and is more accurately described as "Manual with ATP". The Automatic Train Protection may involve wayside signaling or cab signaling. When the term "manual" is used hereto forth in this paper, it inherently assumes a "Manual with ATP" definition, not a Line of Sight definition.

RULES OF THUMB

Numerous other factors can be discussed as transit automation rules of thumb. The following are several major ones.

Dedicated Right-of-Way. A factor of paramount importance in considering automated operation is a dedicated right-of-way (Figure 4). Without a dedicated right-of-way, the transit system will not realize the full benefits of automation. Trains relegated to a shared right-of-way can function no better than buses in terms of maintaining precise, efficient headways and overall operation. Collision avoidance also becomes an issue. In addition to the safety issue of collisions, a shared right-of-way in a roadway environment is typically detrimental to all involved modes of transit. Not only can the over-the-road vehicles interfere with and decrease the efficiency of the trains, the trains will interfere with road-going vehicles and increase the overall roadway congestion.

Figure 4
Dedicated APM Right-of-Way in an
Airport Environment

"Combination" Automatic/Manual Systems. There are transit systems with combinations of both manual and automatic operation resulting primarily from the inability to acquire a fully dedicated right-of-way. An example of a combination system is San Francisco's Municipal Railway (or "MUNI"). MUNI light rail vehicles run in a fully dedicated tunnel right-of-way essentially under full automatic control in certain locations (Previous Figure 3) and revert to manual control with drivers when running on street-level shared right-of-ways (Figure 5). Because the intervals between automatic and manual operation are short, the drivers remain on board the vehicles even when under full automatic control. MUNI is an urban, not airport system.

Figure 5
MUNI LRT under Manual Control

Expansion Considerations. There are several ways to expand the capacity of a transit system. The following two ways are most common. One way is to increase train length by adding cars to the trains which will increase capacity. This requires a modest increase in maintenance staff but no additional drivers for systems with manual train control. A second way is to add trains to the system in order to reduce headways which will increase capacity. This way requires the increase in maintenance staff plus the cost of additional drivers for systems with manual control. Either of these ways can be accommodated by fully automated systems with only the modest increase in maintenance staff. A physical expansion to a transit system (extension of the guideway and/or additional stations) that may be part of phased construction or sequenced implementation of the system usually requires additional trains, not longer trains. Thus, it is generally less expensive to accommodate such phased implementation with automated systems that will not require the increase in labor pool costs for the additional drivers when trains are added.

Costs. A rule of thumb regarding the cost of automation in the context of people mover-scaled transit systems is as follows. In a small shuttle system with only one or two trains, automation may likely increase the cost of the system because the labor pool for drivers is small. As the number of vehicles increases in larger systems, the labor pool for the O&M staff of an automated system grows proportionately less than the required labor pool of drivers for a system with manual train control. Thus, automation may add cost to a small system and save costs for a larger system compared to manual train control. It should be noted that the rule of thumb regarding automation adding costs applies to only the smallest and simplest of APM systems. This is primarily because the major suppliers of APM systems have proprietary designs that inherently include automation and converting these systems to manual train control would require changes and the associated costs for such changes. For example, although a standard Bombardier CX-100 vehicle has physical provisions to manually drive the vehicle on rare occasions, ongoing manual operations would necessitate the addition of a true operator's cab within the vehicle and could incur substantial costs for design and production modifications. The required hours of operation can also have a substantial effect on a transit system's costs, particularly for manually operated systems with a large labor pool of drivers.

MEASUREMENTS OF COST EFFECTIVENESS

The cost differences between transit systems are often measured in comparative fashion in an attempt to establish the cost *effectiveness* of a particular transit type or of a specific transit system compared to another transit system. There are many different widely accepted measures of cost effectiveness and they vary among transit types. Common to Light Rail Transit is the measurement of "dollars per passenger mile" or "cost per revenue mile". Common to the APM industry is "passengers per hour per direction" (pphpd) that can be applied to cost / pphpd. Another commonly used measure is simply the "cost per mile" required to construct the particular system. These and other cost measurements may involve capital costs, operational costs, maintenance costs, or a mix of some or all of these types of costs. Although these various measures of cost effectiveness are each statistically legitimate and widely accepted within the overall transit industry, they are typically not comparable to each other and to do so can result in comparing "apples and oranges". This is because each transit system is designed for a particular set of project-specific parameters. For example, one transit system may move few people over a great distance whereby another transit system may move many people over a very short distance. In this example, the first transit system would likely yield a low "cost per mile" dollar amount yet would provide little capacity whereas the second high capacity system would likely have a high "cost per mile" dollar amount by virtue of the fact that it may not traverse even a single mile. The point of this hypothetical comparison is to propose that there may be nothing comparatively cost effective (or ineffective) between these two systems. They may simply satisfy the particular parameters of site-specific needs and may be, in fact, not usefully comparable to each other. A real-world example of this scenario would be attempting to compare the Clarion Health APM in Indianapolis (Figure 6) with the Narita Tokyo International Airport APM in Japan (Figure 7). The Clarian Health APM consists of two small capacity trains traversing a distance of approximately one and one-half miles, whereas the Narita APM consists of four large capacity trains traversing a distance of approximately 900 feet.

Figure 6
Clarian Health APM
Indianapolis, USA

Figure 7
Narita Tokyo
International Airport APM, Japan

Within the transit industry reside various designers, manufacturers, users, and proponents of various transit types. With the aforementioned range of measurements for cost effectiveness, proponents of various transit types can, either knowingly or unknowingly, readily produce data that may appear to favor a particular transit system type or a particular aspect of a transit system (such as automation or manual control). Such data may be statistically correct but of little practical validity. A litmus test of such data is that various measurements of cost effectiveness must be considered in the context of the system's purpose and such measurements must apply to the most salient parameters of the system being measured. If compared to another system, such system must be of like kind and purpose in order for the comparison to have any meaningful validity.

CLAIMS OF COST EFFECTIVENESS

The internet provides a forum for promotion and discussion of various transit systems and their cost effectiveness via various websites. The content contained in these websites varies from meticulously documented research to factually baseless opinions. The tone of these websites can vary from that of a scholarly academician to that of a sarcastic pundit. Many websites do not offer a point/counterpoint approach and instead are focused on making a particular case in point, or promoting the cost effectiveness of a particular transit system or transit type. Examples include The Light Rail Now Project's website, a well researched and documented, but unabashedly pro-Light Rail, forum. The articles on this website make pro-Light Rail arguments with titles such as "Transit Automation and Operating Cost - Where Are the Huge Savings?" and "Monorail Capital Costs: Reality Check". These articles are well researched, but are often based on "apples and oranges" comparisons. Some use disparaging terms such as "Gadget Transit", particularly when referring to automated Personal Rapid Transit (PRT). Another example is a PRT-oriented website that makes claims of solving modern urban problems with fully automated podcars capable of traveling on a fixed guideway at speeds between 100 and 200 mph at a cost of approximately $2000 USD per vehicle. A common sense reality check for such claims is to consider that India's Tata Motors, a multi-billion dollar industrial giant, is currently engaged in bringing to market what they bill as "the people's car" or as some refer to as "the world's cheapest car", to be produced in mass quantities (Figure 8). It will be capable of neither automated operation, nor cruising speeds of 100 to 200 mph, yet it's projected pricing points already exceed the claimed cost of the aforementioned entrepreneur's podcar.

Figure 8
Tata Nano, "The People's Car"

ACTUAL COSTS

Regardless of the diverse types of measurement for comparative cost effectiveness, and the various claims of cost effectiveness, the actual costs of a transit system can be distilled into two main categories; 1) capital costs and 2) operations and maintenance (O&M) costs. Regardless of the structure of the debt service, capital costs can generally be assumed to be one-time, up-front costs that involve the expenditure of funds during the project's design and construction phases with full expenditure completed at the time of the Certificate of Final Completion. O&M costs can generally be assumed to be ongoing costs. O&M services are typically procured via a multi-year contract, the total value of which is usually stated as annualized costs that are paid monthly to the O&M provider. Because the cost of automation is basically incurred as a capital cost and the cost of a labor pool for drivers is basically incurred as an O&M cost, the most prevalent way to objectively compare the cost differential between fully automated and manual train control is to perform a life cycle cost analysis that considers both capital and O&M costs over the useful life of the system. Although such an analysis is beyond the scope of this paper, prospective owners and planners of transit systems should embrace a life cycle cost analysis as a useful tool in their decision making process.

CONCLUSION AND RECOMMENDATIONS

The general conclusion is that it is best to choose the transit system that best meets project-specific parameters, not to design the project around a particular type of transit system and its characteristics, which includes the aspects of automation or manual operation. A life cycle cost analysis can help in comparing the actual costs involved with automatic or manual operation.

In order to best accomplish this project-specific design, it is recommended that planners and owners temper their design decisions with objectivity gained through self-education and/or the input of outside experts. If consulting experts are engaged, their objectively should be confirmed. If self-education involves internet research, common sense should prevail in discerning which websites offer objective facts, and which offer a bully pulpit for entities advancing a particular agenda.

It is recommended that professionals within the transit industry embrace and advance this same objectivity. Despite competition within the transit industry for market share, funding sources, and political and social acceptance, one transit type need not be championed at the expense of another. Unsubstantiated, skewed, or exaggerated claims made by one damages the credibility of all within the transit industry, and when considering the mobility needs of today's society, it is self evident that no single transit type represents a panacea for fulfilling such needs. Instead, the solution is comprehensive in nature, and requires a multi-modal approach that can best be advanced by cooperation between all factions within the transit industry - and can be inclusive of both automatic and manual operation.

REFERENCES

MonoMobile home. Retrieved 3 Dec. 2008.
http://home.fuse.net/ard/jandress/Prod.html

Monorail capital costs: reality check. Light rail progress. November 2002. Retrieved 3 Dec. 2008. http://www.lightrailnow.org/myths/m_monorail001.htm

Scott, Wade. Personal interview. 12 May 2006.

Tata motors home. Retrieved 3 Dec. 2008.
http://www.tatamotors.com/our_world/profile.php

Transit automation and operating cost - where are the huge savings? Light rail progress. December 2002. Retrieved 3 Dec. 2008.
http://www.lightrailnow.org/facts/fa_monorail002.htm

Rubber Tired APM—A Better Solution for Honolulu Rapid Transit

J. David Mori, P.E.[1]

[1]President, Jakes Associates, Inc., Jakes Plaza, 1940 The Alameda, Suite 200, San Jose, California, 95126, USA, Tel: (408) 249-7200; Fax: (408) 249-7296; E-mail: jakes@jakesassociates.com

Abstract

The right public transportation technology within an urban setting can add significantly to the aesthetic look and functionality of a metropolitan area. With the variety of urban transit systems currently operating in the world today, it's important that cities planning on developing new or extending systems have the right information based on previous projects. The City of Honolulu, in its High-Capacity Transit Corridor Project, has decided to build their new transit system on steel wheel/steel rail technology. An expert panel recently affirmed this decision based upon alleged lower construction costs, operations and maintenance costs, operating noise levels, and providing the most proven mass transit solution. This paper will reconsider whether the official decision to go with a steel wheel system was in fact more beneficial than a rubber tired system.

Introduction

In 2007/2008, the City and County of Honolulu critically evaluated various fixed guideway transit technologies and decided to construct an elevated rail rapid transit line connecting the western portion of Oahu to downtown Honolulu and Ala Moana. Known as the Honolulu High-Capacity Transit Corridor Project (HHCTC), it will comprise 19 stations and more than 20 miles of dual lane trackway and operate trains capable of carrying more than 300 passengers each (6,000 riders per hour). During the planning and feasibility process, the City and County of Honolulu issued an RFI and solicited 11 responses by various technology manufacturers. Of those submitted, 5 were vehicles which have steel wheels operating on steel rails, 4 utilized rubber tires operating on a concrete guideway, 1 was a monorail system, 1 a Maglev system, and the last featured train control systems only. Of these choices, an alleged independent panel of 'experts' selected rail rapid transit as the best long term solution for Honolulu. Eighty percent (80%) of this panel was comprised of professionals having dedicated their entire professional careers to the steel wheel/steel rail rapid transit industry.

Of the numerous characteristics between rail rapid transit (steel wheel) and rubber-tired fixed guideway, 8 were utilized as a basis of comparison. These included: construction costs; cost to maintain and operate; ability to qualify for federal transit funding; passenger capacity; electric propulsion; adverse construction impact on community; relief impact for traffic congestion; operating noise levels; and proven technology. Of these, construction and maintenance and operation costs, noise impacts, and being the most proven mass transit solution apparently tipped the project in favor of rail rapid transit.

Capital Costs

Rubber Tire - APM Transit

A fully automated, rubber-tired APM system requires its own right of way in order to operate. Supporting concrete or steel guideways do not disrupt traffic, especially in urban areas. Full grade separation is also required. Vehicles are fully electrified and non-polluting and can be coupled together in train consists. Train consists can carry from 50 to over 300 passengers.

Steel Wheel – Rail Rapid Transit

Rail rapid transit construction costs can vary depending on where the lines operate and under what conditions. Like APM, vehicles are often coupled together in train consists and are fully electrified. Trains can carry from 100 to over 300 passengers.

Rubber Tire APM and Rail Rapid Transit Comparison

Capital costs per passenger mile for rubber tired APM systems can be seen in the following Table 1, "APM Capital Cost per Passenger Mile". Table 2, "Rail Rapid Transit Capital Cost per Passenger Mile" shows these costs for rapid transit rail systems. On average, the capital costs associated with rail rapid transit solutions are significantly more expensive than comparable APM alternatives.

Table 1
APM Capital Cost per Passenger Mile

Area / City	Cost/Mile*
Jacksonville Skyway	$73
Miami Metromover - Original Line	$80
Miami Metromover – Extension	$91

Costs are in USD Million

Table 2
Rapid Rail Capital Cost per Passenger Mile

Area / City	Cost/Mile*
BART/SFO Extension	$152
WMATA Extension	$174
BART/Santa Clara Extension	$261

Costs are in USD Million

Maintenance & Operations Costs

Rubber Tire - APM Transit

Automated people movers are, by definition, fully automated systems, so their operational costs are kept low. In addition, with today's APM reliability enhancements including redundant control systems and self-diagnosis/monitoring, maintenance costs remain competitive with those of rail rapid transit systems.

Steel Wheel – Rail Rapid Transit

The majority of steel wheel rail rapid transit vehicles can operate either manually or semi-automated from a technology standpoint. The largest single operations and maintenance cost driver in rail rapid transit systems resides with their requirement for having a driver on-board.

Rubber Tire APM and Rail Rapid Transit Comparison

Automated people movers have a life cycle overall cost advantage relative to rail rapid transit because of their ability to operate automatically rather than manually. In addition, APMs running on rubber tires have additional operational advantages which are worth noting. By utilizing rubber tires, vehicles have an opportunity for faster acceleration, reduced braking distances, and the ability to climb or descend steeper slopes than would be feasible with conventional steel rail systems.

Noise Level

According to the Federal Transit Administration, noise from rubber tired or steel wheel vehicles can come from any combination of the following: wheel/guideway interaction, propulsion system, brakes, auxiliary equipment, and/or wheel squeal, cooling fans. In general, noise increases with speed and train length.

Rubber Tire - APM Transit

Rubber tired vehicles are typically known for being quieter during operation in comparison with rail rapid transit technology systems.

Steel Wheel – Rail Rapid Transit

Steel wheel on steel rail rapid transit technologies are often considered one of the noisiest forms of fixed guideway transit. According to Honolulu planning report studies, rail rapid transit is already acknowledged to creating significant noise concerns near planned elevated tracks where trains are expected to operate up to 55 mph about 400 times a day from 4 a.m. to midnight. Excessive noise is also considered to have a potential significant adverse impact on localized real estate values near the planned transit alignment.

Rubber Tire APM and Rail Rapid Transit Comparison

According to the FTA, the introduction of new noise into a community is considered severe if a significant percentage of people would be highly annoyed. A moderate noise impact would be noticeable to most people, but may not cause strong, adverse community reaction. Rubber tired APM technologies are characteristically among the quietest forms of fixed guideway transit systems. Based upon conclusions of the Honolulu 'Expert' technology panel (as referenced above), four of the five panelists rated steel wheel rapid transit technology as the noisiest, though still within acceptable levels.

Proven Mass Transit Solutions

Rubber Tire - APM Transit

Rubber tired APM solutions have a longstanding history of proven service in urban and airport settings throughout the world. Table 3: 'Representative Proven APM Solutions Throughout the World' highlights several of these system applications.

Steel Wheel – Rail Rapid Transit

Similarly, rail rapid transit solutions have established a proven service history in primarily urban settings throughout the U.S. and the world. Table 4: 'Representative Proven Rail Rapid Transit Solutions Throughout the U.S.' highlights several of these system applications successfully operating in the United States.

Rubber Tire APM and Rail Rapid Transit Comparison

Rubber tired APM and rail rapid transit are both widely proven in routine passenger service throughout the world.

Table 3
Representative Proven APM Solutions Throughout the World

#	Country	City / Region	System / Line Name	# Stations	Distance (Miles)
1	France	Laon	Poma 2000	3	0.9
2	US	Jacksonville	Jacksonville Skyway	8	2.5
3	Japan	Kobe	Rokko Island Liner	6	2.8
4	Switzerland	Lausanne	Lausanne Metro - Line M2	14	3.7
5	US	Miami	Miami Metromover	22	4.4
6	Singapore	Singapore	LRT - Bukit Panjang Line	14	4.8
7	France	Paris	Paris Metro - Line 14	9	5.6
8	France	Rennes	Rennes Metro	15	5.8
9	Italy	Turin	Metrotorino - Turin Subway	15	6.0
10	Japan	Tokyo	Nippori - Toneri Liner	13	6.0
11	Singapore	Singapore	LRT - Punggol Line	16	6.4
12	Japan	Kobe	Port Island Liner	12	6.6
13	Japan	Yokohama	Kanazawa Seaside LRT	14	6.6
14	Singapore	Singapore	LRT - Sengkang Line	15	6.6
15	Taiwan	Taipei	Taipei Metro - Muzha Line	12	6.8
16	France	Toulouse	Toulouse Metro - Line A	18	7.7
17	France	Lyon	Lyon Metro - Line D	15	8.1
18	France	Lille	Lille Metro - Line 1	18	8.4
19	Japan	Tokyo	New Transit Yurikamome	16	9.1
20	France	Toulouse	Toulouse Metro - Line B	20	9.3
21	France	Lille	Lille Metro - Line 2	43	19.8

Table 4
Representative Proven Rail Rapid Transit Solutions Throughout the U.S.

#	City, State	Transit Agency Name	# Stations	Distance (Miles)
1	San Francisco, CA	Bay Area Rapid Transit District	43	267.6
2	Washington, DC	Metropolitan Area Transit Authority	83	225.3
3	New York, NY	MTA New York City Transit	468	835.0
4	Boston, MA	Mass. Bay Transportation Authority	53	108.0
5	Chicago, IL	Chicago Transit Authority	144	287.8
6	Atlanta, GA	Metro Rapid Transit Authority	38	103.7
7	Philadelphia, PA	Southeastern Penn. Transportation Authority	75	102.0

Safety

Rubber Tire - APM Transit

Based upon recent data from the United States Department of Transportation, limited information has been compiled on total and average APM fatalities and injuries. Total APM fatalities and injuries can be seen in the following Table 5: 'APM and Rapid Transit Fatalities' and Table 6: 'APM and Rapid Transit Injuries'.

Steel Wheel – Rail Rapid Transit

Rail rapid transit systems are among the safest forms of travel. However, their collective service history is not without injuries and fatalities. Tables 5 and 6 highlight and quantify specific incident histories.

Rubber Tire APM and Rail Rapid Transit Comparison

Based upon statistical service histories for both rubber tired APM and rail rapid transit, the APM solution clearly has a superior safety record in terms system injuries and fatalities.

Table 5
APM and Rapid Transit Fatalities

	2002	2003	2004	2005
APM	0	0	1	3
RT	73	49	59	35

Table 6
APM and Rapid Transit Injuries

	2002	2003	2004	2005
APM	28	29	15	2
RT	4806	4158	4738	3814

Conclusion

Currently, the City of Honolulu has selected steel wheel rail rapid transit as the best option based on reliability, safety, ride quality and cost efficiency. Based upon the aforementioned research and findings, it remains unclear how this conclusion could possibly be supported. That is, until we consider other critical evaluation factors which were not widely known or documented in the 'official' evaluation above. These likely included:

- The political influence of special interest groups (including A/E firms);
- Closed minded 'thinking' in terms of the alleged 'proprietary' nature of APM technologies thereby limiting future procurement options;
- Typical U.S. procurement management differences between the two technology groups;
- Degree of overall project management control retained by the Owner;
- Other.

References

- Panayotova, Tzveta. "People Movers: Systems and case studies." Facilities Planning and Construction, University of Florida. 28 May 2003 <http://www.facilities.ufl.edu/ cp/pdf/PeopleMovers.pdf>.
- 2000 Transit Safety and Security Statistics and Analysis Report. Cambridge, MA: Bureau of Transportation Statistics, Research and Innovative Technology Administration, U.S. Department of Transportation, 2002.
- Honolulu Rail Transit Q&A. Honolulu, HI: City & County of Honolulu, 2008.
- Rohde, Mike. World Metro Database. Metrobits.org. 2008 <http://micro.com/metro/table.html>.
- Heavy Rail Transit Agencies Mileage and Stations Data. Washington, DC: American Public Transportation Association. FY2004 <http://www.apta.com/research/stats/rail/ hrmiles.cfm>.

Guangzhou APM: First Urban APM in China

Rob DeCostro, APM Projects
Bombardier Transportation
Systems Division
Pittsburgh, PA, USA

Abstract

Guangzhou is the 3rd largest city in China. However, it is one of the fastest growing cities as a result of China's continued accelerated urbanization efforts. Located 75 miles northwest of Hong Kong, Guangzhou is the economic center of the Pearl River Delta. The 16th Asian Games will be held in Guangzhou in November 2010 and is considered the "coming out" party for the city. A modern area within Guangzhou City is being developed, known as the Pearl River New City, and will be integrated with primary urban facilities such as international finance, trade, commerce, entertainment, and administrative offices.

In May 2007, Bombardier Transportation signed a contract with the Guangzhou Metro Corporation to provide an Automated People Mover (APM) system that will serve as the central line operating within the Pearl River New City. This paper details how Bombardier will deliver to Guangzhou Metro Corporation a 3.94-km pinched-loop APM system, including fourteen (14) *BOMBARDIER* CX-100** vehicles, the *BOMBARDIER* CITYFLO** 650 signaling system, and an APM Central Control by the end of June 2010. The total underground line, which will travel below the Pearl River, will include nine stations and four interchange stations to lines 1, 3 and 5 of the Guangzhou subway system.

This paper shows how Bombardier and the Guangzhou Metro Corporation (GMC) are working in close cooperation due to the unique nature of the scope-split for this project. GMC is responsible for the civil construction, power supply, platform doors, and other key components required for successful system commissioning, including the installation of all GMC and Bombardier equipment.

The Guangzhou APM will be the first urban driverless APM system in China and is expected to demonstrate the modernization and forward thinking of Guangzhou City.

Introduction

As 2010 approaches, the City of Guangzhou inches closer to becoming the most elite city south of the Yangtze River. The City of Guangzhou is currently undergoing a transformation that is not often seen and happens once in several generations. The transformation is known as the Pearl River New City Project and is a total reconstruction of the downtown area of southeastern Guangzhou. The project includes the development of a business, leisure, and cultural district that the local government hopes will become the perfect host for the 2010 East Asian Games. The area will consist of the City's consulates, several 5 star luxury hotels, luxury residential communities, a new opera house, and an art museum. The area will not only be home to new beautiful sky scrapers, but will also be home to the Guangzhou Television Tower that will be 450 meters tall once complete. The architecture is poised to be world-renowned and the hope by local officials is that this project will place Guangzhou as a globally recognized city. As this area will stretch almost 4 kilometers long, the local transit authority, Guangzhou Metro Corporation (GMC), was commissioned to find a way of connecting the new city that would benefit its residents. GMC selected Bombardier Transportation's Systems Division to implement its *CX-100* technology to connect the Pearl River New City. This quiet, environmentally friendly system can be easily integrated into existing buildings and landscapes. In the case of the Guangzhou Pearl River APM, the system will run totally underground, under the Pearl River, via a north and south direction.

Figure 1: Artistic Rendering of the Pearl River New City

SYSTEM / PROJECT

GMC selected the driverless *CX-100* system and vehicle technology, along with the *CITYFLO* 650 signaling technology, for the 3.94-km Guangzhou Pearl River Automated People Mover system. The system will consist of 9 stations and will initially deploy 14 vehicles but can be expanded to 32 vehicles. The track will have 14 total switches, 12 pivot and 2 turntable switches, with 4 additional pivot switches to be added if a proposed phase 2 of the maintenance depot is completed. Bombardier will also provide its Central Control technology to GMC to be located at station #1. The contract calls for Bombardier to complete the project in a 36-month schedule, beginning June 1, 2007, and ending June 28, 2008. The system will run north and south with the maintenance depot located south of station #1. Figure 2 below illustrates the system from a high-level perspective.

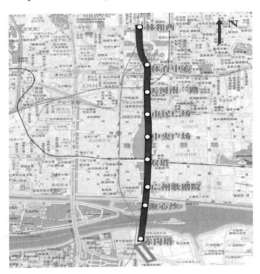

Figure 2: High - Level Map of System

GMC has specified the following performance specification: following testing and commissioning, a Utilization Rate of 99.0% for 14 consecutive days must be completed to successfully complete the demonstration period. Once the demonstration period is completed and a Certificate of Safe Carrier (CSC) is issued, the system must obtain a Utilization Rate of 99.65% for 180 consecutive days. Once the 180 consecutive days has been fulfilled, the system can then undergo a final inspection by GMC. The project is deemed complete once approval on the final inspection from GMC has been granted.

The current project schedule specifies that the project will receive final acceptance from GMC on December 25, 2010. Prior to final acceptance, Bombardier and GMC must collectively manage a very unique scope split to ensure all the proper interfaces are correct and each other's schedules are synchronized to reach the final outcome of final acceptance.

SCOPE SPLIT

In the case of the Guangzhou Pearl River APM, the scope split that was agreed was formulated to assist GMC so that the local government could move forward with the commissioning of the transit project. While this is not unique, it will be the first scope split of its kind for APMs and in the Chinese market.

The agreed scope split stipulated that Bombardier is responsible for the conceptual system design consisting of Civil, Operations Maintenance & Storage Facility, Operations Management, Organized Train Operation, Power Rail, and Power Supply & Distribution. Bombardier also has the responsibility for the vehicles, switches, and signaling equipment, as well as APM system development. Bombardier is also responsible for testing and commissioning the system.

GMC has responsibility for the civil works (tunnel, running surface, and station construction), power distribution system (including power rail), platform screen doors, guide beam, installation of all components, including Bombardier's scope, and the Operations and Maintenance of the system.

In order to ensure that the system will be successful, Bombardier and GMC must work together to seamlessly integrate the complete system.

DETAILED SCOPE SPLIT - BOMBARDIER

VEHICLE

Bombardier is providing the *CX-100* vehicle platform with 14 vehicles to be delivered in summer 2009.

Figure 3: Guangzhou Vehicle Design by Tanghao Designing

The attractive, comfortable *CX-100* vehicle is electrically powered, operates on rubber tires and offers a smooth ride quality on a dedicated guideway. The advanced vehicle design includes a modern end cap, spacious, climate-controlled interiors and large windows with an expansive view. Car interiors are attractive, yet ruggedly resistant to wear.

SIGNALING

GMC required that the system incorporate the latest in signaling technology. In order to accommodate this requirement, Bombardier offered its *CITYFLO* 650 signaling automatic train control. This technology is unlike traditional fixed block systems; *CITYFLO* 650 requires neither standard track circuits nor an on-board operator. Train-to-wayside communication is not transmitted through fixed-track circuits, but through a "contactless" communications medium capable of bi-directional transmission. *CITYFLO* 650 is Bombardier's moving block Communications Based Train Control System (CBTC) technology. Moving block means that the occupancy of the train moves along with the train in a continuous fashion. Communications based means that train control information is transmitted between the train and wayside computers through a RF link. *CITYFLO* 650 is a "contactless" train control system because an RF link exists (i.e. no physical contact) between the trains and the wayside equipment controlling their movements.

SWITCHES

Bombardier is providing 12 guideway pivot switches. The pivot switch is a device used to provide a method of guidance for the *CX-100* vehicle from a one-lane guideway onto an adjacent lane. Guidance of the vehicles through the pivot switch is accomplished by using two pivoting guidebeams, with equivalent cross-section to the main guideway guidebeam, one of which provides a continuous guidance surface for the vehicle. The two switch beams are each pivoted about individual pivot points during the same time interval. Each beam is aligned with a diverging track at the end of the switch. Switching is performed by pivoting the beams such that the desired position is achieved.

Figure 4: Bombardier Guideway Pivot Switch at San Francisco Airport

Bombardier is also required to provide two turntable switches. The turntable switch is a single, rotating beam used to align one of two intersecting guideways. In addition, there are four fixed beams. The rotating beam aligns two of these fixed beams to form a continuous guideway.

DETAILED SCOPE SPLIT - GUANGZHOU METRO CORPORATION

CIVIL

One of the most critical items within GMC's scope is the civil responsibilities on the project. GMC is responsible for the entire guideway system's design and construction, which is comprised of concrete running surfaces, steel guidebeam, guideway switch pits (Bombardier will supply switches to be installed by GMC), and emergency walkways. GMC is also responsible for all system equipment rooms and station construction including the platform build. This undertaking began with the supply of the civil conceptual design for the system as Bombardier's first contractual deliverable to GMC. The guideway interface is the first in many interfaces that show why Bombardier and GMC must work very closely to ensure no mistakes are made. If any part of the running surface is out of specified tolerance, not only could ride quality be affected, but performance requirements specified by the contract could also be impacted.

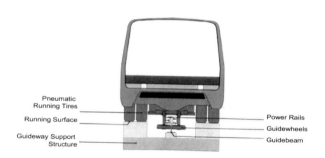

Figure 5: Vehicle to Guideway System Interface

Running Surface - The *CX-100* vehicle rides on a concrete running surface with a compressive strength of 34500 kPa (5000 psi). Outdoor running surfaces have a transverse broom finish. Running surfaces in tunnels have a broom or smooth float finish. In some cases, an alternative running surface design is used for a more aesthetically pleasing structure. This design consists of two separate running surfaces. The Running Surface installation requires careful attention to tolerances.

Figure 6: Running Surface Construction

Guidebeam - GMC is also responsible for the guidebeam that interfaces with the *CX-100* technology. The guidebeam is a critical component of the system and requires GMC to adhere to strict Bombardier provided requirements. Located at the center of the guideway, the guidebeam provides guidance for the vehicle. Four pair of guidewheels on the *CX-100 vehicle* locks the vehicle to a steel guidebeam anchored to the guideway to provide continuous guidance for each moving vehicle. GMC and Bombardier must work in close collaboration to make sure that the fabrication meets the geometry of the system, including vertical and horizontal transitions. The guidebeam installation requires careful attention to tolerances.

Figure 7: Guidebeam Located between Running Surface

Power Distribution System /Power Rail - Per the contractual requirements, Bombardier, with its first deliverable to GMC, is responsible to provide conceptual designs of various systems. Two of those systems are the Power Distribution System and the Power Rail.

The Power Distribution System (PDS) and the Power Rail are the two main components necessary to supply power to the *CX-100* vehicle. To meet the ultimate ridership demand, the system is designed to run thirteen (13) 3-car trains in pinched loop mode. Electrical power to the APM system will be provided from two bulk power substations and transmitted to various propulsion power substations and station power substations located at the passenger station via cables.

The power rail system consists of power rails, ground rails, rail covers, rail joint, connectors for power and ground cables, mounting brackets, on and off ramps and power feed assemblies to route the power cables through the guidebeam. The power rail system is designed to interface smoothly with the power collection system on the APM vehicles. The system is designed to provide continuous power collection to the APM vehicles at all locations on the APM system at operational speeds from a complete stop to the maximum operating speed of the vehicle in both normal and reverse directions of operation. The power rail system must be sized for the system current/voltage drop requirements. The APM system rails and mountings must be of sufficient size to withstand vehicle dynamic loads and electromagnetic and thermal loads due to short circuits. The power rail system will have a design life of 15 years. The power rail system is designed such that maintenance can be performed with standard hand tools. Damaged rail components must be easily replaceable. Field adjustments are not a normal maintenance requirement of the power rail system. The principle components of the power rail system include:

- rails
- rail covers
- mounting brackets
- splices (joint connectors)
- fixed splice
- expansion
- isolation
- cable connectors
- power
- ground
- on/off ramps
- cable feed assemblies

Rail sections, supplied in 30-foot lengths, consist of an aluminum channel type extrusion containing a stainless steel contact surface, on which the *CX-100* vehicle collector shoe rides. Except for that of the rail, which must be open to accommodate the collector shoe, the rail is covered with a protective, insulating cover.

Figure 8: Power Rail

Other key components - GMC is also responsible for the areas of emergency walkways, station construction, and Platform Screen Doors. These systems are also critical that GMC and Bombardier work together to make sure the right dimensions are used, with the utmost importance place on the station dimensions. Figures of these can be found below.

Figure 9: Station

Figure 10: Station Doors

Figure 11: Emergency Walkway

Maintenance Depot - The maintenance and storage facility (M&SF) incorporates all necessary maintenance equipment, and materials for inventory control, maintenance scheduling, maintenance management information processing, servicing, cleaning, inspection, troubleshooting, and repair.

The following items typically installed in the M&SF facility include:
- running surface, where required
- guidebeam, where required
- maintenance pivot switches
- light maintenance bay work platform (either elevated or recess pits)
- heavy bay work areas
- car wash area

INSTALLATION

GMC is also responsible for the installation of the entire system, including the equipment supplied by Bombardier. Bombardier is working very closely with GMC on the installation approach to ensure all equipment is installed to Bombardier specifications. One way Bombardier is assisting with the installation of its responsible scope – vehicles, signaling, and switches – is by providing GMC with detailed installation technical requirement documentation.

For all of GMC scope, Bombardier will send specialists to perform inspections of GMC's work. For example, the guideway will be allocated extra resources by Bombardier to make sure all tolerances are adhered to. Bombardier will be dedicated to working with GMC to consult the overall development of the civil aspects of the project.

Source inspections will be conducted with the typical Bombardier methodology as if the work was being completed within Bombardier's scope. This means that all test procedures normally written for inspection on the items typically in Bombardier's scope will still be written by Bombardier. Bombardier will then send the procedures to GMC for approval. The procedures will then be filled out in close collaboration with GMC, allowing Bombardier to source inspect all GMC work via the internal Bombardier process. The outcome from these reports will indicate if the system is ready to begin testing and commissioning.

TEST & COMMISIONING

The testing and commissioning (T&C) of the Guangzhou Pearl River APM is expected to take 9 months. This includes all inspections, one-time tests, and system integration testing.

Bombardier will take the lead during the field testing program with GMC providing resources and will act as a shadow to supplement its overall training program. Bombardier's focus during this time is threefold. One, Bombardier will conduct T&C of the APM system to ready the line for revenue service by June of 2010. Second, provide the necessary technical information flow to the customer in order that the customer learn the system through a hands on approach. Third, provide adequate training in terms of manual procedures, with an emphasis on safety.

Bombardier's objective of the overall collaboration is to ensure a successful handover to the Guangzhou Metro Corporation.

HANDOVER TO GMC

The Bombardier Division responsible for Automated People Mover Systems has over 35 years of experience in design/build projects and providing a comprehensive range of operations and maintenance services.

This is the first time in the Chinese market that Bombardier is not providing an O&M contract with the supply of an APM system. Due to this fact, the project team, will emphasis training as soon as possible. As mentioned in the T&C portion of this paper, Bombardier will begin formal training throughout this period. Also, during the installation period of the contract, which GMC is responsible for, Bombardier will emphasize maintenance on the guideway, guidebeam, power rail, and PDS.

Through these various activities, the training program will not only be classroom and field, with 6 overall contractual sessions, but will also be a training program throughout the 36-month contract.

Bombardier also plans on keeping the site office in operation for support after project close out to assist GMC and to ensure that the handover to GMC will result in the successful operation of the APM system.

CONCLUSION

To date, Bombardier is currently 17 months into the 36-month schedule. Bombardier's scope is being managed at its West Mifflin, Pennsylvania facility, where the vehicles are being produced. GMC is currently working to complete the tunneling of the system. Once this is complete, GMC will begin the various scope of supply described in this paper. Bombardier and GMC enjoy an excellent relationship and it is Bombardier's goal is to build on this close collaboration over the next 19 months.

---

* *BOMBARDIER, CX-100* and *CITYFLO are* trademarks of Bombardier Inc. or it subsidiaries

Simulation Analysis of APM Systems in Dense Urban Environments – Part 1: Transit User Experience

J. Sam Lott, P.E.[1]
Douglas Gettman, Ph.D.[2]
David S. Tai[3]

[1] Senior Vice President, Kimley-Horn and Associates, Inc., 12012 Wickchester Lane, Suite 500, Houston, Texas 77079; PH 281-597-9300; sam.lott@kimley-horn.com
[2] ALPS Modeling and Simulation Group Leader, Kimley-Horn and Associates, Inc., 7878 N. 16th St. Suite 300 Phoenix, AZ 85068; PH 602-906-1332, doug.gettman@kimley-horn.com
[3] Senior Modeling Software Developer, Kimley-Horn and Associates, Inc, 7878 N. 16th St., Suite 300, Phoenix, Arizona 85020; PH 602-944-5500; David.Tai@kimley-horn.com

Abstract

This paper is the first in a series of two technical papers which define the key issues to be analyzed in the study of dedicated transit systems such as Automated People Mover (APM) systems for circulation within dense urban environments. It is quite complex to accomplish an "apples-to-apples" comparison between alternative transit technologies for urban district circulation systems because of the different travel times, access convenience and walk distances between the alternatives, all of which effect the transit user's choice. These papers describe the suitability of the ALPS ™ simulation tool for such analyses including discussion of the characteristics of the urban environment, the nature of transport systems suitable for circulation system application in urban centers, and trade-offs between at-grade and aerial transit system alignments . Simulation-based case studies are presented that illustrate the analysis of APM transit technologies using ALPS. This first paper addresses the representation of the transit user's complete experience within the simulation model, including the approach to modeling transit ridership and the related pedestrian facilities associated with the transit stations and the means of access to the transit. The second paper describes the capabilities within ALPS to represent transit system operations, including train performance modeling, guideway and alignment configuration, demand-responsive scheduling, and options for dynamic routing of APM systems.

Introduction

The urban renewal that is sweeping through the core of most large American cities is bringing with it new challenges for maintaining access and mobility. To meet this challenge, it is increasingly apparent that the use of dedicated circulation systems will be required in many locations to distribute transit users within the urban district. The circulation system must, therefore, be designed to as an integrated component of the multimodal transportation network.

The resulting opportunities to apply grade-separated, advanced transit technology commonly known as automated people mover (APM) systems are increasingly significant. Furthermore, we are on the threshold of seeing a new subset of APM technologies being applied in the form of demand-responsive, personal rapid transit (PRT) systems. APM systems are important to consider as a viable alternative to more conventional forms of surface transport in dense urban settings due to their flexibility of alignment, frequency of service, and reduction in size of trains and stations. These factors are particularly important when considering connecting regional access systems such as commuter rail or rapid transit metro systems with the heart of urban districts through dedicated circulation systems such as APM. In particular, APM circulation systems that serve an intermodal terminal adjacent to the urban district, and/or remote/intercept parking facilities are being considered more frequently as viable options.

The physical characteristics of dense urban districts and major activity centers impose significant limitations to the way new transit systems can be retrofitted into the built-environment of the urban core. The realities of the district's building density, existing surface transportation infrastructure, and underground utilities – when all are essentially occupying the same space – make any transit installation relatively high in construction impacts and cost. The differences between the design and construction costs of transit alternatives can be readily addressed through typical engineering studies. However, major differences in operational and environmental impacts between alternative circulator system technologies, are too often only assessed with limited qualitative and subjective analysis.

More advanced techniques are therefore needed in the alternatives analysis process to accurately assess the comparative operational / environmental differences between transit technologies for circulation within major urban districts. New simulation-based tools and techniques have been developed that represent all modes and related infrastructure, including traffic operations and street intersection signaling systems, pedestrian systems and conveyances, parking facilities, and transit facilities.

Figure 1 illustrates the multimodal environment that is characteristic of dense urban districts and major activity centers. Each of the modal elements provide an important function within the overall multimodal transportation system. The end result of such an integrated multimodal system will be acceptable access to and mobility within the urban district / activity center.

Alternatives analysis studies of circulator transit systems for urban districts have, in the past, typically focused on the suitability of more conventional, at-grade transit modes such as bus or light rail systems. This conventional technology is usually located with the transit system in mixed operations with street traffic and along alignments that are frequently in conflict with pedestrians. More recent alternatives analysis studies are increasingly beginning to also assess the benefits of grade-separated circulator system(s), such as fully automated APM systems operating on aerial guideways.

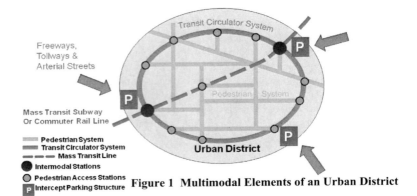

Figure 1 Multimodal Elements of an Urban District

Multimodal Simulation Methodology

For several decades Kimley-Horn has been supporting agencies in analyzing transit systems that operate in multimodal transportation environments using a simulation-based methodology. In order to effectively perform the comparative analysis of alternative transport technologies, the suite of computer programs called the Advanced Land-Transportation Performance Simulation^{TM} (ALPS^{TM}) has been continuously refined and advanced since the early 1980s. Most recently, this tool has been used to study transit technologies as diverse as high speed maglev, urban light rail, street trolleys, and complex APM and PRT systems.

Considering the complexities of a rigorous alternatives analysis process for an urban district circulation system, ALPS is uniquely designed as a simulation tool that can accomplish a true "apples-to-apples" comparative analysis in one analysis package. ALPS simulations can specifically model the following diverse types of circulation systems:

At Grade Modes

- Multiple local bus routes, supplementing a dedicated district-circulation route(s)
- "Smart Bus" circulation bus service operating on the city streets, with variable routing that utilizes a demand-responsive dispatch mode
- Street trolley running in mixed traffic with street vehicles
- Light rail transit (LRT) or bus rapid transit (BRT) systems operating in a dedicated lane with traffic signal priority or signal pre-emption

Aerial Guideway Modes

- Grade separated light rail (LRT) systems
- Fully automated APM systems (scheduled or managed-headway operating mode)
- Personal rapid transit (PRT) systems with pure demand responsive controls
- Group rapid transit (GRT) systems – potentially providing a combination of PRT-type demand responsive and APM-type managed headway operations

The discussion that follows illustrates the most comprehensive study process that can be utilized within the ALPS analysis methodology, specifically with respect to the analysis of the user experience of APM / PRT modes. The methodology is essentially the same for analyzing at-grade transport modes (Gettman and Lott, 2009) and it is possible to only apply partial applications of the methodology are also viable (e.g., modeling a single large station only). .

For purposes of this technical paper that is focused on the user experience aspects of urban district / activity center circulation systems, the discussion below will concentrate on APM technologies that are being increasingly considered for a district's internal circulation functions. Subway lines, commuter rail lines, or any other such line-haul mass transit systems which pass through the district are typically modeled with only a small piece of their total system included in the simulation studies.

Demand Forecasting and Ridership Analysis

One of the important first steps of simulation modeling of transit systems within dense urban settings is forecasting of ridership. Most travel demand forecasting is accomplished using models that are primarily designed and used for regional analysis. These macroscopic models can be very effective at projecting future person-trips for larger segments of the region and over larger periods of time (e.g., daily trips or peak periods of several hours). However, the proper analysis of APM and PRT systems requires a much more detailed disaggregation of trips to more localized sites such as specific buildings or at least segments of a city block.

ALPS trip generation methodology has been developed over a number of years through multiple studies of urban centers and other multimodal activity centers as diverse as Las Colinas Urban Center (Dallas, Tx); Los Angeles, Newark Liberty and Chicago O'Hare International Airports; and the recent models of Downtown Houston.

Figure 2 illustrates the demand model condition where a portion of the model matches the regional model TAZs – configured as the larger spatial area (e.g., a full city block). But a portion of the downtown district model applies the ALPS methodology to break down (i.e., disaggregate) selected TAZ areas into smaller trip production and attraction zones using the demographics of each building / property use. To accomplish this disaggregation, the model applies classic person-trip generation methodology as a function of parameters such as square feet of office space or retail space, number of hotel units or residential units, number of hospital beds, etc.

A key part of modeling a dense urban area in such a detailed manner is the identification of parking facilities and their associated access/egress locations, both for automobiles and pedestrians. If for no other reason than this parking access issue, more detailed modeling within an urban district is required than is possible using the regional model alone. **Figure 3** and **Table 1** give an example of disaggregation subzones, and the corresponding trip generation data tables used to supplement and further disaggregate the regional model's trips for the East End Corridor LRT project in Houston, Texas.

In addition to distributing the spatial aspects of trip generation, ALPS can also provide a time-distribution throughout the day of the daily trips forecast by the regional model for each of the trip production nodes. This temporal distribution is usually applied with time increments as small as 5 minutes. This is necessary in order to properly analyze the ridership demand conditions on an APM circulator system.

Figure 4 illustrates a typical time distribution curve over a 24 hour day for a particular facility (or class of facilities), combined with the particular trip purpose and primary mode. This combination of attributes of a given type of person-trip is called a "travel classification" within ALPS models. For example, a typical travel class that would circulate within an urban district would be the non-home based, non-work trips that occur during the mid-day and evening meal periods. The routes (travel paths) would be from the office to local restaurants, and then back to the office. The corresponding time of day curve would distribute almost all trips between the hours of 10 a.m. and 2 p.m., and between 4 p.m. and 8 p.m.

In addition, if a new development, such as a major new high-rise office building needs to be represented with new person-trips layered over the regional model's data, the ALPS methodology allows these trips to be included as a separate travel classification(s). With this methodology, the analyst is able test multiple scenarios of localized future development much more quickly than is possible with the regional model.

Modal Split Modeling

Under the common analysis approach to transportation "mode choice", the typical regional modeling process performs "mode split to transit" computations as a sequential process that follows after the trip generation and the initial roadway traffic assignment process. Based on the calculated total travel time for the automobile mode, the mode split calculations estimate the portion of the total person trips that would choose to use transit rather than the auto mode. The resulting person-trip assignments to transit constitute the transit ridership, which is typically calculated as a whole day's value. Regional mode split models utilize calculations that include demographic data such as average family income and other such socio-economic data. This approach is reasonable at a regional level, but for focused study of circulation systems such as PRT and APM systems within a urban area , much more specific multi-modal split modeling is required within every time interval of the day.

The ALPS modeling methodology for determining modal split uses the calculation of travel times between "competing routes" to determine the allocation of travelers that use each route. These routes include all the multimodal elements of each travel path (i.e., walking, driving, parking, and transit). Further, such competing paths may or may not contain the same dominant travel mode, so one can analyze the split between various combinations of modes such as park and ride lots that are at different distances between origin and destination. The assignment of trips through the network is then made based on calculations that utilize the "Logit" model to distribute the trips across the possible multi-modal routes. **Figure 4** illustrates this multimodal trip assignment process that has inherent "mode-split" calculations as part of the assignment process.

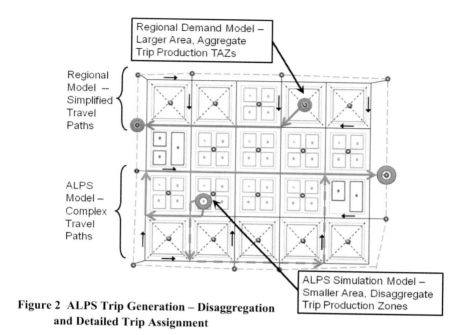

Figure 2 ALPS Trip Generation – Disaggregation and Detailed Trip Assignment

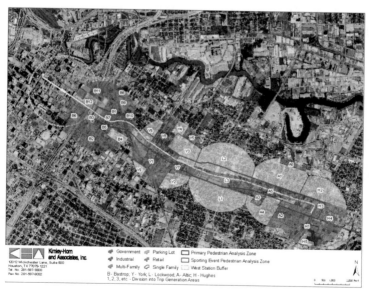

Figure 3 ALPS Trip Generation – Disaggregation and Detailed Trip Assignment

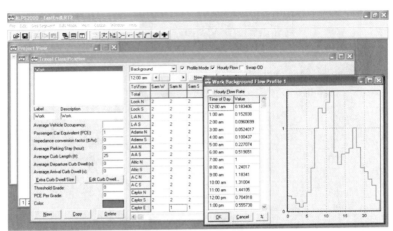

Figure 4 ALPS Trip Generation – Time Distribution of Person-Trips

Table 1 Sample Trip Generation Input and Output Tables

Trip assignments made directly by ALPS from the origin/destination trip tables includes consideration of the diverse time and cost parameters that the typical transit patron will experience, such as:

- Transit travel speed / travel time, including the effect of transit vehicle stops at all stations between the origin station and destination station.
- Out of pocket costs to utilize the transit mode, park a vehicle, and/or pay a toll.
- Headway between successive transit train/vehicle serving the origin station of a patron, or for demand responsive dispatch the wait time for transit service following the demand call placement (e.g. PRT, GRT or demand responsive bus service).
- Transit station proximity to the travel origin and destination nodes or transfer points to other local transit systems, and the corresponding walk distances

These time/cost differences may result in significantly different results for the transit patronage of each case study of a different transit circulator technology alternative. Any such difference in transit ridership scenarios is fundamentally important to quantify for use in the comparative assessment of the transit circulator alternatives.

As illustrated in **Figure 5**, an example of competing travel paths could be:
- Walk to destination utilizing pedestrian mode only
- Walk to APM Circulator system, ride transit to nearest stop, then walk to destination
- Walk to parking, drive to parking nearby the destination, park the vehicle, and walk to destination

In the **Figure 5** illustration, the difference in the travel time and cost "impedance" between the three alternative travel paths through the network would determine the percentage of person-trips assigned to each path according to the Logit algorithm. The driving path alternative would probably be the fastest under normal conditions, but the cost to park represented in the model input data may make this alternative quite unattractive. The "dynamic" aspect of the model means that during times of the day when the traffic builds in the District (as simulated by ALPS), the drive time will increase and the percentage of person-trips assigned to that mode will decrease. But during times of the day when traffic congestion dissipates, a larger percentage will be assigned to drive. And as described above, secondary or tertiary travel path choices can also be included for the drive and park "mode" to represent congestion effects on driving path choice or alternative parking choice with lower parking prices but possibly longer walking distances.

Depending on the number of station stops, the APM transit mode will likely provide the second fastest time for the trip. But the out-of-pocket fare to ride the system would also influence the percentage of person-trips assigned to transit. The APM transit travel time would always be consistent due to its managed headways, unless the APM system is operating near its capacity. In that case not everyone waiting in the station would be able to board the next train entering the station, and simulation would delay the portion of passengers in the station who were forced to wait for the next train. When this occurs, the extension to travel time for some who use the transit mode path would result in fewer transit riders being assigned.

The walking mode path would probably have the slowest travel time of all the modes, but the distance penalty-factor that is automatically applied to walking in the ALPS models (referred to as the "social acceptability factor") would further reduce the attractiveness.

When the distribution of person-trips to competing travel paths is performed at this level of detail (as compared to other methods that are more approximate in nature), the travel time advantages of an APM will be properly reflected in the simulated transit ridership. ALPS performs these trip assignments dynamically, with the travel time calculation representative of simulated conditions at each time of day. This allows the representation of the unique features of PRT systems to deliver passengers more quickly to their destination when sufficient empty vehicles can be delivered by the system to satisfy demand. Since the demand-responsive features of PRT technology are then accurately represented in the ALPS simulation, there is a more accurate forecast between the transit ridership of PRT versus other potential technology applications – including APM, LRT, and bus rapid transit.

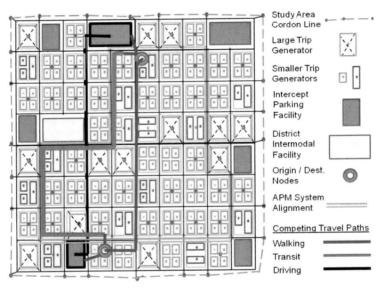

Figure 5 ALPS Trip Assignment and Mode Split – Competing Travel Paths

Pedestrian Components of Transit Simulations

A critically important part of a comprehensive analysis process of transit in dense urban settings is the representation of the pedestrian experience when accessing the station site, circulating vertically to the station platform (for grade separated applications), and then boarding/alighting the vehicles **Figure 6** illustrates the capability of ALPS to model pedestrians taking transit from origins to destination. This example is taken from the East End Corridor light rail study in Houston, Texas.

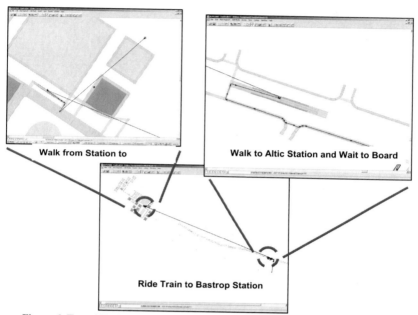

Figure 6 Travel Path Modeled Including Pedestrian and Transit Trip Elements

Figure 7 shows the level of detail of the station facilities that can be represented in the analysis. Although not all potential aspects of station facilities are shown in the example, modeled facilities can include vertical circulation systems (stairways, escalators, and elevators), corridors, ticketing and turnstiles, queue lines, and platforms. This comprehensive modeling of pedestrian activities allows surge effects to be quantified such as conditions where a train arrives at a station and transit patrons alight the train to mix and move through the dense pedestrian group that has accumulated on the platform waiting to board.

ALPS allows this level of pedestrian simulation to be included throughout the complete urban district model when this level of detail is determined to be important to the analysis process. If it is not, simple delays are added to the trip times to represent the time it takes to traverse the platform or station to board or alight the transit system. **Figure 8** depicts a larger scale view of transit platform facilities, based on this hypothetical example of a large urban district.

Transit User Experience Statistics from ALPS Simulations

A wide variety of user experience statistics can be accumulated from ALPS simulations to compare performance of alternative transit systems. Because the trip-table database is common among all the alternatives, APM and PRT systems can be adequately compared to more traditional modal alternatives with respect to the user experience. Using a "pivot-table"-type approach, ALPS can aggregate performance information across essentially any component of the transit system, such as the passenger waiting time in each station.

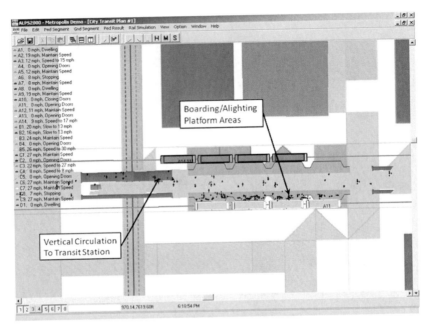

Figure 7 Simulation of Transit Station Pedestrian Facilities

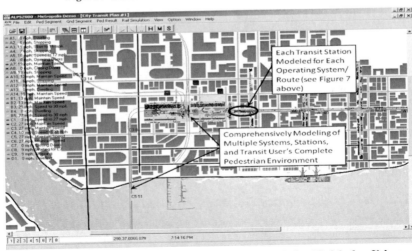

Figure 8 Example of a Multi-System, Multimodal Simulation Model of an Urban District

Figure 9 illustrates an example of such data for multiple stations served by a transit line where the number of transit patrons is given along the vertical axis and the time duration spent waiting for transit service is given along the horizontal axis. These simulation results are also taken from the East End LRT project work in Houston based on a case study of one potential line configuration. The different distributions of waiting time between the different stations form an interesting set of patterns, with one station in particular showing several distinct clusters of different waiting times. The station involved (indicated by the color coding in the legend as the black graph line) was the Hughes Station – a station modeled as a bus transfer point to the LRT system. The pattern of bus service headways throughout the day, when transferring passengers to the train (operating with different headways from the buses) created the unique waiting time distribution.

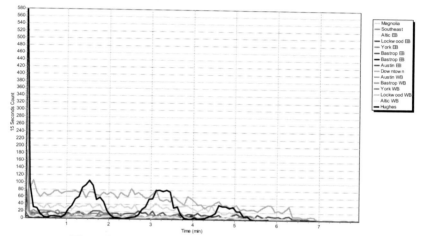

Figure 9. Station Wait Times for the East End LRT System

Other pedestrian and ridership data that is typically extracted for each transit system and its associated stations, such as the ridership occurring on each link, the hourly boarding and alighting activity at each station, and the occupancy and corresponding level-of-service for the pedestrian facilities is available from analysis using ALPS. Other ridership data often useful in the comparative alternatives analysis is the station-to-station transit ridership trip tables, which ALPS can provide with travel classification (i.e. home-to-work, non-work non-home based, retail/restaurant, etc.) breakdowns for all trip totals by hour of the day.

Figure 10 shows simulation results of platform density for a heavily utilized station boarding platforms is common for transit in a dense urban environment. As indicated in the legend, the area directly adjacent to the platform edge where people cluster while waiting for the next train has a comfortable capacity shown in red (typically established in the input data based on 15 square foot per person), whereas the ultimate capacity is much higher and shown as a brown line in the middle of the graph. The accumulation of people reached the comfortable reference capacity multiple times during the day, but did not appreciably exceed this level.

This represents a good level-of-service for such a high level of activity in the station, since alighting passengers can easily pass through the other people waiting to board.

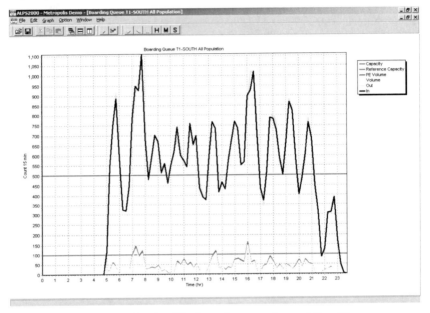

Figure 10 Station Platform Occupancies and Densities

The other type of simulation results that are typically evaluated from simulation models are animations. Animation is frequently used to identify issues with, for example, alternatives that force transit patrons into access and egress patterns that cross and interact with the street vehicular traffic. The activities within an urban district frequently create conditions with high levels of pedestrian flows. This condition can be particularly intense in districts that have large venue sports, entertainment or convention facilities where very large quantities of people exit within a very short period of time. ALPS can represent this aspect of the multimodal operations. **Figure 11** shows a transit access environment in which pedestrian must cross roadway traffic flows, creating a potential conflict zone. The model shown in **Figure 11** is of a future operating condition in downtown Minneapolis, following the conclusion of a baseball game at a new stadium adjacent to the Hiawatha light rail transit line. This case study illustrates the importance of the ability of analyzing the comparison of alternatives where a significant part of the pedestrian activity is moved to grade separated facilities which provide access to aerial APM systems, with respect to at-grade transit alternatives where transit users must cross active traffic lanes.

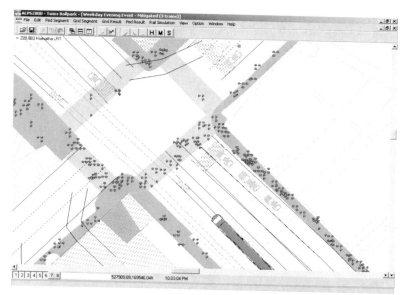

Figure 11. Simulation of Pedestrian, Traffic and Transit Environment

Conclusions

Simulation-based tools that can address the user experience and operational aspects of a complete multimodal environment can provide important capabilities within the alternative analysis process for transit applications within dense urban districts. Comparison of a variety of transit technologies to serve as a circulation system within an urban district requires the analysis of a complex set of scenarios, input parameters, and systems. The ALPS modeling tool has been developed since 1985 to address these types of complex applications in a comprehensive manner. The simulation of pedestrians, transit systems, street traffic, platforms, parking, and a variety of technologies and grade separation aspects is all possible with this specialized modeling tool.

The capability to analyze the complete user experience for alternative transit circulator technologies, when combined with equally rigorous modeling of each transit system's operations, provide the highest possible confidence in the results of an alternatives analysis.

References

1. Gettman, Douglas, Ph.D.; Sam Lott, P.E. (2009); <u>Supporting Renewal and Growth of LRT through use of the Multi-modal ALPS simulation tool</u>; 11th Joint TRB/APTA Light Rail Conference.

Simulation Analysis of APM Systems in Dense Urban Environments – Part 2: System Operations

J. Sam Lott, P.E.[1]
Douglas Gettman, Ph.D.[2]
David S. Tai[3]

[1] Senior Vice President, Kimley-Horn and Associates, Inc., 12012 Wickchester Lane, Suite 500, Houston, Texas 77079; PH 281-597-9300; sam.lott@kimley-horn.com
[2] ALPS Modeling and Simulation Group Leader, Kimley-Horn and Associates, Inc., 7878 N. 16th St. Suite 300 Phoenix, AZ 85068; PH 602-906-1332, doug.gettman@kimley-horn.com
[3] Senior Modeling Software Developer, Kimley-Horn and Associates, Inc, 7878 N. 16th St., Suite 300, Phoenix, Arizona 85020; PH 602-944-5500; David.Tai@kimley-horn.com

Abstract

This paper is the second in a series of two technical papers which define the key issues to be analyzed in the study of automated people mover (APM) systems for circulation within dense urban environments. It is quite complex to accomplish an "apples-to-apples" comparison between alternative transit technologies for urban district circulation systems because of the different travel times, access convenience and walk distances between the alternatives, all of which effect the transit user's choice. These papers describe the suitability of the ALPS ™ simulation tool for such analyses including discussion of the characteristics of the urban environment, the nature of transport systems suitable for circulation system application in urban centers, and trade-offs between at-grade and aerial transit system alignments. Simulation-based case studies are presented that illustrate the analysis of APM transit technologies using ALPS. The first paper addresses the representation of the transit user's complete experience within the simulation model, including the approach to modeling transit ridership and the related pedestrian facilities associated with the transit stations and the means of access to the transit. This second paper describes the capabilities within ALPS to represent transit system operations include train performance modeling, guideway and alignment configuration, demand-responsive scheduling, and options for dynamic routing of APM systems.

Introduction

This paper is a companion document to the paper of the same name, but identified as Part 1: Transit User Experience. This Part 2: Systems Operations paper discusses the operational aspects of an APM system that are important to consider when analyzing the potential for applying such advanced technologies as a transit circulator within a major urban district, large university / medical center campus, or special event or major activity center. Refer to the Part 1 paper (Lott et al, 2009) for a more complete introduction to the topic. In this paper we will assume that the operational models of automated people mover (APM) systems are, by common definition, limited to transit applications with fully automated control systems.

Figure 1 illustrates the multimodal environment that is characteristic of dense urban districts and major activity centers. Each of the modal elements provides an important function within the overall multimodal transportation system. The end result of such an integrated multimodal system will be acceptable service levels provided for access to and mobility within the urban district / activity center.

When a complete urban district or corridor is studied with a dedicated circulation system, it is important to model the entire circulation system. However, other types of line-haul mass transit such as subway lines or commuter rail lines which pass through the district do not need to be modeled in full. Only a small part of these systems need be modeled, particularly when there are associated stations that experience surge flow conditions due to the periodic arrival of large, high capacity mass transit trains.

Alternatives analysis studies of circulator transit systems for urban districts have, in the past, typically focused on the suitability of more conventional, at-grade transit modes such as bus or light rail systems. This conventional technology is usually located with the transit system in mixed operations with street traffic and along alignments that are frequently in conflict with pedestrians. More recent alternatives analysis studies are increasingly beginning to also assess the benefits of grade-separated circulator system(s), such as fully automated APM systems operating on aerial guideways.

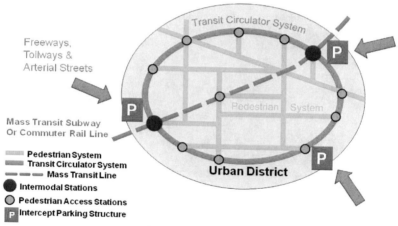

Figure 1. Multimodal Elements of an Urban District

The operational parameters for fully automated APM systems are markedly different from the characteristics of manually-operated trains traveling at-grade, particularly with respect to traffic and pedestrian conflicts. Typical tools used for alternatives analysis are not capable of providing true "apples-to-apples" comparisons that consider technology types across this broad range of very different operating conditions.. The remainder of this paper describes the operational capabilities of the ALPS simulation-based analysis tool that has had successful application in analyses across this broad spectrum of transit technologies and modes.

Methodology for Transit System Analysis

The analysis methodology that has been historically applied in the transit industry involves an initial definition of the system alignment and associated speed constraints, transit vehicle propulsion and braking characteristics, and passenger comfort criteria. Specification of these parameters lead to initial calculations of the train/transit vehicle "performance" along the prescribed route / guideway alignment (note that throughout this paper, reference to "train" will apply generally to both multi-car trains as well as single vehicle operating units). Next, based on the calculated train performance, the operations of the system are then subsequently analyzed based on the estimated station dwell times, headway objectives, system operational capacity calculations, and ridership demand patterns (typically coming from studies completely separate from the performance / operational studies). The end products of this traditional methodology for analyzing the operating equipment and systems are the calculated power consumption, the required train length (for multi-car entrainment systems), the number of trains/vehicles in operation, and the associated size of the total operating fleet.

Kimley-Horn and Associates has sought to advance the state-of-the art in the analysis of transit system operations by developing a more comprehensive approach than that described above. The improved methodology provides a more accurate comparative assessment of alternative transportation technologies for operations over an entire 24 hour day, as compared to the common analysis period of just the peak demand hour. This tool is called the Advanced Land-Transportation Performance Simulation™ (ALPS™). ALPS has been continuously refined and advanced since the early 1980's ALPS can provide simulation analysis of systems operating over multiple routes/lines within the same computational process and even when these systems are different technologies with unique propulsion and control characteristics. ALPS "holistically" models the complete, multimodal transportation environment in which the transit system(s) operate, one of the characteristics that enables "apples to apples" comparisons of advanced transit system alternatives. The methodology is essentially the same for analyzing any transport modes whether operating on alignments that are below ground, at-grade, or aerial. Most recently, this tool has been used to study transit technologies as diverse as high speed maglev, urban LRT (Gettman et al, 2009), street trolleys, as well as complex APM and PRT systems. The remainder of this paper describes various features of the ALPS transit system operational modeling methodology is related to the modeling and simulation of operational characteristics of APM systems.

Vehicle Characteristics

The vehicle characteristics are of fundamental importance to the performance analysis of any transit system. Figure 2 illustrates a screen shot from the ALPS application for configuration of the propulsion system characteristics, transit vehicle size, configuration, performance related comfort limits, capacity and propulsion / braking characteristics for a generic APM system. The propulsion system is quite important to model with reasonable accuracy in order to properly reflect the train travel and round trip times – parameters that determine both the operating fleet required and the headway and capacity of the system. The data entry fields on the right of the data entry table in **Figure 2** allow the mathematical representation of the transit vehicle's "motor curve" through specification of the propulsion force delivered (at the propulsion motor) as a function of operating speed. On the left-hand side of the table, other vehicle characteristics are entered which are important to the calculation of the mechanical losses that result as the propulsion force is transmitted to the point of "traction" between the vehicle and the guideway. Commonly known as the "train resistance" parameters, these parameters principally relate to the mechanical and aerodynamic energy losses for the specific vehicle design (e.g., static resistance per weight in lbs/tons.). These characteristics determine how much power must be delivered to the vehicle at any point in time to meet the control system's request for acceleration, and conversely the braking energy to be removed from the train during slowing and stopping.

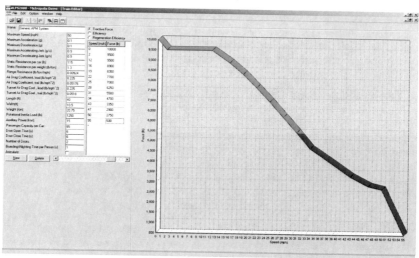

Figure 2. Transit Vehicle Characteristic Data – Generic Large APM Vehicle

Also selectable from this data entry table are the propulsion characteristics of propulsion / braking electrical losses under the label of "Efficiency", and the similar regenerative braking power characteristics labeled as "Regeneration Efficiency". Both can also be used to assess the effects line losses and the benefits of power regeneration. These assessments are useful in sizing traction power substations so that they are neither over nor under designed.

On the left side of the table shown in **Figure 2** are the vehicle configuration parameters such as length and width, number of doors, and maximum passenger load. The door operating time and the maximum rate that passenger board and alight the vehicles are also input in this data entry table. These parameters provide a realistic representation of transit vehicle operational aspects such as station dwell times (affecting round trip times) and related station throughput capacities.

Transit Guideway Features

The modeling of the guideway itself involves the creation of a link-node network representing the alignment on which the transit vehicles will operate. If only the one transit guideway system itself is to be analyzed, the models are fairly simple to create. **Figure 3** shows an example screen shot from ALPS of a simpler, guideway- only model for the fixed route APM system studied for Alameda Island in the Bay Area (Lu et al, 2003). This study compared the existing "traditional" APM system with alternative PRT / GRT technology applications.

Figure 4 shows a data entry screen shot from ALPS where parametric data is entered to represent segments of the guideway that was common for the models of both the traditional APM and the PRT alternatives. In this example, the guideway characteristics are for a very tight radius curve with a constrained operating speed. For automated systems utilizing a fixed block control system, this would represent the command speed to which the vehicle propulsion controls respond at that location on the guideway.

Once the complete guideway network is created, each guideway link is then defined (i.e., the portion of the guideway connecting adjacent passenger stations), including any crossovers, turnouts, and pocket tracks within that link. Vertical curve guideway segments have additional data entry requirements to represent the transition in the vertical alignments, which allows the calculation of related gravitational effects on propulsion and braking power as a function of the train length. If appropriate for the model's purpose, storage guideway can also be included so that vehicle / train dispatches to and from the storage facility are properly represented in the operational simulations. This is a key aspect of empty vehicle management for PRT systems, as discussed in more detail further in this paper.

The modeled values for maximum acceleration, deceleration and jerk rates are of critical importance since they determine passenger comfort levels. The values of 0.01 g shown in the example are common for an APM applied to public transit environments, but are generally an upper limit for vehicles that are designed to primarily accommodate standing passengers.

Train Performance

After the guideway part of an ALPS model has been created and the vehicle types are defined with their characteristic data, the transit models can be used immediately for performance analyses irrespective of the modeling of other transportation modes or transit ridership. **Figure 5** illustrates the link-by-link performance results for a single train operating on the link described **in Figures 3 and 4**.

These performance results for the train / vehicle operating throughout the specific guideway link include the associated power consumption at the current collectors (i.e., not including traction power distribution system losses along the guideway). For the example link shown, the changes in track speed due to the alignment constraints of multiple small-radius curves demonstrate the propulsion and braking aspects of the simulation (refer to the legend within Figure 5 to find the color-coding key for the performance graphs). This type of information is very useful during the preliminary design and even as early as the conceptual planning

AUTOMATED PEOPLE MOVERS 2009 593

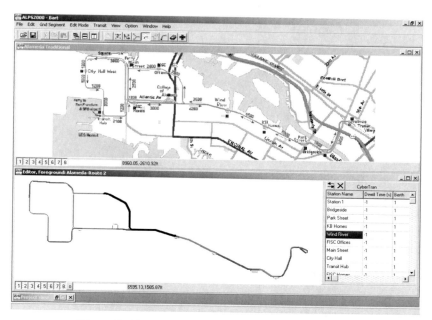

Figure 3. Fixed Route APM System Creation in ALPS

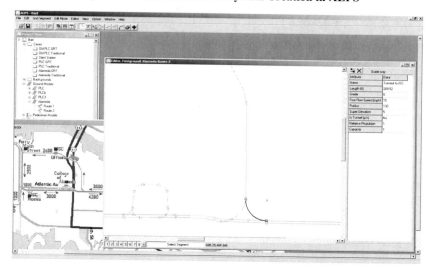

Figure 4. Guideway Segment Characteristic Data

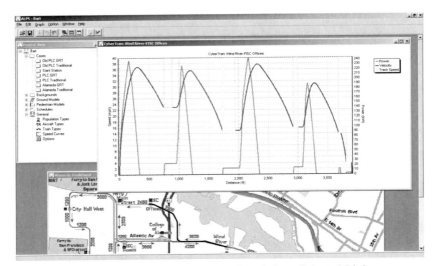

Figure 5. Train Performance Graph for a Selected Link

phases of a transit project, since guideway alignment changes (e.g., enlargement of curve radii) are usually justified by the resulting improvements to operating speeds. All of the graph data for each link can be extracted from ALPS for further assessment in spreadsheets or other tools. The ALPS results also provide a variety of tabular reports, which are described further in discussion that follows.

The particular study depicted in **Figures 3, 4 and 5** was of a proposed APM circulator system for Alameda Island in the San Francisco Bay Area. The speed-graph shown in **Figure 5** indicates that the simulation control mode selected for the ALPS train control simulation provided operations where the train brakes <u>before reaching</u> the speed limited guideway segment. This is a useful simulation mode for planning studies in which no specification of a train control system has yet occurred, and the only speed constraints known at that point are those due to alignment. ALPS can also model the case where the vehicle responds <u>after reaching</u> a change to signaled speed for a given guideway segment, as would be the appropriate analysis when a conventional train control signaling system is well defined.

Figure 6 illustrates a more typical set of train performance curves for a hypothetical high performance transit system with an alignment design for maximum speed, contained within an urban district demonstration model called "Metropolis". In this example, the alignment includes a substantial elevation change and steep grades, and the vehicle characteristics are those of the Generic APM System shown in **Figure 2**. The legend of the figure defines the color coded curves (color code and graph data displayed is selectable by the analyst in ALPS). In this case, the graph includes the guideway grades (left axis gives values in %) and the associated power requirements for such demanding propulsion conditions. The transit system link being represented (highlighted in black) connects the Riverside Station through steep grades to reach the Mt. Vista Resort Station.

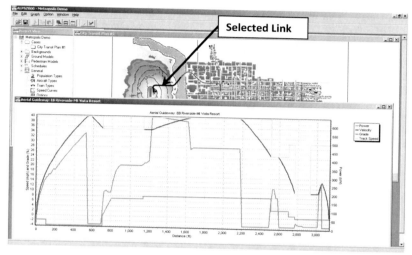

Figure 6. Typical Train Performance Curve with Guideway Grades

Train Operations

The operation of trains or independent transit vehicles (i.e., "single-car trains") along an APM guideway must be modeled based not only on the performance of the trains along the route, but also on other interactions between the train and the automatic train control system. The ALPS transit operations simulation logic can model all types of control systems that are possible for application among transit and APM systems represented in the projects in the industry. These features include modeling the separation from the other operating trains by a suitable "safe-stopping" distance, routing of trains between the sequential stations that each specific train is designated to serve, and optimization of train performance (travel speeds and dwell times) where such modifications are possible. .

<u>Train Progression Management</u> – ALPS can model both approaches used in the transit industry to model operations of trains in fixed route applications where it is important to manage train progress relative to the location of other trains in service. The first approach to vehicle headway management is the feature generally called "Station Ahead Clear". In this operating mode, the following train is not allowed to leave a station until the station ahead is cleared from occupancy by the leading train. This is a feature that fully automated systems often employ to ensure no train becomes stopped on the guideway away from a station during any normal operating mode or common failure / service disruption incident.

In addition to this simple train control logic, ALPS can model active headway management logic. In this case, ALPS models the supervisory control features that continually work to adjust dwell times and rebalance the spacing of trains and their respective operating headways. In addition to analysis of normal operating conditions, ALPS can represent the response of headway management systems in abnormal conditions as well. For example, as trains progress through the assigned route, the representation of real-world conditions can be imposed within the simulation which begin to disrupt the progression of trains in a manner that emulates the way that people interact with the transit system in the real world – such as a passenger holding the train doors open. Another type of real world condition that can be studied with ALPS is of the occurrence of failure-mode incidents which disrupt normal operations for a brief period of time before normal operations resume. In either case, headway management logic is necessary to represent the response of the APM system and corresponding impacts on performance metrics. Headway management logic is also relevant for simulations of at-grade, non-automated transit systems where the train movement is disrupted by traffic, pedestrians, and traffic signal systems.

Fixed Block and Moving Block Control – There are generally two different types of automated train control concepts that are used with trains moving in a guideway network – fixed block control systems and moving block control systems. **Figure 7** depicts a screen shot of one of the pull-down menus in ALPS where the analyst selects between these optional control features. The actual mode that is represented in **Figure 7** is the moving block control mode. In moving block control mode, the safe-stopping distance for each train is continually calculated and train speed is adjusted accordingly. In **Figure 7**, the safe-stopping distance is represented in ALPS as the colored "dot" in advance of each train. This feature was an important aspect of the analysis for the Alameda Island study.

With respect to close-headway operational conditions for PRT and GRT systems, the development of synchronous "moving-slot" control concepts has led to some PRT system developers utilizing a fixed block configuration, and other PRT technologies developing moving block controls to achieve the synchronous operation. The ALPS software allows either control concept to be selected for the operational simulations.

Train and Ridership Status – Also shown in **Figure 7** is an optional display feature of ALPS where the operating status of each train in service is shown along the left-hand edge of the display as the simulation is running. This feature is helpful in observing the over-all status of the transit system and illustrates how the complete performance and operational status of each train / vehicle is being continually calculated within ALPS.

Another simulation feature of ALPS that is represented in **Figure 7** is the passenger loading conditions for each train. The numerical value displayed by each train is the number of passengers on board, which is important for several reasons. By including the actual passenger load on each train, the model can determine at any point in time whether a train/vehicle is full and cannot board additional passengers, or how many of the waiting patrons can board the vehicle at any given station. As described further in (Lott, et al, 2009), these delays are captured on a traveler-by-traveler basis and can be reported and analyzed by station, origin-destination pair, time of day, and other aggregations. The accurate modeling of passenger load on each train also allows the proper dynamic calculations of the power and energy consumed by each train as it progresses along the route.

Service Disruption Incidents – As introduced earlier, with the level of operational detail that the ALPS methodology provides, the tool is useful for studying failure and service disruption conditions. Failure incidents can be interactively induced by the analyst either by using the mouse while the ALPS animation is running, or by entering predetermined failure duration(s) and location(s) at which trains are to be stopped during the simulation run. **Figure 8** depicts a screen shot of the incident scenario data entry menu of ALPS. This example shows a case study for a fixed-route transit system. The animation display is depicting the way trains are queuing during the incident occurrence. The graph display – commonly called a "train graph" or "string line graph" – depicts the train trajectories before, during, and after the incident occurs showing the impact of the incident and the recovery of normal operations.

Demand Responsive Controls

Personal rapid transit (PRT) and group rapid transit (GRT) systems represent the most complex type of automatic train control systems. . These types of transit technologies have two guiding principles of operation:

- Direct ride between the passenger's origin and destination stations at any time of day and between any two stations along the guideway network
- Transit vehicles are dispatched to facilitate the system's response to a "demand-call"

A control system using demand-responsive dispatch features must be continually adjusting operations to provide the most efficient assignments of vehicles whenever a demand call is placed. The automatic train control system must then determine the most efficient route to assign to the transit vehicle to take the rider to their desired destination.

In the discussion below, the term "PRT" will be used as a generic term describing the basic operational principles described above, irrespective of the vehicle size. Also discussed below using the generic "PRT" terminology is the distinction between operational control concepts in which transit vehicles only serve a single travel party onboard a given vehicle vs. a control concept in which transit vehicles serve multiple travel parties together – a concept commonly referred to as a "shared ride" type of service.

In the following sections, we discuss the capabilities in ALPS to represent operational control functions of dynamic routing, empty vehicle management, shared ride service, and hybrid PRT and fixed route service. Throughout the remainder of this paper, reference will be made to "vehicle" rather than "train", since PRT systems operate exclusively with single-car trains.

Dynamic Routing – Dynamic routing is the key operational control feature of PRT systems. The modeling of this type of control system is considerably more complex than the modeling of fixed route service (i.e., the traditional form of transit services where all trains / vehicles operate along a prescribed route that always serves the same stations). The control of PRT operations requires the dynamic determination by the supervisory control system of the suitable minimum path, and the corresponding establishment of a specific route from an origin point to a destination point for a specific vehicle. With this type of dynamic routing the system must also address changing operational conditions which could dictate how vehicle moving along the guideway will have their route assignments changed before the vehicle reaches its destination station.

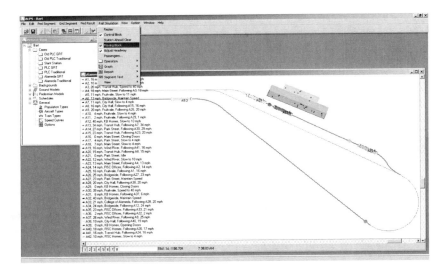

Figure 7. Transit Operating Control Mode Selection Menu, With Moving Block Simulation Displayed

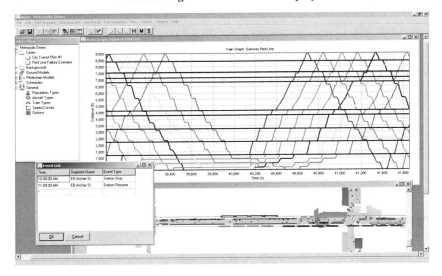

Figure 8. Failure-Incident Data Entry and System Recovery Assessment

ALPS can also model the control system response to conditions where a vehicle encounters an operational delay in an off-line station thereby blocking the path of following vehicles, or conditions where a given off-line station is temporarily overloaded. This situation, even if only momentary, can occur right at the point-in-time when a following vehicle is approaching the destination station, causing the PRT vehicle still in route to be "waved off" from entering the station by the supervisory control system. Under these operational circumstances, the affected vehicle's route must be dynamically changed to allow it to re-circulate back to approach the destination station a second time.

Empty Vehicle Management – The second aspect of PRT operations that is critical to model accurately is the logic for strategically repositioning each empty vehicle to provide the most efficient response to the direct origin-to-destination ridership demand. The operational control (i.e., ATS – automatic train supervision) system must determine the specific empty vehicle to be dispatched in response to the demand and then dynamically assign its route to the proper station. ALPS can also represent variations on the control logic for empty vehicle management involving both the vehicle storage location as well as the timing of the system response for vehicle dispatch – either leading (e.g., typical daily patterns of demand) or lagging a demand call.

Shared Ride Service – For multi-party, shared-ride PRT operations, the demand-responsive operational controls are slightly different from that of single travel party service. For shared-ride service, several demand responsive dispatch modes can be modeled with ALPS for the entire group of transit patrons onboard the shared vehicle, including:

- one common origin and one common destination
- one common origin and multiple destinations
- multiple origins and one common destination

Hybrid PRT and Fixed Route Service – ALPS can also model hybrid PRT service which combines fixed route service operations with demand responsive control features. The simulation studies of Alameda Island and Oakland Airport, performed for Bay Area Rapid Transit (Lu, et al, 2003) using ALPS, concluded that there would be significant benefits in changing the operational mode from demand-responsive to fixed route service, depending on the time of day and/or the station pairs served. With this control mode, some or all stations can be served by pure PRT-type demand-responsive service during much of the day's operations. But during the peak periods of the day when higher vehicle loading efficiencies are needed for capacity purposes, the operating mode can be changed such that the same vehicles operate in a manner that serves some stations, or all stations, with fixed route service.

Analytical Benefits – The simulation of these more complex operational modes allows the potential benefits of non-traditional demand-responsive service to be rigorously compared to traditional, fixed-route APM service while the urban district's circulation system is still in the conceptual planning phase. As the system design progresses, more refined simulation studies can continually evaluate the operational impacts of any design changes that are made to station and guideway configurations. Combining dynamic routing APM service with the complete pedestrian and passenger travel path representation and origin-destination assignments provides a powerful and realistic representation of the operational characteristics of PRT systems. Refer to Part 1 of this two part paper for more detailed discussion of the trip assignment modeling capabilities of ALPS.

System Performance Statistics from ALPS Simulations

Essential to the analysis of any transit system's performance and operations is the generation of text-based reports, and associated data tables and graphs.

By way of definition, an ALPS case study is defined as an assembly of the model components comprising:

- a specific trip generation / transit ridership demand scenario
- a specific configuration of transportation facilities for all modal elements
- specific operational conditions imposed by the analyst, including failure incidents

The ALPS reporting methodology produces a uniform presentation of a complete set of text reports, graphs and data tables for each transit system technology "case study" performed. This allows a direct comparison to be made between all of the alternatives that might be analyzed for an urban district circulator system.

Some of the reporting features of ALPS have been illustrated in the figures above. Other simulation results can be reported through formatted reports, most of which are in tabular form and comprising ridership, train performance and operational statistics.

The Transit Ridership reports quantify parameters such as station boarding and alighting volumes, link volumes, and station waiting time. The Transit Performance reports quantify parameters such as travel distances, propulsion power demands, and station-to-station in-route travel times. The Transit Operations Report quantifies parameters such as trains / vehicles in service, vehicle-miles and vehicle-hours, and total energy consumption. Several examples of these types of reports are shown in **Figure 9**.

When the alternatives for an urban district's circulator system are all tested using an appropriately comparable set of case studies for defined scenarios of demand, system / facility configuration, and operational conditions, a meaningful "apples-to-apples" comparison can be made between the various technology alternatives. Further, by using the ALPS holistic methodology that combines all of the public transit and personal transportation modes, many of which share common infrastructure and facilities, a complete picture of the performance of the multimodal transportation system can be obtained.

The integration of the pedestrian models with the ground transportation and transit models allows the comprehensive analysis of transit user mode choice within the context of capacity limitations and operating conditions. This methodology allows a more accurate representation of the real world in which transit users make their decisions of personal mode preference based on time, cost and convenience. Although this methodology could be applied using more traditional tools, the analyst would be required to apply a variety of conventional modeling and analysis tools involving a number of data-handoffs within the process. In the end, the level of complexity of the comprehensive methodology described in the paper requires a versatile tool designed for this purpose, such as ALPS.

Conclusions

Simulation-based tools which can address both the user experience and operational aspects of a complete multimodal environment provide important capabilities within the alternative analysis process for transit applications within dense urban districts. The potential application of a variety of transit technologies to serve as a circulator system within a district creates a complex set of parameters that cannot be compared adequately without detailed simulation. The ALPS model has been developed to address this type of analysis in a comprehensive manner by allowing simulation of pedestrians, street traffic, parking, and of course transit system operations with a variety of control technologies and grade separation / alignment configurations. The capability to analyze the complete user experience for alternative transit circulator technologies, when combined with equally rigorous modeling of each transit system's operations, provide the highest possible confidence in the alternatives analysis results.

References

Lott, J. Sam, P.E. (2009); Simulation Analysis of APM Systems For Application in Dense Urban Environments – Part 1 User Experience; 12th International Conference on Automated People Movers; American Society of Civil Engineers.

Gettman, Douglas, Ph.D.; Sam Lott, P.E.; David Tai (2009); Supporting Renewal and Growth of LRT through use of the Multi-modal ALPS simulation tool; 11th Joint TRB/APTA Light Rail Conference; Transportation Research Board.

Lu, Richard; David Hathaway, P.E.; J. Sam Lott (2003), P.E.; BART's Investigative Study of the Group Rapid Transit Concept: The Technical Feasibility of GRT Operations; 9th International Conference on Automated People Movers; American Society of Civil Engineers.

ALPS Transit Ridership Report
Transit Station Boardings

Transit System: Med Distr Circulator, C-CW Route,
Vehicle Technology: 40 Ft Generic APM
Operating Configuration: 2-Car Trains

Project: Metropolis Hospital District
Case Study: Traditional APM, Option A, 2030 Design Year, Typ. Monday
Incidents: None

Population Types Included: Hospital Employees, Med School Staff, Students, Patients and Visitors, Office Employees, Service/Support Staff, Other Transit

Ending Time-of-Day: 7:45 a.m.
Simulation Time Interval: 15 Minutes

To Station \ From Station	St. Josephs Hospital	University Hopital	Sheraton Hotel	H-Inn/ Hilton Hotels	SW Childrens Hosp.	Baker Burn Center	South Parking Garage	Metro Blue Line	Veteran's Hospital	Mitchell Schl. Of Nursing	North Parking Garage	Med. Center Intermodal Sta.	Total Boardings
St. Josephs Hospital	0	4	26	17	18	4	102	26	9	7	143	36	392
University Hopital	2	0	3	1	3	7	53	26	6	2	45	75	223
Sheraton Hotel	3	7	0	0	0	21	4	17	33	6	15	26	167
H-Inn/Hilton Hotels	7	11	2		34								
SW Childrens Hosp.	2	1	2										
Baker Burn Center	0	23	14										
South Parking Garage	195	108	5										
Metro Blue Line	22	45	20										
Veteran's Hospital	6	25	22										
Mitchell Schl. Of Nursing	24	22	5										
North Parking Garage	204	112	6										
Med. Cntr Intermodal Sta.	75	68	25										
Total Boardings	540	426	130										

ALPS Transit Performance Results
Transit System Power Report

Transit System: Mt. Vista Aerial Guideway
Vehicle Technology: 40 Ft Generic APM
Operating Configuration: 2-Car Trains

Project: Metropolis Demo
Case Study: Traditional APM, Start of Passenger Service, 2010 Year, Typ. Weekend Day
Incidents: None

Round Trip Time	10:36.0	min:sec
Trip Distance	14903.8	ft
	2.82268	miles

Auxiliary Power Input	15.000	KW/veh.
Total System Max Power	1,421.000	KW
Total System R.M.S Power	497.594	KW
Total System Avg. Power	395.019	KW

Maximum Speed	40.01	mph
Average Trip Speed (incl. dwells)	16.00	mph
Average Travel Speed	23.06	mph

Maximum Regenerative Power	462.945	KW
R.M.S. Regenerative Power	137.753	KW
Average Regenerative Power	95.420	KW

Figure 9 Example ALPS Ridership and Performance Reports

An Enhanced Bombardier CX-100 APM Vehicle

Jack Galanko, PE[1] and Scott Moore[2]

[1]Project Engineer, Bombardier Transportation, Systems Division, Pittsburgh, PA, 15236; (412) 655-5278; email: jack.galanko@us.transport.bombardier.com
[2]Senior Vehicle Engineer, Bombardier Transportation, Systems Division, Pittsburgh, PA, 15236 PH (412) 655-5917; email: scott.a.moore@us.transport.bombardier.com

ABSTRACT

Bombardier Transportation's Systems Division has developed an enhanced version of the *BOMBARDIER* CX-100** APM vehicle to replace the original fleet of C-100 Miami-Dade Transit (MDT) Metromover downtown people mover vehicles.

The paper details significant new designs and enhancements including:

- Enhanced styling featuring new highly sloped end cap design and end interior,
- End-equipment packaging configuration featuring modular framing,
- Manual controller configuration with improved human interface,
- Revised door opening configuration,
- Undercar package Heating, Ventilation, and Air Conditioning (HVAC) unit sized for subtropical climates and optimized for maintainability,
- Air distribution system with optimized air flow characteristics,
- Trainline design with universal coupling for improved operational flexibility,
- Expanded use of anti-corrosion materials and details,
- A new Vehicle Monitoring and Control System (VMCS) component of the on-board Automatic Train Control (ATC) system,
- Communications subsystems including voice and data radio, CCTV with digital video recorder, passenger info system, and mobile wireless router.

3D design and analysis software tools were effectively applied in developing the design and tooling for the new carbody end cap and end interior, end-equipment packaging configuration, and air distribution system design. The result was an enhanced CX-100 vehicle with clean modern styling and improved functionality.

BACKGROUND

Bombardier Transportation was awarded a contract by Miami-Dade Transit (MDT) in 1980 to supply the Metromover System – an automated people mover system in downtown Miami using Bombardier's C-100 technology. Bombardier was awarded a contract in 1989 to add two extensions to the original loop system: one to the north (the Omni Extension) and one to the south (the Brickell Extension). These projects resulted in a large APM system which circulates people within downtown Miami's central business district and interfaces with mass transit modes including MDT's bus (Metrobus) and heavy rail (Metrorail) systems. See Figure 1. The Metromover System includes 4.4 miles of dual-lane elevated guideway with grades up to 10%, twenty-one (21) open –platform stations, twenty-four (24) guideway switches, an off-line maintenance facility, and associated power distribution, control, and communications systems. The Metromover system is a driverless system with fixed-block ATC and a passenger flow rate capacity of 7200 passengers per hour per direction (pphpd). MDT performs the operations and maintenance with in-house staff.

Figure 1: Metromover System Map

Twelve (12) C-100 vehicles were supplied with the original loop system. An additional seventeen (17) C-100 vehicles were supplied with the extensions, resulting in a total fleet size of twenty-nine (29) vehicles. The Metromover System operates with a mix of one and two-vehicle trains.

MDT awarded Bombardier Transportation a contract in 2006 to replace the original fleet of twelve (12) C-100 vehicles with a fleet of enhanced and customized *CX-100* APM vehicles. These twelve (12) vehicles were manufactured and commissioned in 2008. In 2008 MDT exercised an option for Bombardier to provide an additional seventeen of the enhanced and customized *CX-100* APM vehicles. Manufacturing of the additional seventeen (17) new vehicles is scheduled for 2010.

INTRODUCTION

The new *CX-100* APM vehicles for the Metromover include enhancements and customized features based on requirements from MDT and their consultant Washington Group International (WGI). The new Metromover vehicles also include various enhancements introduced into the extensively service-proven *CX-100* APM vehicle platform in recent years. This paper cites and discusses the new and enhanced design configuration of the *CX-100* APM vehicles for the Metromover, outlines the design and decision making process applied, and summarizes the resulting functionality and benefits achieved.

VEHICLE BODY STYLING AND DESIGN

The new Metromover vehicles utilize a re-styled end cap configuration on a modified *CX-100* shell design. The new Metromover vehicles provide the same proven structural and dependable aspects found in previous C-100 and *CX-100* vehicles, while the new end caps give the vehicles an innovative, futuristic appearance. Figures 2, 3 and 4 show the new vehicles installed on the Metromover system.

Figure 2: New Vehicle in Maintenance

Figure 3: New Vehicle Leaving Maintenance

Figure 4: Two Car Train on the Metromover System

The most significant change to the vehicle was the streamlined styling of the vehicle exterior. MDT management required that the vehicles not only be new, but also look dramatically different from the existing fleet. This new and modern look would demonstrate to the public the value of the new and improved vehicles. Early in the proposal stages, the customer selected a unique approach and greatly increased the slope of the end caps. Figure 5 shows an initial comparison layout showing the exterior differences between the existing vehicle and the new vehicle concept. Figure 6 shows the vehicle interior portion of the initial layout.

Figure 5: Initial Exterior Layout

Figure 6: Initial Interior Layout

This approach offered a challenge, since the car could not be lengthened to achieve additional slope. The position of the doors was also fixed and therefore offered the other limit on the extent of end cap slope. Initially the customer proposed a slope of 40 degrees from vertical. This slope proved to be problematic in that it would require half of the door panel to be unsupported in the open position. During the preliminary design process it was agreed that this overhang of the door panel should be eliminated. To achieve this goal, the slope of the end cap was changed to 35 degrees and the door openings were reduced to 75 inches wide from 84 inches. Figures 7 and 8 show the outline of the existing vehicle compared with the new vehicle.

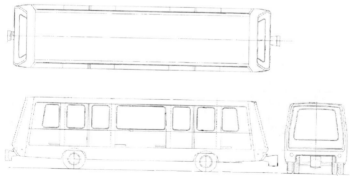

Figure 7: Original Metromover Vehicle Outline

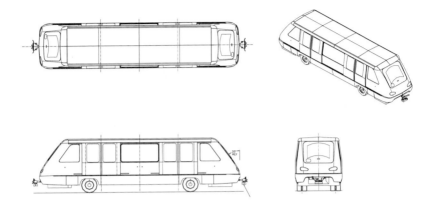

Figure 8: New Metromover Vehicle Outline

The end construction was changed to eliminate aluminum skins between the door posts and the end posts. A fiberglass end cap was then designed to wrap around the end of the car from door post to door post. See Figure 9 below for a view of the car structure. A supplementary structural analysis was performed to confirm that stresses with the revised end construction do not exceed stresses in the baseline analysis.

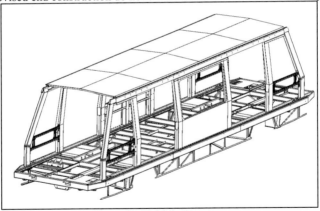

Figure 9: Vehicle Structure

Design software was employed to develop the shape of the new end cap design. The software that was primarily used was Pro Engineer Wildfire 2.0. All concept models were generated using this 3D design tool. These models were then imported into other software to create renderings, animations, and scale models. Once the basic shape was decided, the detail parts were created. Once all of the details of these parts were finalized, 3D data was transferred to the fiberglass part supplier for tooling production. The use of 3D CAD/CAM ensured that the final parts were representative of renderings and animations that were produced for the customer at various stages of the design process. It also resulted in very good fit of all body components the first time, minimizing schedule delays and costly retooling.

Once the basic shape of the vehicle end was developed, then the details of the windows and headlights had to be approved by the customer. A number of concepts were developed quickly and the customer chose the configuration shown in Figure 10.

Figure 10: Final End Configuration

Once the basic shape of the end and the window and headlight details were agreed upon with the costumer, then the detail parts were designed. Figure 11 shows the final end cap part. After developing the end cap part, the master plug part was developed to be used for tooling production, see Figure 12.

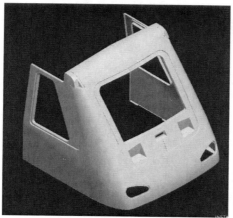

Figure 11: End Cap Part

Figure 12: End Cap Plug

The fiberglass supplier was able to take 3D CAD data and use 3 Axis CNC machines to make a full scale master plug part to be used for tooling production. Figure 13 shows the machining process, and Figure 14 shows the finished plug part. The plug was then used to make fiberglass tooling, see Figure 15 for a photo of this process.

Figure 13: Plug Machining

Figure 14: Finished Plug

Figure 15: Tooling Process

The new Metromover vehicles employ the latest in light emitting diode (LED) technology for both headlights and taillights. This gives the vehicle a unique look and greatly reduces maintenance cost associated with lamp replacement.

The tropical environment of Miami offered another challenge relative to the new vehicle design. Stainless steel plymetal and floor decking were installed on the new vehicles to help prevent sub floor deterioration during the life of the vehicle. Marine grade plywood was also used in the plymetal for this purpose. Some of the undercar hardware was also changed to stainless steel in order to prevent corrosion. The hardware that could not be changed to stainless steel was coated with a rust preventative to further protect the vehicle from damaging and unsightly corrosion.

The vehicle door system remained mostly unchanged except for the size and construction of the door panels. The door panel construction was changed to an aluminum skin and honeycomb core composite panel with minimal framework. Some benefits of this new panel are reduced weight and fewer exposed fasteners.

VEHICLE INTERIOR DESIGN

The redesign of the exterior end of the vehicle impacted equipment space allocation and end interior design. This change also offered an opportunity to improve many aspects of the end interior design relative to aesthetics and subsystem performance. The end interior for the new Metromover vehicles includes a higher end compartment than the standard *CX-100* vehicle. This design adds needed volume to the end equipment space that was reduced by the slope of the end cap. Figure 16 shows the floor plan for the new Metromover vehicle. The length of the floor area was reduced slightly in order to accommodate the highly sloped end design. Figure 17 shows the location of the manual control panel in the end interior. Figure 18 shows an interior side elevation of the new Metromover vehicle.

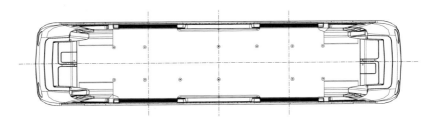

Figure 16: Interior Floor Plan

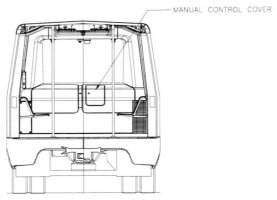

Figure 17: End Interior View

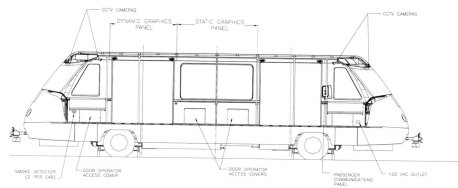

Figure 18: Interior Side Elevation

The vehicles have been designed to comply with applicable portions of the Americans with Disabilities Act (ADA). Door openings and stanchion layout allow for easy maneuvering of a wheel chair in and out of the vehicle. The emergency communication panels have been repositioned for easier access to a wheelchair bound passenger. Stanchions adjacent to door openings have a contrasting color to aid the visually impaired. Remaining stanchions have a contrasting color band at eye level.

The interior of the vehicles utilizes resilient finish materials. All soft trim has been eliminated to prevent premature wear, soiling, and tearing. Wainscot panels are covered with a durable laminate material with texture and color to complement the vehicle interior design. Floor covering is slip resistant safety flooring that will resist water and help prevent sub-floor rot. These materials combined with other standard finishes in the interior result in a clean appearance that can be easily maintained over the life of the vehicles. Figures 19 and 20 show interior photos of completed vehicles.

Figure 19: End Interior

Figure 20: End Interior with Cover Open

INTERIOR EQUIPMENT

Pro Engineer Wildfire 2.0 design software was used to develop virtual solid models of the new end interior equipment compartment packaging in conjunction with the new vehicle exterior and interior designs in advance of manufacturing. See Figure 21.

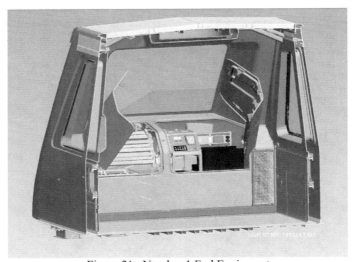

Figure 21: Number 1 End Equipment

Equipment within the number 1 end interior equipment compartment includes the ATC cradles, relay logic, brake and propulsion controls, tire pressure monitoring system (TPMS) receiver, and manual control panel designs provided by Bombardier, along with the associated wiring, connectors, and termination panels. A modular aluminum equipment frame system set up to receive 19-inch rack mounted equipment, and readily adaptable to special sizes of equipment, was applied to provide ease of assembly and facilitate modification to accommodate different equipment applied on other projects. A modular wiring approach was applied to facilitate wiring at the bench assembly level and thereby maximize consistency and quality. See Figure 21.

Communications equipment within the number 2 end interior equipment compartment includes the following equipment specified by MDT for the Metromover application and integrated by Bombardier into the new *CX-100* APM vehicles for the Metromover application: CCTV system including eight (8) cameras and digital video recorder, voice and data radio systems, Passenger Information System (PIS) for triggering of voice and dynamic graphics announcements, Mobile Access Router for wireless transmission of CCTV camera video and Vehicle Monitoring and Control System (VMCS) data, and associated power supplies. A relay panel assembly designed by Bombardier provides the interface between MDT's specified voice radio and the passenger intercom, maintenance microphone, and voice announcement system. Other equipment within the number 2 end interior equipment compartment includes a manual control panel similar to the one in the number 1 end compartment, an auxiliary control panel, operator information panel including indicator lights, gauges, and HVAC unit remote controls, and a pneumatic equipment panel. The end equipment packaging configuration with modular framing allows for a great deal of flexibility in accommodating packaging and installation of different equipment required for other projects.

The manual control panels are located at the center of the end equipment compartments. The design maximizes the field of vision for the manual operator.

AUTOMATIC TRAIN CONTROL (ATC), VEHICLE MONITORING AND CONTROL SYSTEM (VMCS), AND VEHICLE OPERATION

A modern *BOMBARDIER* CITYFLO** 550 on-board fixed block ATC hardware and software platform, similar to that applied on several new *CX-100* in recent years was included on the new Metromover vehicles to interface with the existing legacy fixed-block ATC system on the Metromover System wayside. The on-board ATC system performs required automatic train operation and automatic train protection functions.

A new on-board Vehicle Monitoring and Control System (VMCS) was also included. The VMCS hardware is based on a commercially available PowerPC board. Bombardier developed the custom software package.

The VMCS provides the capability to monitor and store on board data including alarm information and operational data, resulting in improved diagnostics. The VMCS stores 72 hours of data via a 4GB industrial grade flash memory device. The VMCS receives the data to be logged from the ATC via a high speed serial interface.

A Portable Test Unit (PTU) based on a ruggedized laptop computer interfaces with the VMCS. The PTU provides the capability to monitor, manipulate, graph and chart VMCS data, and to download all or part of the event and operational data stored in flash memory. The PTU software application was developed by Bombardier and includes pre-programmed screens, user-configurable screens with capability to display up to 40 variables, and alarm/ event screens. The VMCS also interfaces with an on-board Mobile Access Router to provide the capability for a wireless link to transmit VMCS event data from the vehicle to a wayside network being developed by MDT. Figure 22 is a block diagram of the VMCS.

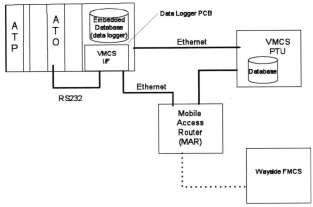

Figure 22: VMCS Block Diagram

For the Metromover project, MDT specified logic units for the passenger information system and the data radio based on similar hardware already in use within their transit properties. Where not specified by the customer, Bombardier has in-house designs to apply to achieve these functions. Bombardier developed a serial interface between the *CITYFLO* 550 ATC and the passenger information system logic unit to trigger audio announcements and dynamic sign messages. Bombardier also developed a serial interface with the data radio logic unit to communicate vehicle alarms to the MDT's wayside radio network.

A fluorescent display on the manual controller provides speed, door status, a subset of the alarm list, and other operational data for quick access by technician operating a train manually.

The trainline design provides for universal coupling functionality, i.e. the capability to couple any end-to-end combinations (e.g. number 1 to number 2 end, number 1 to number 1 end, and number 2 to number 2 end). This is an important operational feature on systems where vehicles turn around end-for-end with each loop as on the outer loops of the Metromover. The new vehicles also allow for mechanical coupling to the original vehicles for recovery purposes during the few years when the new and original fleets overlap.

MDT specified that the new Metromover vehicle design be forward compatible to convert to *BOMBARDIER* CITYFLO** 650 moving block ATC operation, in the event that MDT upgrades their wayside by overlaying a CITYFLO 650 moving block automatic train control system. Design provisions for such a conversion include modular framing applied for end interior equipment packaging, spare circuits, and equipment space allocations. The *CITYFLO* 650 equipment can easily be installed at the original build of this new version of the *CX-100* vehicle for future applications.

UNDERFRAME EQUIPMENT

Undercar equipment consists mostly of the existing typical standard package of *CX-100* APM vehicle bogies, propulsion and braking, control, pneumatics, and auxiliary equipment. Corrosion protective features on undercar equipment assemblies include powder-coated enclosures and stainless steel hinges and latches.

As discussed separately, the HVAC units (one at each end of the undercar) are a new design configuration.

The automatic couplers include mechanical couplers and electric heads and are similar to those applied on several other contracts in recent years. The configurations of the trainline hardware including the coupler electric heads, trainline junction boxes, and associated hardware within the end interior equipment compartments were designed to accommodate universal coupling functionality.

HEATING, VENTILATION AND AIR CONDITIONING (HVAC)

New HVAC unit and air distribution design configurations were developed in conjunction with the new vehicle end design. Ductwork design was developed and optimized using 3D CAD and computation fluid dynamic (CFD) analysis tools. The required air flow was thereby achieved while avoiding undue pressure losses and acoustical noise and also avoiding the use of costly full scale prototypes. Ductwork and equipment packaging design efforts occurred concurrently. Adjustments in equipment packaging were made during the process as required to allow optimization of the adjacent return and supply air ductwork configurations. The final design utilized molded duct work that was manufactured using CAD/CAM similar to the end cap process. The ducts are sealed to prevent leakage of conditioned air and ingress of unfiltered air. Figure 23 shows the HVAC system layout including the HVAC unit, the supply duct, and return duct. Figure 24 shows a CFD output plot.

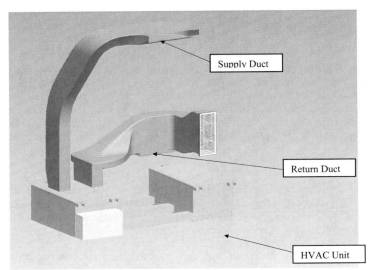

Figure 23: HVAC Layout

Figure 24: CFD Output Plot

The HVAC units are an undercar package unit configuration. Each vehicle contains two (2) identical undercar HVAC units (one at each end of the vehicle).

The HVAC units apply service-proven components, include a rigid support frame with resilient mounts and operate from 575 VAC, 60 Hz, 3 phase rail power.

Refrigerant is R22 single component HCFC. Blended refrigerants such as 407C were considered; however, such refrigerants have certain disadvantages. Given that there was no clear industry consensus on a long-term replacement for R22 at the time, the joint decision between Bombardier and MDT was to apply the extensively service proven and thermodynamically efficient R22 single component refrigerant.

Each HVAC unit provides cooling, heating, and ventilation modes of operation. The operation of each HVAC unit is fully automatic and includes an independent remote control panel with power on/off switch, set point selector, and temperature display.

The HVAC unit control panels are located within the number 2 end interior equipment compartment.

Each unit has a scroll compressor. The condenser and evaporator coils have copper tubes and fins. The condenser coils are cooled by air drawn through the coils by an axial fan. The evaporator blower (single speed) draws air through the evaporator coils and forces the supply air up through sidewall duct to the air distribution system in the ceiling of the car. Ceiling diffusers distribute the supply air throughout the length of the passenger compartment to maintain temperature uniformity in the car. Heaters within the HVAC units are direct resistance type with over-temperature control.

The HVAC system is designed for 90°F dry bulb and 77°F wet bulb temperatures per ASHRAE 1% summer design conditions for Miami, Florida. Each unit has 15.8 kW (4.5 tons or 54,000 BTU/hr) cooling capacity. The HVAC unit was qualification tested in a climate chamber in an independent test lab.

Battery powered emergency fresh air fans separate from the HVAC units and integrated into the duct system provide outside ventilation air if primary power is lost. The HVAC unit provides alarms indicating "Hot or Cold" vehicle condition and "HVAC Failure" conditions to the VMCS system.

CONCLUSION

As illustrated by this enhanced version of the *BOMBARDIER* CX-100** APM vehicle, Bombardier Transportation's System Division has the capability and flexibility to develop new and improved APM vehicle designs while working closely with customers to satisfy their unique requirements. Bombardier has the comprehensive engineering resources and tools to facilitate the significant amount of design optimization and integration required for such projects.

*BOMBARDIER, CX-100 and *CITYFLO* are trademarks of Bombardier Inc. or its subsidiaries.

Author Index

Page number refers to the first page of paper

Adams, Brian K., 428
Adams, Gregory A., 91
Al-Fadala, Hassan Eisa M., 179
Al-Ibrahim, Ghanim Hassan, 179
Anderson, J. Edward, 436
Andreasson, Ingmar J., 343
Angelov, Elisabet Idar, 257

Baumgartner, Jenny, 34
Baxandall, Frank W., 219
Beebe, Steve, 11
Bhattacharjee, Sambit, 80, 492
Bortolini, E., 245
Brackpool, Jon, 116

Castaneda, Steven M., 471
Coleman, Larry, 141, 201
Cottrell, Wayne D., 164
Culver, Frank, 511

Davenport, Nick, 321
de Leão, A. Gehlen, 245
DeCostro, Rob, 562

El-Aasar, Moni, 190

Galanko, Jack, 603
Gary, Dennis, 56
Gettman, Douglas, 574, 588
Green, James W., 484
Griebenow, Bob, 1
Guala, Luca, 361
Gurol, Husam (Sam), 233
Gustafsson, Jörgen, 389

Hammersley, John, 321
Heath, John A., 428
Hoshii, Tsutomu, 525, 534

Inada, Takaomi, 525

Jeffers, Frank, 91
Joy, John E., 546

Kapala, John, 69, 80
Kawai, Masafumi, 534
Kerr, A. D., 450
Kim, Dae-Hwan, 525
Kim, Yong-Kyu, 333
Koike, Yuji, 534
Krappinger, Heimo, 151
Kutchins, Scott, 20
Kuwabara, Masaaki, 525, 534

Landman, Dean, 190
Leder, William H., 219
Lee, Jun-Ho, 333
Lee, Seung-Cheol, 525
Lees-Miller, John, 321
Lennartsson, Svante, 389
Lindau, L. A., 245
Lindsey, Hal, 403
Lineback, Gary B., 428
Little, David D., 105
Lohmann, Robbert, 361
Lott, J. Sam, 574, 588

MacDonald, Raymond, 379
McDonald, Shannon Sanders, 297
McFadden, Dan, 492
Mitake, Masaya, 507
Mitchell, Terry, 20
Mokhtech, Kamel-Eddine, 179
Moore, Harley L., 1, 11, 34
Moore, Scott, 603
Morgan, Glenn, 116
Mori, J. David, 212, 555

Morita, Katsuaki, 507
Moss, Margaret Hawkins, 141
Muller, Peter J., 190, 309, 350

Nagahara, Genichirou, 525
Nguyen, Tuan, 492
Nuevo, Mario, 511

Oates, R. J., 450
Ogawa, Hiroshi, 507

Page, Jerome, 50
Pereira, B. M., 245
Picard, Margaret, 105
Piltingsrud, Mark, 56

Qaddoura, Hassan, 179

Redd, Melvin, 50
Riester, Thomas E., 233
Rincon, Diego, 20

Schrock, Steven, 190
Schroeder, B. M., 128
Shah, Sanjeev N., 141, 179, 201
Sheakley, Thomas, 91
Sproule, William J., 413, 219
Stirrup, Franklin, 141
Sturgell, Matthew, 499

Tai, David S., 574, 588
Taliaferro, David, 44
Tanaka, Motoaki, 534
Tatecho, Kunihiro, 534
Tegnér, Göran, 257
Todt, E., 245
Tomber, Dave, 1
Troy, Eric, 212

Williams, Mike, 80
Woodley, Russell, 50

Yang, Jackie, 11
Young, Stanley E., 190

Zuber, Marc, 425

Subject Index

Page number refers to the first page of paper

Airport terminals, 11, 116
Airports and airfields, 1, 20, 34, 44, 50, 56, 69, 80, 91, 105, 128, 141, 164, 428, 450, 492, 525
Austria, 151
Automation, 546

Business districts, 179

California, 34
Case reports, 190
Chicago, 56
China, 562
Colorado, 309
Comparative studies, 245
Composite materials, 525
Computer aided simulation, 333
Computer software, 603
Construction, 428
Control systems, 546

Design, 425, 428, 436, 603
Design/build, 128

Economic factors, 245, 546
Emissions, 379
Energy efficiency, 425, 507

Facility expansion, 69, 91
Failures, 484
Financial management, 201, 257
Florida, 128, 141, 492, 511

Georgia, 50, 69, 80, 91, 428
Guideways, 350, 361, 403, 413, 428, 436, 450

Hawaii, 555

History, 413

Illinois, 56
Implementation, 511
Industrial plants, 425
Integrated systems, 80, 164, 297, 471, 534

Kansas, 190
Korea, South, 389

Liability, 212

Magnetic levitation trains, 233
Michigan, 219
Middle East, 11, 179
Military engineering, 309
Monorail, 471

Nevada, 164, 212

Optimization, 484
Owners, 201

Parameters, 245
Parking facilities, 1, 297
Partnerships, 201
Passengers, 1, 321, 343, 379, 403, 534
Pennsylvania, 233
People movers, 1, 11, 20, 34, 44, 50, 56, 69, 80, 91, 105, 116, 128, 141, 151, 164, 179, 212, 245, 389, 413, 471, 484, 492, 499, 507, 511, 525, 534, 555, 562, 574, 588, 603
Private sector, 201
Procurement, 141, 179, 484

Railroad stations, 151, 350

Rapid transit systems, 190, 321, 333, 343, 350, 361, 379, 389, 403, 436, 450, 499, 525, 546, 555
Real estate properties, 105
Rehabilitation, 492
Ridership, 257, 321
Rubber, 555

Simulation, 574, 588
Standards and codes, 212, 511
Sustainable development, 297, 309, 361, 379, 425
Sweden, 257, 389
System analysis, 574, 588

Technology, 471
Telecommunication, 333
Tests, 389, 499, 20, 44
Transportation management, 201, 297
Transportation networks, 219, 257

United Kingdom, 116, 450
Universities, 219, 233
Urban areas, 257, 562, 574, 588

Vehicles, 343, 603

Washington, 1